the politics of SPACE

a comparison
of the Soviet
and American
Space Programs

William H. Schauer

HOLMES & MEIER, PUBLISHERS

New York and London

Published in the United States of America 1976 by
Holmes & Meier Publishers, Inc.
101 Fifth Avenue
New York, N.Y. 10003

Great Britain:
Holmes & Meier Publishers, Ltd.
Hillview House
1, Hallswelle Parade, Finchley Road
London NW11

LIBRARY OF CONGRESS CATALOGING IN PUBLICATION DATA

Schauer, William H.   1943—
    The politics of space.

    Bibliography: p.
    Includes index.
    1. Astronautics—Russia. 2. Astronautics and state—Russia. 3. Astro-
nautics—United States. 4. Astronautics and state—United States.
I. Title.
TL789.8.R9S32 .   387.8'0947   74-84657
ISBN 0-8419-0185-6

PRINTED IN THE UNITED STATES OF AMERICA

CONTENTS

Acknowledgments

The author especially wishes to express his gratitude to Professor Roger E. Kanet, who guided the preparation of the original version of this book, and to Professors Roy D. Laird and Clifford P. Ketzel, who provided helpful advice and many suggestions incorporated in the work.

To his wife, Pamela, the author owes an immense debt of gratitude for her encouragement, inspiration and invaluable assistance in preparation and completion of the present work. Thanks are also due the Committee on Astronautics and Space Sciences of the United States Senate for the wealth of information related to the author's topic collected by the committee and by the Library of Congress which was made available to the author.

Despite the assistance of the aforementioned persons and of many others, the author must bear exclusive responsibility for the generalizations and conclusions made in the study.

the
politics
of SPACE

1 Sputnik and Before

This study will be primarily concerned with the period dating from the launch of Sputnik I on October 4, 1957. This chapter, however, deals with the extended period of Russian and Soviet interest in rocketry and space that forms a heritage to Sputnik.

REACTIONS TO SPUTNIK

In the countries outside the "socialist camp" the reactions to the launch of Sputnik I included shock, alarm, bewilderment, dismay, and disbelief. The Eisenhower Administration was inclined to dismiss the Soviet achievement, Defense Secretary Charles F. Wilson calling it a "neat scientific trick." Others took Sputnik more seriously. Adlai Stevenson thought, "Not just our pride but our security is at stake." Averell Harriman saw it as "shocking" that a backward nation like the USSR could perform such a feat. Senator Henry Jackson viewed the Sputnik as "a devastating blow to US scientific, industrial and technological prestige" which plunged the US into "a week of shame and danger." Senator Mike Mansfield called for a "new Manhattan project" to wrest missile superiority from the Soviets, and Senator Symington demanded a special session of Congress and an investigation by the Senate Armed Services Committee. The Democratic National Committee accused the Eisenhower Administration of "complacency" and "failure." Senator Styles Bridges, a Republican, thought the nation had to change its values and priorities and had "to be more prepared to shed blood, sweat and tears if this country and the free world are to survive."[1]

Reactions from other countries included one unidentified Western ambassador who thought that "on October 4 the balance of political and diplomatic power shifted from Washington to Moscow." In France the Sputnik was viewed as a "rude awakening," and in Italy as a warning to the West. A correspondent for the British *Sunday Express* thought that it had thrown

1

the US into "frantic, fearful and angry confusion." One French general, formerly with NATO, thought that if the US did not pool its brains and its resources, it would be "condemned in advance." Japan's leading rocket expert saw the Sputnik as the most significant scientific success since Newton discovered the law of gravity.[2]

In the third world Sputnik was greeted by Nehru as a "great scientific advancement." Cairo radio thought it "will make countries think twice before tying themselves to the imperialist policy led by the US" and minimized "all kinds of pacts and military bases" which "are no longer secure from guided missiles."[3]

In Peking *Ta Kung Pao* considered the USSR "the most advanced country in the world in the field of science and technique," and Khrushchev, in one of the first of a long series of boasts and threats based on the Soviet ICBM, proclaimed the age of bombers over, describing the ICBM as a "merciless weapon" and promising "more things up our sleeve."[4]

Examining the background to Sputnik, one must conclude that the reactions of shock, alarm, bewilderment and disbelief were ill-made in the light of substantial achievements and firm predictions of Russian space scientists and rocketeers.

DEVELOPMENT OF RUSSIAN ROCKETRY

The Chinese were probably the first to use rockets as far back as pre-Christian antiquity. The first recorded military use of rockets as an incendiary weapon was also by the Chinese in 1232 A.D., with the introduction of rockets to Europe shortly afterward traced to the English monk Roger Bacon. Occasional use of rockets for show and as military weapons are noted in Europe in the thirteenth through the seventeenth century, with the first purely scientific work in rocketry credited to a German colonel in 1668. It was probably not until the seventeenth century that rockets were used in Russia, where history records a fireworks display in the town of Ustiuga in 1675. Peter the Great is reported to have established a Rocket Works in Moscow in 1680 for the fabrication of military signal and illumination flares.[5]

The Russian artillery officer Alexander D. Zasiadko (1779–1837) pioneered Russian use of rockets as an artillery weapon, and another artillery officer, Konstantin Konstantinov (1817–1871), applied mathematics to the scientific study of rocketry, eventually suggesting the use of rocket propulsion for travel. Yet another artillery officer, I. I. Tretskii (1821–1895), proposed rocket propulsion for dirigibles in 1849.[6]

A more interesting figure in the annals of Russian rocketry is the explosives expert of the revolutionary organization *Narodnaia Volia*, Nikolai Kibalchich (1853–1881). It was Kibalchich who built the bomb which mortally wounded Alexander II on March 13, 1881. (March 1 by the old Russian calendar).[7] Kibalchich, arrested shortly after the event, spent his jail time designing a flying platform powered by a gunpowder rocket. At his trial he presented the authorities with his proposal asking that he be allowed to discuss it with a government expert. The prosecution falsely informed him that his proposal was under study by a panel of experts. Kibalchich was convicted and hanged, and his crude designs remained sealed in an envelope in the police archives for thirty-six years.[8]

Besides an anonymous Chinese of dim antiquity, the fathers of modern rocketry and astronautics are three: they are a Russian,* Konstantin E. Tsiolkovskii (1857–1935), an American, Robert H. Goddard (1882–1945), and a German, Hermann Oberth (b. 1884).

Tsiolkovskii was a half-deaf, self-taught Russian schoolteacher who, inspired by Jules Verne's stories, went on to become acknowledged as the first great theoretician of interplanetary flight. Among other ideas, Tsiolkovskii conceived the laws of motion of bodies in cosmic space, the velocities required for earth orbit and escape, the use of multistage rockets for space travel, the use of liquid hydrogen and liquid oxygen as rocket fuels, the use of liquid fuels to cool rocket engines, the need for heat shields in reentry from space, and the creation of permanent space stations in orbit with self-contained regenerative environments.[9] In 1954 the Soviet Academy of Sciences paid tribute to this pioneer by establishing the K. E. Tsiolkovskii Gold Medal for outstanding work in the field of space travel.[10]

In 1881 Tsiolkovskii began his research in three areas: airplanes, interplanetary rockets, and metal dirigibles. In his diary in 1883 he described the motions of bodies in outer space. In 1890 Tsiolkovskii built the first Russian wind tunnel for conducting experiments in aerodynamics and in 1894 wrote an article describing an airplane not dissimilar to those subsequently built. In 1893 and 1895 he published science fiction stories about a flight to the moon and an artificial earth satellite.[11]

It was after reading an 1896 pamphlet by A. P. Fedorov on a jet flying device that Tsiolkovskii became exclusively interested in rocket flight. In 1897 he derived "Tsiolkovskii's formula" establishing the relationship between the velocity of a rocket and the velocity of its expelled gases. Further work led in 1903 to his publication of an article in a Russian scientific review on "Exploring Cosmic Space with Reactive Devices." An enlarged

---

*Leonid Vladimirov, a journalist for *Znanie-Sila* who defected to the West, states that Tsiolkovskii was Polish. See *The Russian Space Bluff*, pp. 26-27.

version of the article was reprinted in 1911 in the *Herald of Aeronautics*. In 1919 Tsiolkovskii became a member of the newly organized Soviet Academy of Sciences, receiving in 1921 a life pension from the Council of the People's Commissars. In 1924 Tsiolkovskii published a major work, *Cosmic Rocket Trains*, on the multistage rockets necessary to reach cosmic velocities. Subsequent studies delved into nuclear and electric propulsion (1925 and 1926), acceleration of space rockets by land vehicles, airplanes and dirigibles (1926), jet aircraft (1930), and in 1935, the year of his death, a paper on multistage rockets in tandem and in parallel configurations (similar to the clustering of rocket boosters employed by the standard Soviet launch vehicle used since Sputnik). In a letter to the Central Committee of the Communist Party, Tsiolkovskii bequeathed his work "to the Bolshevik Party and the Soviet Government."[12]

Tsiolkovskii was by no means the only man investigating rocketry and space travel in Russia or elsewhere. In St. Petersburg in 1904 Ivan V. Meshcherskii completed his doctoral thesis on *The Dynamics of a Point of Variable Mass*, which examined the acceleration of a rocket under condition of loss of mass through consumption of fuel, considering the effects of thrust, gravity, and air resistance.[13] In Riga in 1907 the German-Russian Fridrikh A. Tsander (1887–1933) began a "Spaceship Notebook" of ideas on building a spacecraft. In 1908 Tsander organized the Riga Student Society of Space Navigation and Technology of Flight. As a student at Riga Politechnic Institute, Tsander's early efforts compare with those of Goddard, who in 1909 began his own calculations of rocket flight at Clark University in Massachusetts. Also like Goddard and unlike Tsiolkovskii, Tsander's work extended to the actual fabrication of working rockets.[14]

In 1919, the year Goddard published the scholarly treatise *A Method of Reaching Extremely High Altitude*, which introduced the concept of multistage rockets, the Soviet Government made the first investment in the space age with the founding of the Zhukovskii Academy of Astronautics with its advanced facilities for teaching, research and experiment.[15] The Academy was named for the Russian aviation pioneer Nikolai E. Zhukovskii, who founded in 1918 the companion *Tsentralnii Aero-Gidrodinamicheskii Institut* (TsAGI), which corresponds roughly to the National Advisory Committee for Aeronautics (the forerunner of NASA) established by President Wilson in 1915.[16]

By the mid-1920s a pattern emerged in which the Soviet Government as indicated above took an active interest in rocket travel. Abroad Goddard and Oberth were working largely as self-supported individuals apart from any organized framework or government funding.* Tsiolkovskii also continued to work this way in Russia, although he corresponded with Tsander and others. Another Russian researcher, Iurii V. Kondratiuk (1897–1942), who in 1929 published *The Conquest of Interplanetary Space and Rockets*, also evidently worked in relative isolation.[17]

The Soviet Government in 1924 created and supported the Central Bureau for the Study of the Problems of Rockets (TsBIRP) with the objectives of:

1. Bringing rocket researchers together;
2. Obtaining the results of research abroad;
3. Disseminating and publishing correct information on the current position of interplanetary travel;
4. Engaging in research and study, particularly of the military application of rockets.[18]

A group more similar in nature to the voluntary American Rocket Society, but founded six years before it and three years before the German Society for Space Travel, was the All-Union Society to Study Interplanetary Communications ("interplanetary communications" is the literal translation of the Russian expression for "space travel": *mezhplanetarie sviazy*). The group, OIMS, had goals similar to TsBIRP and was divided into sections for research, popularization and propaganda. Among its 150 members were Tsiolkovskii, Tsander, and the popularizer of science and astronautics, Ia. I. Perelman.[19]

The extent to which popular and governmental interest in space and rocket research in the Soviet Union exceeded that of other countries in this period may be accounted for by three factors. First, the atmosphere of experimentalism, modernism and futurism in the new land of "scientific socialism" favored this kind of interest in science as it did in the arts. Second, as the program of TsBIRP shows, the government, considering itself with some justification to be surrounded by enemies, would be interested in any expedient that could bolster its military strength. Last, Tsiolkovskii's works were undoubtedly a stimulus to Russian interest in space.

Following the publication of Oberth's *The Rocket into Interplanetary Space* (1923) and the first successful launching of a liquid-fueled rocket by Goddard in 1926, a space exhibition was held by the Moscow Association of Inventors in 1927 featuring designs of Tsiolkovskii, Tsander and others.[20] In 1928 the Soviet Government took over the sponsorship of the Moscow

---

*Oberth was employed as a professor. Goddard was able to secure some help from private foundations. See Daniloff, pp. 18, 20.

laboratory set up by the rocket scientist Nikolai I. Tikhomirov in 1921, renaming it the Gas Dynamics Laboratory (GDL) of the Military Scientific Research Commission.[21] A companion to GDL for solid fuel engines was established by Tikhomirov's assistant, Vladimir A. Artemev, in Leningrad the same year. In 1929–1930, at a time when the suspicious atmosphere of the Stalin purges was already rife, OIMS and TsBIRP were disbanded and some of their members accused of crimes against the state.[22] GDL with its military connections survived the purges, although it was under the command at that time of the ill-fated General (later Marshal) M. N. Tukashevskii. Also in 1929 Tsander, Ia. I. Perelman and others established the Group for the Study of Reactive Motion (GIRD). GIRD was divided into branches: *Mosgird, Lengird* and affiliated GIRDs throughout the Soviet Union. They became more politically secure successors to OIMS and TsBIRP by virtue of their ties to OSOAVIAKHIM, the paramilitary society for the promotion of air defense.[23] While GDL focused on the development of engines, including the first liquid-fueled Russian rocket motors, *Mosgird* worked on design, construction, and flight testing of rocket vehicles.[24] The man in charge of the development of liquid-fueled engines for GDL has been identified as V. P. Glushko (b. 1906), the "Chief Designer of Engines" for the Soviet space program and referred to only by this title or by an alias in published documents to this day.[25] At the same time the Ukrainian Sergei P. Korolev (1907–1966), later the "Chief Designer of Spacecraft" and the chief of Soviet space personnel until his death, was working in *Mosgird* and became its chairman upon Tsander's death in 1933. Another member of *Mosgird* was Mikhail K. Tikhonravov (b. 1900), later the nameless "Chief Theoretician" of the Soviet space program. *Lengird* had about forty members. Tukashevskii secured it a grant of R14,000 (ca. $2,000) in 1931 and a promise of more.[26]

With the approval of Tukashevskii and Stalin's deputy Grigorii Ordzhonikidze, sections of GIRD and GDL were merged in 1934 to form the Jet Scientific Research Institute (RNII) aimed at developing solid and liquid-fueled rockets for military use.[27] In 1935 the GIRDs, which had flown several experimental rockets as high as 16,000 feet, were disbanded, and all work on rocketry was clamped behind the bars of military secrecy. Many previously published studies were withdrawn from open circulation.[28] Dating the inception of a high-priority military rocket program in the Soviet Union from the formation of RNII in early 1934 would put the Soviets five years behind the Germans and eight years ahead of the United States in this regard.[29]* During the late 1930s Soviet rocketry suffered the effects of the

---

*Perhaps only two years behind the Germans in that while the *Verein für Raumschiffahrt* was founded in 1929, it was not until 1932 that the German Army took an active interest. See Stoiko, p. 68.

purges as a result of the general terror and probably in part as a result of its association with Tukashevskii. Many rocket pioneers were arrested or shot, and some, probably including Korolev, continued their work in special prisons.[30] The Soviet Armed Forces made extensive use of rocket weapons in the subsequent war, including the "Katiusha" artillery rockets, antiaircraft rockets, rocket-assisted takeoff of airplanes, and a flying bomb purportedly designed by S. P. Korolev and similar to the German V-1.[31] During World War II rocketry was unquestionably developed furthest in Germany.

Working at Peenemünde, Korolev's German counterpart, Wernher von Braun (b. 1912), and Major General Walter Dornberger led the group that developed the A-4 rocket, better known as the V-2. The A-4, first successfully flown in 1942, is a landmark of the space age comparable to Goddard's first liquid-fuel rocket, Sputnik and Apollo. In thrust and range the A-4 was in a class beyond any comparable weapon developed during the war and the immediate period after. Some 3,300 were launched during the war. As successors to the A-4, the Germans had plans for the A-9/10 ICBM and outlines of long-range plans for earth satellites, moon rockets and interplanetary travel.[32]

The German Scientists

In the wake of the launching of Sputnik there were some in the United States who attributed the Soviet success to the work of "German scientists" captured by the Russians at the war's end. While the USSR was clearly ahead of the United States in rocket theory and technology in 1945, both countries were far enough behind Germany so as to make the difference academic if one side was in a position to exploit German efforts while the other did not. This did not happen. When the Russians occupied Peenemünde, the production and experimental facilities had been dismantled and removed, and the leading German scientists and technicians had left. About 150 of the German experts, including Dornberger and von Braun, arranged their surrender to the Americans through von Braun's brother, Magnus. These men, along with much of the German equipment and documentation, were rounded up by the American "Operation Paperclip." Meanwhile, the Russian and British armies were conducting similar operations of their own.[33] Perhaps the biggest Soviet catch was the German scientist "Mikhail" Iangel, who later became an important figure in the Soviet space program after the other German scientists had left the USSR.[34] "Paperclip" succeeded in removing the central A-4 underground production center from the Soviet zone before the area was occupied by Red Army troops.[35]

While the US probably got more in terms of brainpower and equipment, both sides captured A-4s in various stages of construction, sufficient men and blueprints to build more, and German plans for more advanced vehicles.[36] Both were then in a position to equal the German efforts and to carry them forward as they saw fit. The "German scientists" undoubtedly contributed more to the United States, which got the best of them and had less of its own expertise and organization in advanced rocketry.

POSTWAR DEVELOPMENTS

As evidenced later by the detonation of an atomic bomb by the Soviets in 1949, they were anxious in 1945 and after, despite the devastation of their country, to press ahead with the development of strategic weapons. Friction had developed among the erstwhile Allies even before the war's end. The Soviet leaders were evidently unhappy about a situation in which the Americans had monopolies on the atomic bomb and the long-range bomber (B-29) capable of delivering it. Moreover, they may have been concerned that the US would unite the further development of German rocket technology with American nuclear technology, making the USSR extremely vulnerable to attack or the threat of attack with no means of retaliation.[37] The A-4 itself was of little use in this regard since it had a range of only about 200 miles with a warhead considerably lighter than the primitive A-bomb.

In 1945* the Russians rebuilt the A-4 production line at Peenemünde and set their Germans to working on it. In 1946 they moved the whole outfit to the USSR.[38] Evidently the Soviet leaders drew little comfort from the testimony before a Senate committee of the US wartime director of the Office of Scientific Research and Development, Dr. Vannevar Bush, that the building of an ICBM was so far beyond the range of current capabilities that even thinking about it would be undesirable.[39] This author does not know what the reaction of Soviet leaders and rocketeers was to American Air Force and Rand Corporation proposals in 1945 and 1946 to develop A-4s into intercontinental missiles and employ them to launch earth satellites or to the lack of action on these proposals.[40] The Rand study was remarkably prescient in its prediction that the first artificial satellite to orbit the earth would "inflame the imagination of mankind" and that the launching state would be "acknowledged as the world leader in both military and scientific techniques." "[O]ne can imagine the consternation and admiration," Rand said, "that would be felt if the US were to discover some other nation had already put up a successful satellite."[41] The US Air Force in the spring of 1946 did provide $1.4 million to Convair Inc. for the first work

*Vladimirov reports that in 1945 Korolev was released from prison. See *The Russian Space Bluff*, p. 40.

on what later became the Atlas ICBM, but the project was canceled in an economy move that summer and not renewed until 1952.[42]

The USSR was acting with dispatch. In 1946 the Council of Ministers established an Academy of Artillery Sciences (later the Academy of Rocket and Artillery Sciences) under the Soviet Ministry of Defense. The director of the new Academy was Lt. General Anatolii A. Blagonravov, later the official spokesman for the USSR in its foreign space relations.[43] At this time the Russians must have set to work on improving the A-4, while continuing to produce unimproved models and launching them in tests as was then the practice in the United States. Working under Iurii A. Pobedonostsev, a former member of TsAGI, GIRD and RNII, a German-Russian team had developed by 1948 an improved rocket known as the T-1, or *Pobeda*, a rocket with a range of 560-700 miles comparable to the US Redstone developed by the von Braun team for the Army in 1951. *Pobedas* were in production by 1949 and deployed with A-4s to the first Soviet Rocket Division as operational weapons in 1950–1951.[44]

Some information on the organization and goals of the Soviet rocket program has come from Colonel Grigorii A. Tokaty-Tokaev, the Red Air Force officer-scientist and chief of the aerodynamics laboratory of the Moscow Military Air Academy who defected to Britain in 1948.[45] Tokaty-Tokaev reports two high-level Kremlin meetings of April 14 and 16, 1947, the first attended by Georgii M. Malenkov, Politburo member and vice-chairman of the Council of Ministers in charge of aircraft production, chief of GOSPLAN Nikolai A. Voznesenskii, Air Marshal T. F. Kutsevalov, Armaments Minister Dmitrii F. Ustinov, aircraft designers Artem I. Mikoian and A. S. Iakovlev, Tokaty-Tokaev himself and others. Discussions included the possibilities of developing intercontinental nuclear missiles and rocket-boosted long-range bombers on the pattern of the German Sanger Project. Tokaty reports he suggested that the missile could also be used to launch satellites.[46]

The second meeting was a joint session of the Politburo and Council of Ministers attended, among others, by Stalin, Molotov, Anastas Mikoian, Zhdanov, Beria, Beria's deputy Colonel General Ivan A. Serov, Voznesenskii and Voroshilov. This meeting resulted in a secret Council of Ministers' decree of April 17 to investigate and begin work on the ICBM and Sanger projects. A commission was formed to do this, chaired by Serov, with Tokaty-Tokaev as deputy chairman and members including Mstislav Keldysh of the Armaments Industry (later president of the USSR Academy of Sciences and an important spokesman on space), M. A. Kishkin of the Ministry of Aircraft Production, and Stalin's son, Major General Vasilii I. Stalin. The commission was titled the Governmental Commission for Long-

Range Rockets and ordered to return a feasibility study to the Council of Ministers by August 1, 1947.[47]

It is likely that the report and direction of this commission facilitated subsequent developments in Soviet rocketry. A 2,000-mile-range IRBM, the T-2 or model 103, was authorized in 1949 and went into production in 1952.[48] Tokaty reports that he made a proposal for a three-stage satellite-launching vehicle in September, 1947, but his proposal was not acted on. He says that this and outside interference with his group contributed to the disaffection among rocket cadres which in his case led him to defect. He believes that if his proposals had been accepted, a Soviet satellite could have been launched sometime during the period 1950–1952.[49] This would probably have been accomplished by a multistage missile similar to the Viking rocket that launched the US Vanguard and much smaller than the ICBM used to launch Sputnik. Proposals by the US Navy and Air Force to develop satellite programs were likewise rejected by a Department of Defense evaluation group in 1948 on the grounds that their military utility would not be commensurate with their costs. One may presume that the same reasons had been given Tokaty.[50]

In spite of Colonel Tokaty's difficulties, the Soviet leaders' focus on rocket development for presumably military purposes did not extend to the total exclusion of the application of rocket technology to scientific pursuits. Smaller rockets were used for geophysical and meteorological probes beginning in 1946. In 1947 A-4s were used for experiments in the upper atmosphere, and in 1949 Soviet scientists were permitted use of the large *Pobeda* rockets. The first Soviet experiments in bioastronautics were evidently carried out in 1949–1952, consisting of the launching of dogs to altitudes of up to 400 kilometers.[51]

It was probably sometime in the 1949–1952 period that designs were settled on for the Soviet ICBM and the earliest plans made for its possible use as a launch vehicle for satellites and manned vehicles. Some indication of this came in the October 2, 1951, issue of the children's newspaper *Pionerskaia Pravda*. There former *Mosgird* member and member of the Academy of Artillery Sciences, Mikhail K. Tikhonravov, who at some point became the "Chief Theoretician" of the Soviet space program, wrote that the Soviet Union would be carrying out manned rocket flights in the period 1961–1966.* The next major indication of Soviet intentions came at the World Peace Council in Vienna, where, on November 27, 1953, A. N. Nesmeianov, then president of the USSR Academy of Sciences, said, "Science has reached a state when it is feasible to send a stratoplane to the

_____
*The first manned orbital flight was Iurii Gagarin's on April 12, 1961.

moon, to create an artificial satellite of the earth . . ."[52] Nesmeianov would almost certainly have been aware of the state of Soviet space technology and plans for the future.

In the United States the Eisenhower Administration did give top priority to developing an ICBM in 1954,[53] and the Army had started work on Project Orbiter, a satellite to be launched by the Redstone missile developed by the von Braun group (Army Ballistic Missile Agency, hereinafter referred to as ABMA). The Naval Research Laboratory was also working on what later became Project Vanguard, the first official US satellite program. The Secretary of Defense, Charles Wilson, put little stock in orbiting satellites and said he would not be alarmed if the Soviets did it first.[54]

The USSR in 1954 saw the publication of several articles on space flight, including a section of the *Bolshaia Sovetskaia Entsiklopediia* by M. K. Tikhonravov entitled "Interplanetary Communications" (Volume 27). As noted above, 1954 also saw the establishment of the K. E. Tsiolkovskii Gold Medal. Another article in the July *Tekhnika Molodezhi* included a section by Iu. Khlebsevich describing exploration of the moon by an automatic tankette laboratory in communication with earth. This is precisely what *Lunakhod* did in 1971. About this time the astronautics section of the Central Chkalov Aeroclub of the USSR was formed to facilitate the realization of cosmic flights for peaceful purposes. Khlebsevich was among its charter members.[55]

It was also in 1954 that the design of the Soviet ICBM, submitted by a group led by S. P. Korolev and his deputy L. A. Voskresenskii, was accepted and authorized for production. The engines, probably developed at the Leningrad GDL under V. P. Glushko, were then in their final stages of work.[56] In April, 1955, the *Vestnik* (*Herald*) of the Academy of Sciences announced the formation of the permanent Interdepartmental Commission for Interplanetary Communication (ICIC) under the Academy's Astronomy Council. Among the members were leading space scientists and prominent academicians from other fields. Leonid I. Sedov was the chairman and Tikhonravov his deputy. Among the members were physicist Piotr L. Kapitsa, mathematician N. N. Bogoliubov, General Blagonravov, astronomer V. A. Ambartsumian, Iurii Pobedonostsev, the rocket pioneer, and Major General Georgii I. Pokrovskii, an explosives expert and space publicist.[57] Tokaty-Tokaev claims that the commission was in fact formed years earlier but only announced at this time.[58] The commission serves in part as the public face of the Soviet space program and may have been formed partly in response to the International Geophysical Year (1957–1958). The IGY developed out of discussions in 1950 at the home of the American physicist James Van Allen. The Special Committee for the IGY (CSAGI) invited the Soviets to participate. The Soviet Union agreed to participate after the deadline for submitting plans in May, 1954. Meeting in Rome in September-October, 1954, CSAGI suggested the launching of satellites as part of the scientific investigations of the IGY.[59] This was also prominently identified in the April, 1954 *Vestnik* as "one of the immediate tasks" of ICIC.[60]

One may surmise that by this time the Soviet satellite program was firmly under way. The launch vehicle made available would be the first Soviet ICBM which was already in its initial stages of development. Discussions of what scientific use to make of the vehicle were probably in progress, including the plans for Sputnik I and several of its successors. This became clear in 1955 with the publication in the USSR of several articles on the subject of artificial earth satellites. In *Krasnaia Zvezda* (August 7, 1955), the organ of the Red Army, Professor K. P. Staniukovich described an "Artificial Earth Satellite" very like Sputnik I.[61] Similar articles were written by A. G. Karpenko (*Moskovskaia Pravda*, August 14), G. I. Pokrovskii (*Izvestiia*, August 19), and Leonid Sedov (*Pravda*, September 26).[62] Sedov, the only member of the first Soviet delegation to an annual Congress of the International Astronautical Federation (IAF) in August, 1955, announced that "it will be possible" to launch an artificial earth satellite "within the next two years."[63] Sputnik I was launched two years and two months later.

Sedov's comments closely followed the first American announcement (July 29) of the intention to launch a satellite as part of IGY.[64] The CIA had warned the Eisenhower Administration that the Soviets might achieve this first and recommended haste.[65] In September a Defense Department group met to consider proposals by the Army, Navy and Air Force to conduct the launch. The Army proposal, Project Orbiter, was based on the efforts of the von Braun team to create a military ballistic missile and would employ the operational Redstone rocket with upper staging added to achieve orbital velocities. Use could also be made of the larger Jupiter rocket nearing final stages of development. The Army promised a launch date of January, 1957, saying this could match or beat the Russian program. In opting instead for the Vanguard program of the Naval Research Laboratory based on the nonmilitary Viking rocket (smaller than Redstone and Jupiter and *much* smaller than the Soviet ICBM), the Eisenhower Administration put little stock in competing with the USSR for an early launch date, choosing instead to make a rigid separation between its IGY program, a civilian-scientific effort, and its ICBM program, a military weapons technology effort.[66]* Work began on Vanguard in October.[67] The

---

*This rationale lost some of its cogency when the Vanguard program was later classified secret. See Shelton, p. 50.

target launch date was 1958. Evidently the announcements from the White House and the National Academy of Sciences, along with stories in the Western press, gave the Soviets the opportunity to compare their own capabilities with those of the Americans and to take these into account in formulating their plans. In his remarks at the IAF Congress in August, L. I. Sedov had implied that the Soviet satellite would be larger than the American.[68] Moreover, Korolev evidently informed the Soviet political leaders that the Soviet satellite could be larger and launched sooner as part of his proposal for Sputnik.[69] This very probably helped to secure the support of Khrushchev and other leaders for the project.

This consideration, along with the effort to develop *some* kind of ICBM capable of carrying nuclear warheads to the United States, probably encouraged Soviet political, military and economic leaders to go ahead with the development of the missile now variously known as "A," SS-6, Sapwood and *Vostok,* which for strictly military purposes is something of a "turkey." First, the Sapwood (*NATO designation*) is a larger vehicle than was necessary to launch the smaller thermo-nuclear warheads made possible by advances in 1952–1953.[70] Second, its reliance on nonstorable cryogenic fuels makes launching it a long and awkward process during which it is especially vulnerable in conditions of war. Third, its clustered configuration makes it inefficient in thrust-lift ratio and difficult to fit in a silo, and finally, its radio guidance makes it subject to electronic "spoofing" by an opponent.[71]

Undoubtedly these considerations played a part in the decision by Soviet leaders not to deploy the Sapwood extensively as an ICBM, which gave rise to the imaginary "missile gap" of 1959–1962. None of these factors, however, constitute significant disadvantages for the use of the Sapwood as a space booster, and it continues to be used as such today (e.g., to launch *Soiuz* manned vehicles). The bomb-carrying versions have been dismantled or transferred to space uses.[72] It was only after 1964 that the United States had available more advanced space boosters of thrust ranges comparable to the Sapwood (Saturn I and Titan III, which use this thrust much more efficiently).

The Soviet military planners may have been satisfied to have this bulky vehicle at an early date, whatever its drawbacks as an operational weapon. Certainly the men who looked forward to scientific investigation of space must have been delighted to be able to make use of it. One may imagine of these men that if they were anything like Tsiolkovskii, Goddard, Oberth and von Braun, they were similarly infected by dreams of space travel. Korolev's biographers indicate this was true of him,[73] and the same could probably be said of other Soviet space scientists, especially those whose interest in the field went back to the days of GIRD, OIMS and TsBIRP, when their own interests, not lavish government projects, spurred progress in space science.

Apparently there was not such a happy confluence of wills in the American space program at this time. The Army and the Air Force, having been shut out of Project Vanguard, continued to press for satellites and more advanced space programs. The Army, which had successfully tested the Jupiter missile in 1956, pressed on two occasions for permission to launch a satellite.[74] The rocket men (ABMA) and the military men evidently were in agreement (subject to rivalry between the services actually encouraged by the Department of Defense). The political branch took a different view. Lt. General James M. Gavin, who was responsible for Brig. General Medaris' ABMA team in Huntsville, Alabama, was given written orders by Defense *not* to attempt a satellite launch with the Redstone or Jupiter rockets.[75]

In 1956 the Soviet Academy of Sciences applied for membership in IAF. Sedov went to the annual Congress in September and was elected a vice-president of the Federation. That same month at the fourth general meeting of CSAGI in Barcelona, Vice-President of the Academy of Sciences and President of the Soviet IGY Committee I. P. Bardin announced in terse general phrases that the USSR was preparing launch of a satellite as part of its investigations during the IGY. The satellite would measure temperature, pressure, cosmic rays, micrometeorites, geomagnetism and solar radiation.[76] A more comprehensive report of the USSR's plans was submitted by Bardin to the CSAGI the following June.[77]

In May of 1957 the astronomer A. A. Mikhailov disclosed the satellite's orbital period of 90 minutes and gave instructions on visual tracking. In June A. N. Nesmeianov, who had announced the Soviet plan for a satellite in 1953 (see above), indicated that all the "technical difficulties" involved in launching a satellite had been overcome and that the necessary "apparatus" had been created.[78] In August the Soviet Government announced the successful completion of a series of tests of a 6,000-mile-range ICBM.[79] (This was the "apparatus" of which Nesmeianov spoke). Days before the launch the frequencies at which the satellite would broadcast were announced. Then, on October 4, 1957, the world's first artificial satellite was launched. A pattern was set for subsequent Soviet space firsts, including the first rocket to the moon (1959), the first interplanetary rocket (1960), the first man in space (1961), the first woman in space and group flight (1963), the first multimanned flight (1964), the first space walk (1965), and the first orbital space station (1971).

Perhaps no space achievement has had the impact of Sputnik I, and it is from October 4, 1957, that the age of space and the space race can be dated.

NOTES

[1]Quotations from *New York Times*, October 6, 8, 9, 1957; *Time*, October 21, 1957, and *Newsweek*, October 21, 1957.

[2]Quotations from *Newsweek, op. cit.,* and *New York Times*, October 6, 7, 8, 9, 1957.

[3]*New York Times*, October 6, 8, 1957.

[4]*New York Times*, October 7, 1957, and *US News and World Report*, October 18, 1957.

[5]Michael Stoiko, *Soviet Rocketry: Past, Present and Future* (New York: Holt, Rinehart & Winston, 1970), pp. 1-3, 5-6.

[6]*Ibid.,* pp. 8-10.

[7]*Ibid.,* p. 11.

[8]Avrahm Yarmolinsky, *Road to Revolution, A Century of Russian Radicalism* (New York: Collier Books, 1962), pp. 274, 276.

[9]Tsiolkovskii's work is described in Stoiko, *op. cit.,* pp. 17-24; and in Nicholas Daniloff, *The Kremlin and the Cosmos* (New York: Alfred A. Knopf, 1972), pp. 11-19. His biography is A. A. Kosmodemianskii's *Konstantin Tsiolkovskii, His Life and Work* (Moscow: 1956 and 1960, in Russian).

[10]"Gold Medal Established for Outstanding Work in the Field of Interplanetary Communications," *Pravda*, September 25, 1964, in F. J. Krieger, *Behind the Sputniks, A Survey of Soviet Space Science* (Washington, D.C.: Rand Corporation, Public Affairs Press, 1958), Appendix A, p. 329.

[11]Stoiko, *op. cit.,* pp. 18, 19, 20.

[12]*Ibid.,* pp. 20-24; and Daniloff, *op. cit.,* pp. 17, 19.

[13]Stoiko, *op. cit.,* p. 22; and F. J. Krieger, *Soviet Space Experiments and Astronautics*, The Rand Corporation, Paper P-2261, March 31, 1961, p. 1.

[14]Daniloff, *op. cit.,* pp. 17, 21.

[15]*Ibid.,* p. 17; and Stoiko, *op. cit.,* p. 30.

[16]Stoiko, *op. cit.,* pp. 27, 30; and John M. Logsdon, *The Decision to Go to the Moon: Project Apollo and the National Interest* (Cambridge, Mass.: MIT Press, 1970), p. 53.

[17]Stoiko, *op. cit.,* pp. 33, 36-40.

[18]*Ibid.,* p. 31.

[19]*Ibid.,* pp. 31, 33.

[20]Daniloff, *op. cit.,* p. 24.

[21]*Ibid.,* p. 29; and Evgeny Riabchikov, *Russians in Space*, edited by Col. General Nikolai P. Kamanin, translated by Guy Daniels, prepared by the Novosti Press Agency Publishing House, Moscow (Garden City, N.Y.: Doubleday & Co., Inc., 1971), p. 122.

[22]Daniloff, *op. cit.,* p. 29; and Stoiko, *op. cit.,* pp. 32, 41, 43.

[23]Krieger, *Behind the Sputniks . . . , op. cit.,* p. 1; and Daniloff, *op. cit.,* p. 26.

[24]Stoiko, *op. cit.,* p. 46.

[25]Riabchikov, *op. cit.,* p. 123.

[26]Daniloff, *op. cit.,* pp. 26-29, 58; and Stoiko, *op. cit.,* pp. 62-64. See also Leonid Vladimirov, *The Russian Space Bluff* (New York: Dial Press, 1973), which identifies the "Chief Theoretician" as Mstislav Keldysh, now president of the Soviet Academy of Sciences (p. 67).

[27]Daniloff, *op. cit.,* p. 30.

[28]Stoiko, *op. cit.,* p. 54.

[29]Krieger, *Soviet Space Experiments and Astronautics, op. cit.*, p. 2.

[30]Vladimirov, *op. cit.*, pp. 24-40.

[31]Daniloff, *op. cit.*, p. 33.

[32]Stoiko, *op. cit.*, pp. 70-71.

[33]*Ibid.*, pp. 70-72.

[34]Vladimirov, *op. cit.*, pp. 43, 47, 49.

[35]Martin Caidin, *Red Star in Space* (New York: Crowell-Collier Press, 1963), p. 94.

[36]*Ibid.*, pp. 95-96; William Shelton, *Soviet Space Exploration* (New York: Washington Square Press, Inc., 1968), p. 5; von Braun in Jerry and Vivian Grey, eds., *Space Flight Report to the Nation* (New York: Basic Books, Inc., 1962), p. 170; Robert W. Buchheim, *New Space Handbook: Astronautics and Its Applications* (New York: Vintage Books, 1963), pp. 285-286.

[37]See Shelton, *supra*, pp. 3, 39; and Tokaty-Tokaev in Daniloff, *op. cit.*, pp. 219-221.

[38]Stoiko, *op. cit.*, p. 72.

[39]US Congress, Senate, Special Committee on Atomic Energy, *Atomic Energy, Hearings . . . Pursuant to Senate Resolution 179*, 79th Congress, 1st Session, 1945, Part 1, p. 179 in Vernon Van Dyke, *Pride and Power, the Rationale of the Space Program* (Urbana: University of Illinois, 1964), p. 10.

[40]Theodore von Karman as quoted in *ibid.*, p. 10; and Joseph M. Goldsen, "Outer Space in World Politics" in Goldsen, editor, *Outer Space in World Politics* (New York: Frederick A. Praeger, 1963), p. 5.

[41]Quoted by R. Cargill Hall, "Early US Satellite Proposals," in Eugene M. Emme, editor, *The History of Rocket Technology* (Detroit: Wayne State University Press, 1964), p. 70 as cited in John M. Logsdon, *The Decision To Go to the Moon: Project Apollo and the National Interest* (Cambridge, Mass.: MIT Press, 1970), p. 44.

[42]Daniloff, *op. cit.*, pp. 42-43.

[43]Philip J. Klass, *Secret Sentries in Space* (New York: Random House, 1971), p. 26.

[44]Daniloff, *op. cit.*, p. 44; Stoiko, *op. cit.*, pp. 59, 75; Krieger, *Behind the Sputniks . . . , op. cit.*, p. 2; and Buchheim, *op. cit.*, p. 286.

[45]For his published writings, see Daniloff, *supra.*, p. 258.

[46]*Ibid.*, p. 49; Krieger, *Behind the Sputniks . . . , op. cit.*, p. 2; and Merle Fainsod, *How Russia Is Ruled* (Cambridge, Mass.: Harvard University Press, 1963), pp. 317-318.

[47]Daniloff, *supra*, pp. 49-50; and Stoiko, *op. cit.*, p. 73. See Tokaty-Tokaev, *Comrade X*, trans. by Alec Brown, (London: Harvill Press, 1956).

[48]Stoiko, *op. cit.*, p. 59; and Krieger, *Behind the Sputniks . . . , op. cit.*, p. 2.

[49]Stoiko, *supra*, p. 74; and Arthur C. Clarke in Grey, *op. cit.*, pp. 179-180.

[50]Van Dyke, *op. cit.*, p. 12.

[51]Daniloff, *op. cit.*, p. 52; Stoiko, *op. cit.*, p. 77; and Krieger, *Behind the Sputniks . . . , op. cit.*, p. 9.

[52]*Pravda*, November 28, 1953, p. 2.

[53]General Bernard A. Schriever, "Does the Military Have a Role in Space?" in Lillian Levy, editor, *Space: Its Impact on Man and Society* (New York: W. W. Norton & Co., Inc., 1965), p. 66; and Edwin Diamond, *The Rise and Fall of the Space Age* (Garden City, N.Y.: Doubleday & Co., Inc., 1964), p. 19.

[54]Van Dyke, *op. cit.*, p. 13. See also Jay Holmes, *America on the Moon: The Enterprise of the Sixties* (Philadelphia: Lippincott, 1962), p. 49; and Shelton, *Soviet Space Exploration, op. cit.*, p. 50.

[55]Krieger, *Behind the Sputniks . . . , op. cit.*, pp. 3, 4, 17. The two articles appear in

English in *ibid.,* pp. 28-34, 35-49; Khlebsevich on pp. 46-49.

[56]Stoiko, *op. cit.,* p. 59; Daniloff, *op. cit.,* p. 55; and Vladimirov, *op. cit.,* pp. 34, 49, 68, 95.

[57]"Commission on Interplanetary Communications," *Vecherniaia Moskva,* April 16, 1955, in Krieger, *Behind the Sputniks . . . , op. cit.,* Appendix B, pp. 329-330; and Daniloff, *supra,* p. 56.

[58]Tokaty-Tokaev in Daniloff, *supra,* p. 224.

[59]*Ibid.,* pp. 54-55. See also International Council of Scientific Unions, *Annals of the International Geophysical Year* (London: 1958), p. 456.

[60]"Commission on Interplanetary Communications," *op. cit.,* pp. 329-330.

[61]Krieger, *Behind the Sputniks . . . , op. cit.,* p. 16. See K. Staniukovich, "Artificial Earth Satellite Principles," *Krasnaia Zvezda,* August 7, 1955, in *ibid.,* pp. 237-239. See also Shelton, *Soviet Space Exploration, op. cit.,* p. 47.

[62]Krieger, *supra,* p. 16; "On Flights into Space," an interview with L. I. Sedov, *Pravda,* September 26, 1955; A. G. Karpenko, "Cosmic Laboratory," *Moskovskaia Pravda,* August 14, 1955; and G. I. Pokrovskii, "Artificial Earth Satellite Problems," *Izvestiia,* August 19, 1955, all in *ibid.,* pp. 112-115, 240-242, 243-245.

[63]Shelton, *Soviet Space Exploration, op. cit.,* p. 47.

[64]Krieger, *Behind the Sputniks . . . , op. cit.,* p. 16.

[65]Daniloff, *op. cit.,* p. 57.

[66]Diamond, *op. cit.,* pp. 22-23; Van Dyke, *op. cit.,* p. 14. See *Project Vanguard, A Scientific Earth Satellite Program for the IGY,* Report to the Committee on Appropriations by Surveys and Investigations Staff, US Congress, House, Committee on Appropriations, Subcommittee on Department of Defense Appropriations for 1960, Hearings, 86th Congress, 1st session, 1959, Part 6, pp. 69-70. See also US President, *Public Papers of the Presidents of the US, Dwight D. Eisenhower, January 1 to December 31, 1957* (Washington, D.C., 1958), pp. 734-735.

[67]Frank Gibney and George J. Feldman, *The Reluctant Spacefarers: A Study in the Politics of Discovery* (New York: New American Library, 1965), p. 46.

[68]Shelton, *Soviet Space Exploration, op. cit.,* pp. 47, 50.

[69]P. T. Astashenkov, *Akademik S. P. Korolev* (Moscow, 1969), p. 119. See also Daniloff, *op. cit.,* pp. 101-102; and Vladimirov, *op. cit.,* pp. 52-58.

[70]Diamond, *op. cit.,* p. 18.

[71]*Soviet Space Programs, 1966-1970, Goals and Purposes, Organization, Resources, Facilities and Hardware, Manned and Unmanned Flight Programs, Bioastronautics, Civil and Military Applications, Projections of Future Plans, Attitudes Toward International Cooperation and Space Law,* Staff Report prepared for the Use of the Committee on Aeronautics and Space Sciences, US Senate, 92nd Congress, 1st session, Senate Document #92-51, December 9, 1971, pp. 160 and 559 (illustration). Afterward referred to as *Soviet Space Programs, 1966-1970.*

[72]*Ibid.,* p. 160.

[73]Astashenkov, *op. cit.;* and Alexander P. Romanov, *Designer of Cosmic Ships* (Moscow, 1969, in Russian). See Daniloff, *op. cit.,* pp. 257, 258. There is also much information about Korolev in Leonid Vladimirov's *The Russian Space Bluff.*

[74]*Project Vanguard . . . ,* Hearings, 86th Congress, House, *op. cit.,* p. 60.

[75]Shelton, *Soviet Space Exploration, op. cit.,* p. 54.

[76]Krieger, *Behind the Sputniks . . . , op. cit.,* pp. 5-7.

[77]I. P. Bardin (President, USSR National IGY Committee: Vice-President, USSR Academy of Sciences), "USSR Rocket and Earth Satellite Program for the IGY," submitted to CSAGI at Brussels, June 10, 1957, in *ibid.,* pp. 282-287.

[78]Krieger, *supra*, p. 10; and A. N. Nesmeianov (President, USSR Academy of Sciences), "The Problems of Creating an Artificial Earth Satellite," *Pravda*, June 1, 1957, in *ibid.*, pp. 276-281.

[79]TASS, "Report on Intercontinental Ballistic Report," *Pravda*, August 27, 1957, in Krieger, *supra*, pp. 233-234.

2   Organization of the Soviet Space Program

In contrast to the United States, the Soviet Union has never made public its space budgets, the organizations involved in the space programs and the names of its leading space officials. Two of its major launch sites have never been publicly acknowledged, and the location of the third site has been falsely identified.* In spite of this aura of secrecy, estimations of the level of effort of the Soviet space program and its organization can be made by putting together bits and pieces of public information and by making inferences from known facts.

LEVELS OF EFFORT

From October, 1957, through 1973 the USSR made some 708 successful space launches of 853 payloads to earth orbit and 47 payloads to the moon and beyond. The US totals are 599 launches, 770 earth satellites and 61 earth escapes. The Soviets have probably had in addition to this a number of failures on the order of the United States, which in this period suffered 96 launch failures of 120 satellites and 12 escape payloads, although they have never admitted such failures. No other country succeeded in launching more than eight satellites in this period. Each year after 1957 the US led the Soviet Union in total payloads until 1968, when the rising Soviet total surpassed the falling American total.[1] In comparing the numbers of payloads launched by the two countries, one should keep in mind that the US has generally been able to accomplish more per satellite than the USSR because of its superiority in electronics, rocket-engine technology and other fields. So, for example, American satellites often carry more equipment, more sophisticated equipment, and have longer useful lives in terms of data return than their Soviet counterparts.

*Recently the existence and general vicinity of a second launch site has been identified.

19

TABLE I
SCIENTIFIC SPACE PAYLOADS BY MISSION CATEGORY, 1966-1973

	1966		1967		1968		1969		1970		1971		1972		1973		Totals*	
	USSR	US	USSR	US	USSR	US	USSR	US	USSR	US	USSR	US	USSR	US	USSR	US	USSR	US
Earth Orbital Science	4	17	5	12	10	16	3	14	9	4	7	11	13	9	14	2	86	148
Earth Orbital Engineering		1		6		1		5		1		10	1	1	0	0	1	51
Earth Orbital Man-related	2	6	3	1	4	0		1	1	0	2	0	1	0	4	0	25	11
Earth Orbital Manned	0	5	1	0	1	0	5	0	1	0	3	0	0	0	2	4	21	18
Lunar Man-related	0	1	2	1	3	2	1	0	2	1	0	3	0	3	0	0	8	16
Lunar Manned						2		8		2		4		4		0		20
Lunar Unmanned	6	4		8	1	1	3	0	3	0	2	0	2	0	1	1	28	22
Mercury																1		1
Venus	0	0	2	1	0	0	2	0	2	0	0	0	2	0	0	0	19	2
Mars	0	0	0	0	0	0	0	2	0	0	5	1	0	0	4	0	14	5
Outer Planets														1		1		2
Interplanetary		1		1		1					0	0	0	0	0	0	5	5
Vehicle Tests		2		0		0					0	0	0	0	0	0		13
Totals including other categories	44	100	66	87	77	64	71	66	91	39	102	53	97	40	115	27	814	831

*Totals include previous years. Military and applications flights are accounted for in Chapters 3 and 10. Eighty-six Soviet orbital launch platforms put in orbit to fire earth-escape vehicles are not counted.

Sources: C. S. Sheldon II, "Overview, Supporting Facilities and Launch Vehicles of the Soviet Space Program," p. 120; *Soviet Space Programs, 1971*, p. 5; and Sheldon, *United States and Soviet Progress in Space, 1972*, p. 42, and *United States and Soviet Progress in Space*, 1973, p. 42.

As Table I reveals, the United States has mounted a more ambitious program in space science and engineering; the Soviets have engaged in a larger planetary effort and a rather larger effort in manned orbital missions. As is well known, the USSR has made no manned lunar flights like the American Apollo missions, although the number of unmanned and man-related lunar flights has been about equal.

The total Soviet space budget has been estimated at 1.4 to 2.0 per cent of GNP, or the equivalent of about $5 billion annually.[2] The 1964 Soviet space budget was estimated at R1.3 billion, with the value of the defense ruble estimated at 2.50 to 3.50 in US equivalent dollars.[3] The more neutral International Institute of Strategic Studies in London pegs the defense ruble at $2.00 to $2.50.[4] This would put the 1964 1.3 billion ruble space budget at $2.60 to $3.25 billion. The 1966 Soviet space budget has also been estimated at R1.3 billion,[5] which would lead to the same figures. It is probably safe to say that the total Soviet space budget in recent years has been the equivalent of between $3.0 billion and $5.0 billion a year. NASA's budget in recent years has declined from about $6.0 billion to around $3.2 billion, while the DoD space budget has climbed from about $1.6 billion to over $2.0 billion for a total of around $5.5 to $7.5 billion a year.[6] The NASA budget has been split approximately as follows: manned missions, 70-75 per cent; space research, 15 per cent; space applications, five per cent, and aeronautics, five per cent.[7] About 90 per cent of the budget goes to private contractors.[8] Clearly the US spends more, perhaps twice as much, but the Soviet space budget probably consumes a similar proportion of GNP, the Soviet GNP being about half the American.

Pegging the defense ruble at a higher ratio than that of the consumer ruble or the official exchange rate is the result of complex economic calculations which are based on the relatively greater productivity of the Soviet defense sector compared with the civilian sector. This results from diversion of the best men and techniques into the defense sector as well as from many subsidies and fiscal preferences. One result is that while the ruble cost of Soviet defense and space spending appears rather low, the *opportunity cost* is considerably higher.

This author is inclined to adopt the lower figures for Soviet space spending. To avoid complex questions relating to dollar or ruble figures, one may look at the physical output of the space programs, keeping in mind that the US bears a substantially lower opportunity cost for an equivalent output because of its greater wealth and productivity. The Soviets have not mounted anything comparable to Project Apollo, which absorbed some 75 per cent of NASA's budget in 1964–1968.[9] Moreover, their space successes have a "Potemkin village" aspect about them since most have involved the same obsolescent booster developed in 1957.[10] Of 708 launches through 1973 over three-fifths have used the old SS-6 Sapwood ICBM as a first stage. The Sapwood employs a crude but effective design with four rocket bodies strapped onto a central core. The total effect is rather like bundling five obsolescent US Redstone missiles together. One hundred and twenty-seven launches involved the much smaller and cheaper SS-4 Sandal MRBM (dispatched to Cuba in the fall of 1962), and 80 launches have used the intermediate SS-5 Skean IRBM. Only 31 launches employed the large "D" or *Proton* rocket and 37 the second-generation SS-9 Scarp ICBM as first stages. NASA officials have for years indicated that the Soviets are developing a superbooster with a lifting capacity in the Saturn V class, but it has never flown. Although the *Protons* and SS-9s began to be used in 1965 and 1966, the presumably cheaper SS-6 is still the mainstay of the Soviet space program. By contrast the US has in recent years made greater use of the more expensive Titan and very expensive Saturn I and V launch vehicles.[11]

Some comparison of effort of the US and the USSR can be gleaned from Table II, which matches launch vehicles and gives launch totals through 1973.

TABLE II
US AND USSR LAUNCH VEHICLES AND LAUNCH TOTALS THROUGH 1973*

LAUNCH VEHICLES	US	TOTAL	COMPARABLE SOVIET	TOTAL
Small, nonmilitary	Vanguard	3	None	
	Scout	56		
MRBM	Redstone	4	Sandal	127
IRBM	Jupiter	4	Skean	80
	Thor	294		
First-generation ICBM	Atlas	137	Sapwood	433
Second-generation large ICBM	Titan	79	Scarp	37
Large nonmilitary	Saturn I-B	12	*Proton*	31
Very large nonmilitary	Saturn V	13	None	
Totals		602		708

*Sources: Sheldon, "Overview, Supporting Facilities and Launch Vehicles of the Soviet Space Program," in *Soviet Space Programs, 1966-1970*, pp. 130-160, 511; *Soviet Space Programs, 1971*, p. 9; Sheldon, *United States and Soviet Progress in Space* (1973), p. 21. The 1973 figures are: Scout (1), Sandal (10), Thor (6), Skean (15), Atlas (4), Sapwood (53), Titan (8), Scarp (1), *Proton* (7), Saturn I (3) and Saturn V (1).

The strapping together of smaller rockets and serial production of the Sapwood ICBM probably make it comparatively cheap. Although this booster was very inefficiently employed to launch the first Sputniks, the addition of upper stages has improved its efficiency without substantially affecting the cost. As newer orbital and escape missions increased weight and velocity requirements, improved upper stages were merely piled on top of the old booster.[12]

LAUNCH SITES AND SUPPORT FACILITIES

The United States has two major spacedromes, and the USSR has three. The original "Baikonur" cosmodrome is the largest and, like Cape Kennedy, launches all the Soviet-manned, lunar and deep-space vehicles. It is located near Tiuratam in Kazakhstan about 230 miles from Baikonur. In 1966 the USSR began launches from near Plesetsk in northwest Russia. The Plesetsk launch site has never been officially acknowledged by the Soviets. It launches mostly military flights including the bulk of all Soviet flights in recent years and compares with the American Vandenberg Air Force Base. In 1962 smaller launch vehicles began flying from near the town of Kapustin Iar in the vicinity of Volgograd. This launch site has only recently been acknowledged. It is the base for small scientific and some military launches. It compares with lesser American facilities at White Sands, New Mexico, and Wallops Island, Virginia.[13] Without a superbooster such as the Saturn V, the Tiuratam site probably does not match Cape Kennedy in launch facilities for such a vehicle.

The USSR does not have as extensive a ground-based, worldwide tracking system as the US. It supplements its foreign-based tracking stations with a fleet of tracking ships operated in the Atlantic and Indian oceans by the Soviet Academy of Sciences. The three largest ships in the tracking fleet are the 11,000-ton *Kosmonavt Vladmir Komarov*, named for the late pilot of *Soiuz* 1, the 21,000-ton *Akademik Sergei Korolev*, and the giant 45,000-ton *Kosmonavt Iurii Gagarin*.[14] Military tracking ships in the Soviet Pacific missile range also track Soviet spacecraft. Unlike the US, the USSR does not have deep-space tracking facilities outside its borders except for the *Komarov*, *Korolev* and *Gagarin*. NASA maintains its own tracking stations and cooperates with those of several foreign countries. Overall, the American tracking capability is superior.[15]

At its peak the US space program employed some 600,000 people, most of them working for the more than 800 private contractors under contract to NASA and the military.[16] As of June, 1964, NASA itself employed approximately 11,300 scientists and engineers and its contractors 62,000, together about five per cent of the nation's total supply of engineers and scientists.[17] This gave rise to concern that the space program was putting too great a strain on the nation's scientific and technical talent, threatening an imbalance in the economy.[18] Such worries have been submerged by recent NASA budget cuts and the recession in the aerospace industry. It is not openly known how many people are employed in the Soviet space program, though the total has been estimated at 600,000. There is probably less concern in the Soviet Union about manpower shortages. The USSR graduates about three times as many technicians each year as the US.[19]

In the resources of money and productive capacity the Soviet space program is probably carried on at a greater relative cost than the American because the USSR is only about half as productive as the US and because the Soviet technological base is quite inferior to the American. One authority thinks that space expenditures resulted in a general lag in the Soviet economy from 1958 to 1965, especially in the civilian sector.[20] Another has suggested that agricultural problems have forced limits to the number of Soviet flights.[21] Indeed agricultural difficulties in the Soviet Union can be compared to the Vietnam War, which has had an inhibiting effect on NASA budgets; they have been lower every year since 1966. Perhaps these difficulties have kept the USSR from successfully completing development of a superbooster and joining the race to the moon.

It is hard to estimate the commitment of Soviet leaders to their space program except by noting the extent of what the USSR has accomplished in space at a substantial cost. As with other policies, the space program is not subject to the public criticism of the kind that goes on constantly in the United States. However, the Soviet leaders have been lavish with their praises and prizes for Soviet space workers and cosmonauts, as exemplified by the burial of Gagarin and Komarov in that holy of holies, the Kremlin Wall, and by the celebration of Cosmonautics Day, April 12, the anniversary of Gagarin's first manned flight.[22] Compared to the US, Soviet leaders are more willing to impose long-term sacrifices on a population which is more habituated and willing to accept them and certainly less free to do anything about it. Moreover, there is something of a cult of science in the Soviet Union which is associated with "scientific socialism" and an older Russian tradition in science. This cult favors such activities as space exploration and has no direct counterpart in the United States.[23]* The long-term economic plans in the Soviet Union, usually five or seven years, may be more favorable to long-term space plans than the annual budget struggles of NASA and DoD in the US.

*In a further justification of science spending, *Pravda* of February 4, 1969, claimed that "a ruble of expenditure on science and development gives an increase in national income greater than an investment in expanding production capital in the absence of technical progress."

ADMINISTRATION OF THE SOVIET SPACE PROGRAM

It will be recalled from the first chapter that the Gas Dynamics Laboratory (GDL) and the Interdepartmental Commission on Interplanetary Communications (ICIC) were important organizations in the pre-Sputnik Soviet space program. The Leningrad branch of the older GDL (1928) was and is under the direction of Academician Valentin Petrovich Glushko, who is the man now identified as the Chief Designer of Engines of the Soviet space program.[24] GDL can be compared to the Jet Propulsion Laboratory at Caltech.

As noted previously, ICIC was announced by the Academy of Sciences in April, 1955, and formed in 1954 or earlier. The first chairman of ICIC was Academician Leonid I. Sedov until his replacement by Evgenii Fedorov and later by Academician (and Lt. General) Anatolii Arkadevich Blagonravov, former president of the Academy of Artillery Sciences.[25] ICIC's members, as noted, included the most prominent Soviet academicians. ICIC was established under the Astronomy Council of the Academy.[26] The Soviet Academy of Sciences has approximately 600 full members and has been presided over by Mstislav Keldysh since 1961. Keldysh appears to have played some role in mathematical calculations for the Soviet space program. Blagonravov was the head of the Academy's Department of Technical Sciences.[27] The Soviet Academy of Sciences wields considerably more political influence than the American National Academy of Sciences. This applies generally and also as regards the space programs of the two countries. Within the limits of the Soviet power structure (of which it is a part) the Soviet Academy performs an important role in the planning and execution of the Soviet space program. By comparison the National Academy of Sciences is limited to an informal advisory function for NASA and the President.[28]

The name of ICIC was subsequently changed to the Commission on the Exploration and Use of Space (CEUS).[29] ICIC-CEUS was originally divided into five sections: (1) rocket technology, (2) astrophysics, (3) space navigation, (4) radio guidance, and (5) space flight biology.[30] The original vice-chairman was Mikhail K. Tikhonravov, the Chief Theoretician of the Soviet space program. ICIC-CEUS has been composed to a large extent of engineers with defense backgrounds who are now important managers and political leaders. This is clear with Academician Blagonravov, a three-star general and rocket expert; G. I. Pokrovskii, a major general,* nuclear physicist and explosives expert, V. F. Bolkovitinov, also a major

---

*Now, like Blagoravov, a Lt. General. See Beck, "Soviet Space Plans . . . ," in *Soviet Space Program, 1962-1966*, p. 355.

general and professor of Aeronautics, and Iu. A. Pobedonostsev, a colonel, aerodynamics professor and Hero of the Soviet Union.[31] Since the early 1960s 20 of the 31 members of the Academy's Department of Technical Sciences chaired by Blagonravov were generals and admirals.[32]

Besides CEUS, the Academy of Sciences also maintains a Space Research Institute previously directed by Sedov, later by Academician Georgii I. Petrov, and now by Academician Roald Sagdeev, which has responsibility for the design and construction of space vehicles; a Commission on Space Law chaired by the international law specialist E. A. Korovin and set up in 1960; a Council on Space Cooperation, or "Intercosmos" Council, chaired by Boris N. Petrov, and an Orbit Calculations Institute as well as other institutes which make partial or occasional contributions to the space effort.[33] As noted previously, the Academy is responsible for the tracking of Soviet space flights and also operates a space data collection and coordinating center comparable to NASA's Manned Spacecraft Center in Houston.[34]

Although accounts differ, there are some indications that the overall day-to-day direction of the Soviet space program is under the direction of a national commission or state committee at the ministerial level of the Soviet Government. Such committees proliferated under Khrushchev as a means of centralizing control, thus taking away with the left hand what the right hand had conceded to decentralization in the formation of the *sovnarkhozy*, or regional councils of the national economy.[35] This organization is possibly called the State Commission for Space Exploration, or State Commission for the Organization and Execution of Space Flight.[36] The commission is not an operational organization like NASA, but very likely one more like the National Aeronautics and Space Council (NASC), which draws in top leaders from the organizations involved in making and launching spacecraft as well as high officials of the government and the party and has its own staff.[37]* If such a commission did exist, it would be characteristic of the Soviets to keep it secret, just as they have done with the names of leading space scientists such as Korolev and Glushko.

The first chairman of this state commission is believed to have been Konstantin N. Rudnev, who held various managerial posts in defense production from 1948 to 1958, served as chairman of the State Committee for Defense Industries in 1958–1961, and then became chairman of the State Committee for the Coordination of Scientific Research. He is now Minister of Instrument-Making, Automation and Control Systems. The commission is said to include, besides the chairman, a deputy chairman, a launch director, the Chief Theoretician (Tikhonravov), the Chief Designer (formerly

---

*For a brief description of NASC, see below. NASC was dissolved in 1973.

Korolev) and several military representatives.[38] Evidently three of these positions were held by Korolev, who was Rudnev's deputy, the launch director and the Chief Designer.[39] Rudnev was among those named who received the Order of Lenin for his part in the Gagarin flight in 1961. Academy President Keldysh was also honored on this occasion.[40] The most prominent guests at the wedding of the cosmonauts Nikolaev and Tereshkova in November were Khrushchev, Rudnev, two marshals and the unnamed designers of spaceships (Korolev and Glushko).[41]

Rudnev was possibly replaced later by Academician Vladimir A. Kirillin, a former power engineer, who is also the former vice-president of the Academy of Sciences, a deputy prime minister and full member of the Central Committee since 1966.[42] Under Kirillin the State Committee for the Coordination of Scientific Research was abolished and reformed as the high-level State Committee for Science and Technology. If Kirillin did replace Rudnev on the Space Commission, then the Commission (if it exists) is probably organized under the State Committee in its own right. As chairman of the State Committee, Kirillin could be the ex officio chairman of the Commission. Kirillin's appointment and the reform of the State Committee were the result of the change of leadership in 1964–1965. The tasks of the State Committee are the financing and planning of science and the supplying of scientific institutions. It also attempts to foster the development and diffusion of new techniques in the economy and has charge over international scientific and technical exchange. Kirillin has contrasted the Committee to the Academy of Sciences by stressing its involvement in applied rather than basic research. President Keldysh is a member of the Committee, and Kosygin's son-in-law is the member of the Committee's Collegium, or executive council in charge of negotiations with foreign governments. Statements by the chairman and his deputy, Academician Vadim A. Trapeznikov, indicate that the Committee lobbies actively within the Council of Ministers for increased funding of science. The Committee since 1965 has played a role in coordinating work in atomic energy and developing computer technology. If the Committee does not now play the leading role in the space program, it is quite likely to do so in the future.[43]

In 1963 the Aerospace Technology Division of the Library of Congress attempted to identify the possible organization of the administration of the Soviet space program* and the organization of agencies responsible to the State Commission for Space Exploration. At the top stand the leading

*See *Management of the Soviet Space Program* (Washington, 1963); and **Holmfeld**, "Organization of the Soviet Space Program," in *Soviet Space Programs, 1966-1970* (US Senate document), pp. 87-90.

bodies of the Communist Party and Government, i.e., the Central Committee (and its Secretariat and Politburo) and the Council of Ministers (and its Presidium). Various organizations responsible to the State Commission would also be under the authority of the Supreme Council of the National Economy (since abolished), the State Commission for the Coordination of Scientific and Research Work (later reorganized), the Ministries of Defense, Communications and the Red Army, Air Force and Rocket Troops as well as various ministries associated with the service industry and machine-building industry. The Space Commission is also probably responsible to the State Planning Commission (*Gosplan*) for its expenses and procurements and is linked through its subordinate operational organizations to various state committees such as those for aviation technology, defense technology, radioelectronics, electronics technology, and automation and machine-building.[44] Since 1965 many of these state committees have been abolished in favor of ministries with the same or similar responsibilities. As indicated previously, the Academy of Sciences and many of its committees and research institutes are closely linked with the Soviet space program. Some review of space activities and appropriations is also probably carried out by the relevent committees of the Soviet parliament or Supreme Soviet. Five Soviet cosmonauts have been honored by nomination and election to the Supreme Soviet, and Gagarin served in each of its two chambers. Further support of the space programs comes from the universities, research institutes, and design bureaus attached to them and to other agencies such as the ministries and the Academy of Sciences.[45]

The membership of the State Commission for Space Exploration is said to include, in addition to the chairman of the State Committee for Science and Technology, the Chief Designer (formerly Korolev and subsequently Mikhail Iangel, an academician and aviation engineer with a background in the machine-building industry, who died October 25, 1971,* and was purportedly one of the German scientists who came to the USSR in 1945). Other members indicated are the Chief Theoretician (Tikhonravov), the director of the Academy's Orbit Calculations Institute (also Tikhonravov?), a Red Air Force representative (probably General Kamanin), Petr Ermolaevich and others.[46] V. P. Glushko would almost certainly be a member as well. The key role played by Korolev and Glushko is indicated by their treatment in published materials. After their election to the Academy as full members shortly after the first Sputnik, their names along with Tikhonravov's vanished from press reports and later editions of Soviet books on

*See *Soviet Space Programs, 1971*, p. 61.

space.* Glushko was formally listed as holding jobs at a provincial university and the Logging Institute (!) in Voronezh.[47] It was only after Korolev died that he (Korolev) became the beneficiary of lavish public praise and official recognition, although his identity had been previously discovered by the Western press.[48]

Other persons with leading roles in the Soviet space program may include Dmitrii F. Ustinov, a member of the Party Central Committee Secretariat, candidate member of the Politburo and important in space affairs by virtue of his past responsibilities for defense production.[49] Ustinov was purportedly the party's choice to replace Marshal Malinovskii (deceased) as Minister of Defense, but the military prevailed in forcing the choice of their own candidate, Marshal Grechko.[50] Among those publicly honored after the Gagarin flight were Khrushchev; Rudnev; Keldysh; Ustinov; Khrushchev's heir-apparent, Frol Kozlov, who died in early 1964; the chairman of the State Committee for Radioelectronics, Valerii D. Kalmykov, and the then President of the USSR, Leonid Brezhnev.[51] The space scientists and technicians themselves were decorated anonymously. Among the prominent academicians who have published works on space science and plans for the future are two not previously mentioned: Iu. S. Khlebsevich, a radio-astronomer with an interest in automatic space vehicles, and N. A. Varvarov, astronautical consultant to the USSR Armed Forces with a special interest in the moon.[52]

The cosmonaut Iurii Gagarin may have played a role in the leadership of the space program before his accidental death in 1968. Gagarin purportedly served as Kamanin's deputy in the training of cosmonauts and had charge of projects associated with lunar exploration, possibly being assigned himself to command the Soviet lunar landing crew.[53] Whatever Gagarin's responsibilities, they may likely have passed to cosmonaut-engineer Konstantin P. Feoktistov (a doctor of technical sciences) whom NASA has identified as the leader of Soviet man-in-space efforts.[54]

Other academicians who have been leading spokesmen for the Soviet space program are Georgii I. Petrov, who succeeded Leonid Sedov as director of the Academy's Institute for Space Research; Kiril Kondratev, who has written for *Pravda* and *Izvestiia* on economic applications of space activities; Boris N. Petrov, who has been a frequent commentator in the popular press; two vice-presidents of the Academy, Boris Konstantinov and A. P. Vinogradov, and G. V. Petrovich, specialist in rocket engines, who is really Glushko, writing and being interviewed under a pseudonym.[55]

*For example, see V. I. Feodosev and G. B. Siniarev, *Vvedenie v raketnuiu tekhniku* (Moscow, 1956, 1960), pp. 13, 14.

Whatever the details of organization of the Soviet space effort, there is no question that in the USSR, as in the USA, it is the highest political leaders who are responsible for approving the major programs and immense costs entailed. Khrushchev, as noted previously, was honored for his role in the space program, receiving the Order of Lenin for his efforts.[56] *Pravda* credited him with the "triumph of Soviet rocket technology."[57] The overall responsibility previously held by Khrushchev is now presumably shared between the general secretary of the Party Central Committee Secretariat, Brezhnev, and the chairman of the Council of Ministers, or prime minister, Kosygin.

ORGANIZATION OF THE AMERICAN SPACE PROGRAM COMPARED WITH THE SOVIET ORGANIZATION

As is the case in the USSR, the highest American officials have overall responsibility for the American space program and ultimate responsibility is concentrated in the person of the President of the United States. If it was Khrushchev who deserves the credit for the rapid progress in Soviet space technology and achievements from 1957 to 1964, a similar measure of credit goes to President John F. Kennedy and to Lyndon B. Johnson in his roles as Senate majority leader, chairman of the Senate Preparedness Subcommittee, the first chairman of the Senate Aeronautics and Space Committee, Vice-President and President.[58] Johnson took Sputnik I as a challenge to the leading position of the US in world affairs and led his Preparedness Subcommittee in 1958 through an extensive investigation of the reasons and remedies for what he saw as the failure of political leadership in the US in science policy.[59] To a large extent the National Aeronautics and Space Act, which gave birth to NASA (National Aeronautics and Space Agency), grew out of the hearings of Johnson's subcommittee and the efforts of LBJ and like-minded Congressmen. Even President Eisenhower, who refused to be perturbed by Sputnik and whose commitment to a "scientific" space program had influenced the rejection of Army proposals to launch a satellite before Sputnik, was moved to appoint a full-time science advisor to the President.[60]

If Eisenhower left a permanent mark on the Space Agency, it was in the delineation made by the Space Act between civilian space activities (a responsibility of NASA) and military space (the competence of the Department of Defense, here abbreviated DoD).[61] Space and science committees were established in the House and Senate, and the State Department opened a separate office to deal with the international implications of space.[62] In the USSR it is the Soviet Academy of Sciences, not the Ministry of Foreign Affairs, which is responsible for international aspects of space both as a whole and through its *Interkosmos* Council.

As will be detailed in the next chapter, the separation of military and civilian space activities by the Space Act was unpopular in military and some congressional circles, and Vice-President Johnson even went so far as to declare that it was not national policy to have separate programs. Nevertheless, an important difference with Soviet policy is that the United States has maintained what is *at the very least* a pretense of separate civilian and military programs, while the Soviets have, on the contrary, maintained an obviously false pretense to operating a wholly scientific and civilian program.

In comparison with the organization of the Soviets in which the State Commission for Space Exploration (if it exists at all) acts as an almost ad hoc body for *coordinating* the work of existing agencies in the scientific, industrial and military fields, NASA from the start acquired a very substantial in-house capability which extended to its own launch, tracking, testing, training and communications facilities as well as to its own international programs and design functions, even though most of its work is contracted out to industry, universities, and the armed services.[63]

The American agency which does correspond to the Soviet Space Commission is the former National Aeronautics and Space Council (NASC). As established under the Space Act in 1958, NASC was to advise the President on space policy. Its members included the President (as chairman), the Secretaries of State and Defense, the administrator of NASA and the chairman of the Atomic Energy Commission. Eisenhower made little use of NASC and recommended to Congress in 1960 that it be abolished. This proposal was blocked in the Senate by the man who became the chairman of NASC in 1961, Lyndon B. Johnson. Instead of abolishing NASC, the Congress reorganized it, replacing the President with the Vice-President as chairman.[64] On taking office, the Kennedy Administration resolved to dramatically step up the pace of the space program. Part of this effort included the resignation of NASA Administrator Glennan, who shared Eisenhower's belief in a limited space effort. Johnson played the leading role in choosing James E. Webb, an executive in the oil empire of his closest Senate colleague, Robert Kerr.[65] The 1960 reorganization of NASC eliminated its nonstatuary members, including the president of the National Academy of Sciences and the director of the National Science Foundation.[66] NASC was dissolved upon the recommendation of President Nixon in 1973.

As part of a policy which emphasized the broadly political importance of space (foreign and domestic) against the scientific and military aspects focused on by Eisenhower, the Kennedy-Johnson Administrations promoted the role of politicians and government and business administrators as opposed to scientists and military personnel. Compared to the Soviet Union, where the Academy of Sciences plays a leading role, the American National Academy was limited to a purely advisory function and came to be the source of criticism for the low priority accorded science in the American space effort.[67]

Although all four members representing the scientific community were removed from NASC in 1961 to help clear the way for politically determined policies,[68] it would be an exaggeration to say that the scientific community plays the leading policy role in the Soviet space program and no such role in the American space program. As shown in this and in the next chapter, the USSR Academy of Sciences is composed in large part of people whose careers have often involved high administrative posts in the military and industrial bureaucracies, and as such they cannot be identified as clearly as "spokesmen for science" in the sense that most members of the American National Academy of Science can. The USSR Academy is itself part of the governing structure of the Soviet Union and part of the military-industrial-party-governmental complex. As such it serves as a conduit for the advancement of the points of view of the scientific community more than as a spokesman for science. Because no such conduit exists into the highest reaches of the American Government, it is easier for the President to effectively shut science out of policy implementation and space decision-making than for the Soviet leadership to do so. By coopting leading scientists into the party, government and defense elites, the Soviet system permits greater political control of science and scientists, but also increases the possibilities for science to influence politics. It should also be mentioned in this connection that while most members of the Soviet political elite began their careers as engineers and scientists, most American political leaders began as lawyers and businessmen. One might expect from this that the Soviet leaders would be more open to considering the opinions of scientists in policy than the American leaders would be. The early blending of scientific, political and military objectives in the Soviet space program accounts in large measure for the Soviet head start in space.[69] In his account of the Soviet space program, the defector Vladimirov contends that the political leaders have effectively excluded scientific purposes from the program, but he does suggest that occasionally scientists like Korolev were able to manipulate the politicians for their own purposes.[70]

Such support of the scientific community as the Kennedy-Johnson Administrations got for its all-out man-in-space effort aimed at beating the Soviet Union to the moon came largely from the Space Science Board of the National Academy, established in 1958 to advise the government on space policy.[71] The original decision to give the highest priority to space emerged out of Kennedy's criticisms of Eisenhower's space policies during his election campaign.[72] The details for the new commitment to space were worked out by an ad hoc committee of shifting membership led by Vice-President Johnson and Secretary of Defense McNamara. At various times members of this committee included Administrator Webb; his deputy Hugh Dryden; Director of Defense Engineering John Rubel; the President's science advisor, Jerome Wiesner; Kenneth Hansen of the Bureau of the Budget; von Braun; Air Force General Schriever; Senators Kerr and Bridges; Edward Welsh, executive secretary of NASC and others.[73] It is very likely that in areas of the most important and broadest formulation of policy (as in this case) the Soviet Union also makes use of the ad hoc committee form bringing together high officials from government, party, military, scientific and fiscal bodies. Day-to-day operations in both cases are evidently left in the hands of permanent organizations with overall management accomplished in the USA by NASC, NASA and the House and Senate space committees, and in the USSR to the Space Commission and the various agencies in the defense industry, Academy of Sciences, etc. By making use of the technique where most production and operational work is contracted out, both countries enjoy considerably more flexibility than if they entrusted the entire space program to a single agency.

The next chapter details the close relationship between NASA, the DoD and the various military services. Some 200 officers of the armed services are on permanent duty with the Space Agency,[74] and many projects are under joint NASA-USAF management.[75] The situation is probably not dissimilar in the USSR. In the USA and the USSR the science academies provide quasiofficial representation of their governments to organizations like the International Geophysical Year Committee (CSAGI) which are part of the International Council of Scientific Unions and under the aegis of its Committee on Space Research (COSPAR).[76] A similar role is played by organizations in the US, such as the American Rocket Society, and sections of the USSR Academy of Sciences which are affiliated with the International Astronomical Federation (IAF).[77] While all the groups attached to the Soviet Academy are quasigovernmental and rather conservative in international dealings, the US members of the IAF are more independent and free-wheeling and generally support more extensive international programs than either the US Government or the National Academy of Sciences.[78]

The Soviet counterparts to the American congressional committees are the two committees of the Supreme Soviet on Education, Science and Culture. It is not known what impact these committees have had on the Soviet space program, but it is a safe bet that their influence is considerably less than that of their American counterparts, whose responsibility is largely confined to space, whose staff facilities and activities are more extensive, and whose political power is considerably greater.[79] Nonetheless, the congressional committees are to some extent the clients rather than the controllers of NASA, as indicated by their willingness to let NASA do its own self-investigation and house-cleaning in the wake of the Apollo fire in 1966. An anonymous NASA source has claimed that it is not the space committees but the Bureau of the Budget that decides on the allocation of resources to NASA.[80]

The involvement of industry in the two countries is conditioned by the nature of the capitalist and socialist economic systems. The competition for NASA and DoD contracts among American aerospace firms far exceeds any such competition in the USSR.[81] This competition certainly gives rise to greater duplication of effort and technological competence in the US. However, the enthusiasm and efficiency bred by competition and the prestige associated with space (by companies whose more destructive products lack the same popular appeal) probably make up for the inefficiency of duplication.[82]

Another American space organization which has no Soviet counterpart is the Communications Satellite Corporation (Comsat), created by the Communications Satellite Act of 1962.[83] Comsat is a privately owned and managed commercial corporation designed to make profits. Needless to say, such an enterprise is unthinkable in the USSR. In providing for the creation of Comsat in 1962, one may surmise that it was the intention of the President and Congress to secure a wider role in the space program for nonaerospace firms, to substitute the market mechanism for political decision-making in allocation of satellite communications channels, and to provide a flexible mechanism to encourage *international* use of an American satellite communications system. Unlike the USSR, in the USA most of the users of satellite communications channels are likely to be private corporations and individuals, whether American or foreign. Soviet clients are more likely to be government agencies, at least in the USSR and the other socialist countries. Thus, while the responsibility for communications decisions may be irksome and controversial for American political leaders, it would be less so for their Soviet counterparts, who have always jealously guarded their control over the communications media.

In spite of these distinctions, Comsat is regulated as a common carrier subject to oversight by the Congress, the President, NASA, the Federal Communications Commission, the Department of State, and the Anti-Trust Division of the Department of Justice. The ownership of Comsat is evenly divided between private US common carriers and the investing public at large. Three of its directors are appointed by the President, six by US carriers, and six by the other investors.[84] Launch vehicles and launching are purchased from NASA. It will be seen in Chapter 9 that the USSR has sought to exclude private entities like Comsat from space by international law.

Comsat is the manager and majority stockholder of Intelsat, the International Telecommunications Satellite Consortium of some 80 countries. Like Comsat, Intelsat purchases its launch vehicles and launching from NASA and its satellites from US firms. It has secured a healthy profit (14 per cent on investments) for Comsat and its other owners.[85] Through Intelsat (as well as myriad other business connections) the United States maintains ties to many nations throughout the world which underlie and in some cases undercut formal political relations in a manner not wholly dissimilar to the ties the USSR retains through the CPSU to the other ruling and nonruling Communist parties of the world. *Intersputnik*, the Soviet international satellite organization, is almost entirely a Communist-bloc organization. As will be further detailed in a subsequent chapter, *Intersputnik* does not carry on programs nearly so extensive as those of Intelsat in the communications field.

Despite obvious differences between the USSR and the USA in the space role allotted to private entities, the space programs of both countries, like similar efforts in atomic energy, health and other fields of scientific and technological research, have relied primarily on government initiative and expenditure. For the present period it is difficult to see how the immense costs of space programs and the degree of coordination of manpower required can be provided by means other than direct government action.[86] This will tend to limit the differences between the organization of the space programs in the "capitalist" USA and "socialist" USSR, the only two nations which have seen fit to mount (and perhaps the only nations which can afford) massive efforts in scientific and technological research. Perhaps the only comparable programs of other countries were the war efforts of Japan, Britain and Germany in 1939–1945.

Programs on the scale of efforts in space and defense have the effect of creating structures in American political and economic life which *approach* the mobilized organization of Soviet society with the creation of a military-industrial complex and scientific-technological elite.[87] Whereas such efforts have been a feature of Soviet society at least since the collectivization and industrialization of 1928–1941, they largely took form in the US only during the Second World War. In the USSR the ideology of socialism and planning and the cult of science in Marxism are congenial to these developments, as is the relatively monolithic and hierarchical organization of all political, economic and cultural life. In the USA these developments go against the grain of an ideological commitment to limited government and "subsystems autonomy" in the spheres of politics, economics and culture. Phenomena like the space and defense efforts in the USA tend to minimize the differences between capitalism and socialism, especially in certain sectors. To the extent that private enterprise and the politics of pluralism are givens in the environment of the American space program, it remains importantly different from the Soviet program. It is not clear to this author whether the advantages of the Soviet system such as unified and central control and the relative absence of short-run subsystem commitments outweigh the American advantages of flexibility and national and international openness. Both countries have scored notable space successes already, but the conquest of space will unfold only in the future.

NOTES

[1]Charles S. Sheldon II, *United States and Soviet Progress in Space: Summary Data Through 1973 and a Forward Look* (Library of Congress, Congressional Research Service, January 8, 1974), p. 11.

[2]John D. Holmfeld, "Resource Allocation and the Soviet Space Program," in *Soviet Space Programs, 1966-1970*, p. 114. See also Sheldon, *Review of the Soviet Space Program With Comparative US Data* (New York: McGraw-Hill Book Co., Inc., 1968), pp. 83-84, which holds to the figure of two per cent of GNP.

[3]Leon M. Herman, "Soviet Economic Capabilities for Scientific Research," in *Soviet Space Programs, 1962-1965, Goals and Purposes, Achievements, Plan and International Implications*, Staff Report to Committee on Aeronautics and Space Sciences, US Senate December 30, 1966, 89th Congress, 2d Session (hereinafter *Soviet Space Programs, 1962-1965*), p. 419; and *The Research and Development Effort in Western Europe, North America and the Soviet Union* (OECD, Paris, 1965), p. 128.

[4]International Institute of Strategic Studies, *The Military Balance, 1971-1972* (London: 1971), p. 5.

[5]*Soviet Space Programs, 1962-1965, op. cit.*, p. 13, which converts this to $4.5 billion.

[6]See Odishaw's introduction in Hugh Odishaw, editor, *The Challenges of Space* (Chicago: University of Chicago Press, 1963), p. 158; and Edward Kokum, "Budget Steadies Aerospace Outlook," *Aviation Week and Space Technology, 100* (6), February 11, 1974, pp. 12-13.

[7]B. W. Augenstein, *Policy Analysis in the National Space Program*, Rand Corporation, Paper #P-4137, July 1969, p. 17.

[8]Seth T. Payne and Leonard S. Silk, "The Impact on the American Economy," in Lincoln P. Bloomfield, editor, *Outer Space, Prospects for Man and Society* (New York: Frederick A. Praeger, Inc., 1968), p. 97.

[9]Hugh Odishaw, "Science and Space," in Bloomfield, *supra*, p. 76.

[10]See Edwin Diamond, *The Rise and Fall of the Space Age* (Garden City, N.Y.: Doubleday & Co., Inc., 1964), p. 10.

[11]Sheldon, *United States and Soviet Progress in Space* (1973), *op. cit.*, p. 21; Sheldon, "Overview, Supporting Facilities and Launch Vehicles of the Soviet Space Program," in *Soviet Space Programs, 1966-1970, op. cit.*, p. 131; and *Soviet Space Programs, 1971, A Supplement to the Corresponding Report Covering the Period 1966-1970*, Staff Report prepared for the Use of the Committee on Aeronautics and Space Sciences, US Senate, by the Science Policy Research Division, Congressional Research Service, Library of Congress, April 1972, 92nd Congress, 2d Session (hereinafter *Soviet Space Programs, 1971*), p. 9.

[12]Sheldon, "Overview . . . ," in *Soviet Space Programs, 1966-1970, op. cit.*, pp. 135-139.

[13]*Ibid.*, pp. 126-129.

[14]*Ibid.*, pp. 148-152; and *Soviet Space Programs, 1971, op. cit.*, p. 60.

[15]Charles S. Sheldon II, "Projections of Soviet Space Plans," in *Soviet Space Programs, 1966-1970, op. cit.*, p. 354.

[16]Erin B. Jones, *Earth Satellite Telecommunications Systems and International Law* (Austin: The University of Texas, 1970), pp. 20-21; and Sheldon, *United States and Soviet Progress in Space* (1973), *op. cit.*, p. 15.

[17]James R. Killian, Jr., "The Crisis in Research," *Atlantic Monthly, 211* (3), March 1963, pp. 69-72.

[18]*Loc. cit.*; and Amitai Etzioni, *The Moon-Doggle, Domestic and International Implications of the Space Race* (Garden City, N.Y.: Doubleday & Co., Inc., 1964), pp. 28-30.

[19]*Soviet Space Programs: Organization, Plans, Goals and International Implications*, US Senate, Committee on Aeronautics and Space Sciences, 87th Congress, 2d Session, May 31, 1962 (hereinafter *Soviet Space Programs, 1962*), p. 327; R. W. Retterer, "Career Opportunities in the Space Age," in Lillian Levy, editor, *Space: Its Impact on Man and Society* (New York: W. W. Norton & Co., Inc., 1965), p. 94; and Sheldon, *supra*, p. 16.

[20]Herman, *op. cit.*, pp. 420-423. See also Stanley H. Cohn, "Economic Burden of Defense Expenditures," *Soviet Economic Prospects for the Seventies*, US Congress, Joint Economic Committee, 93rd Congress, 1st Session, pp. 150-155; and Herbert Block, "Value and Burden of Soviet Defense," *ibid.*, p. 195.

[21]E. H. Kokum, "USSR Space Effort Hits Economic Snag," *Aviation Week, 80*, March 16, 1964, p. 137.

[22]Joseph G. Whelan, "Political Goals and Purposes of the USSR in Space," in *Soviet Space Programs, 1966-1970, op. cit.*, pp. 27-29.

[23]Sheldon, "Projections of Soviet Space Plans," *op. cit.*, pp. 396-398.

[24]Evgeny Riabchikov, *Russians in Space,* edited by Col. General Nikolai P. Kamanin, translated by Guy Daniels, prepared by the Novosti Press Agency Publishing House, Moscow (Garden City, N.Y.: Doubleday & Co., Inc., 1971), p. 123; and Michael Stoiko, *Soviet Rocketry: Past, Present and Future* (New York: Holt, Rinehart & Winston, 1970), p. 101.

[25]Nicholas Daniloff, *The Kremlin and the Cosmos* (New York: Alfred A. Knopf, 1972), p. 77.

[26]Fermin J. Krieger, *Behind the Sputniks, A Survey of Soviet Space Science* (Washington, D.C.: The Rand Corporation, Public Affairs Press, 1958), Appendix B, pp. 329-330.

[27]John D. Holmfeld, "Organization of the Soviet Space Program," in *Soviet Space Programs, 1966-1970, op. cit.*, pp. 82-84; and Leonard N. Beck, "Recent Developments in the Soviet Space Program: A Survey of Space Activities and Space Science," in

Soviet Space Programs, 1962-1965, op. cit., p. 184. The Department of Technical Sciences was abolished in 1963. On Keldysh, see Leonid Vladimirov, *The Russian Space Bluff* (New York: Dial Press, 1973), pp. 49, 67, 102, 108.

[28]Holmfeld, *loc. cit.*; Sheldon, "Soviet Military Space Activities," in *Soviet Space Programs, 1966-1970, op. cit.,* p. 396; and Hugh Dryden in Jerry and Vivian Grey, eds., *Space Flight Report to the Nation* (New York: Basic Books, Inc., 1962), p. 181.

[29]Daniloff, *op. cit.,* p. 77.

[30]M. Vasilev, *Putashestviia v kosmose* (Moscow, 1958), pp. 239-241 as cited in J. Baritz, "The Military Value of the Soviet Sputniks," *Bulletin of the Institute for the Study of the USSR, 7* (12), December 1960, p. 35.

[31]*Soviet Space Programs, 1962, op. cit.,* pp. 64-65; and Nicholas DeWitt, "Reorganization of Science and Research in the USSR," *Science, 133,* June 23, 1961, p. 1990.

[32]Leon Trilling, "Soviet Astronautical Scientists: How They Work and Where They Publish," *Aerospace Engineering, 20,* July 1961, p. 38.

[33]Whelan, "Political Goals and Purposes of the USSR in Space," in *Soviet Space Programs, 1966-1970, op. cit.,* p. 59; Holmfeld, "Organization of the Soviet Space Program," *op. cit.,* pp. 81, 83-84; E. A. Korovin, "Outer Space and International Law," *New Times, 17,* April 25, 1962, p. 13; Stoiko, *op. cit.,* p. 192; and *Aviation Week and Space Technology, 99* (14), October 1, 1973, p. 20.

[34]*Krasnaia Zvezda,* May 16, 1968, p. 4; *Pravda,* April 12, 1969, p. 6; and *Izvestiia,* June 4, 1970, pp. 1, 4.

[35]Leonard Shapiro, *The Government and Politics of the Soviet Union* (New York: Vintage Books, 1965), pp. 119-121.

[36]Stoiko, *op. cit.,* pp. 192, 194; and Daniloff, *op. cit.,* pp. 77, 78, who refers to Aerospace Information Division, Library of Congress, *Management of the Soviet Space Program* (Washington, D.C., October 24, 1963); A. N. Kiselev, M. F. Pebrov, *Ships Leave for Space* (Moscow, 1967), p. 318; Alexander P. Romanov, *Kosmodrom, Kosmonavty, Kosmos, dnevnik spetsialnogo korrespondenta TASS* (Moscow, Izd. DOSAFF, 1966). See also Beck, "Recent Developments in the Soviet Space Program . . .," *op. cit.,* pp. 152-153.

[37]Krieger in Greys, *op. cit.,* pp. 181-182; and Holmfeld, "Organization of the Soviet Space Program," *op. cit.,* pp. 70-71, 83-84.

[38]Daniloff, *op. cit.,* pp. 78, 80.

[39]Alexander P. Romanov, *Designer of Cosmic Ships* (Moscow, 1969, in Russian), p. 64. Vladimirov identifies as Korolev's administrative superior a man with close ties to Khrushchev, one V. N. Chalomei. See *The Russian Space Bluff, op. cit.,* pp. 41-42, 46-47, 49, 53-54, 82, 119.

[40]*Izvestiia,* June 20, 1961, p. 1.

[41]Theodore Shabad in *New York Times,* November 12, 1963, p. 2.

[42]Daniloff, *op. cit.,* p. 81.

[43]Holmfeld, "Organization of the Soviet Space Program," *op. cit.,* pp. 94-105.

[44]Beck, "Recent Developments in the Soviet Space Program . . . ," *op. cit.,* p. 152.

[45]Holmfeld, "Organization of the Soviet Space Program," *op. cit.,* pp. 72-88. Most research institutes and design bureaus in the USSR are attached to the various government ministries and state committees rather than to the Academy of Sciences or the universities.

[46]*Ibid.,* pp. 70-71; Beck, "Recent Developments in the Soviet Space Program . . . ," *op. cit.,* pp. 147, 153; and Vladimirov, *op. cit.,* pp. 43, 47, 49.

[47]Beck, *supra,* pp. 144, 149-150.

[48]Shabad in *New York Times,* November 12, 1963, p. 2.

[49]Daniloff, *op. cit.*, p. 81.

[50]T. W. Wolfe, "Are the Generals Taking Over?" *Problems of Communism, 18,* July-October 1969, p. 107.

[51]*Izvestiia*, June 20, 1961, p. 1.

[52]Sergei Gouschev and Mikhail Vassiliev, editors, *Russian Science in the 21st Century* (New York: McGraw-Hill Book Co., Inc., 1960), ix.

[53]According to Shelton, *Soviet Space Exploration* (New York: Washington Square Press, Inc., 1968), p. 276, who surely errs in identifying Kamanin as head of the Space Commission.

[54]Daniloff, *op. cit.*, p. 87, "conversation" with NASA officials. See also Vladimirov, *op. cit.*, pp. 127-128. It appears that former Cosmonaut Major General Vladimir Shatalov has replaced General Kamanin as Chief of cosmonaut training (see *Time* May 5, 1975, p. 54).

[55]See Sheldon, "Projections of Soviet Space Plans," *Soviet Space Programs, 1966-1970, op. cit.,* p. 358 and the index to that volume. Glushko has recently published under his own name.

[56]*Izvestiia*, June 20, 1961, p. 1.

[57]Victor Orlov in *Pravda*, August 26, 1961, p. 2.

[58]See Levy, *op. cit.*, ix; and Lyndon B. Johnson, "The Politics of the Space Age," in *ibid.*, pp. 3-9.

[59]See Johnson, *loc. cit.*; Philip J. Klass, *Secret Sentries in Space* (New York: Random House, 1971), pp. 26-27; Charles S. Sheldon II, "An American 'Sputnik' for the Russians?" in Eugene Rabinowitch and Richard S. Lewis, editors, *Man on the Moon —The Impact on Science, Technology and International Cooperation* (New York: Basic Books, Inc., 1969), p. 54.

[60]Joseph M. Goldsen, "Outer Space in World Politics," in Goldsen, editor, *Outer Space in World Politics* (New York: Frederick A. Praeger, 1963), p. 8.

[61]See Sheldon, *Review of the Soviet Space Program . . . , op. cit.,* p. 80; and *Review of the Soviet Space Program*, House Committee on Science and Astronautics, 90th Congress, 1st Session, November 10, 1967, p. 80.

[62]Goldsen, "Outer Space in World Politics," *op. cit.*, p. 8; and Paul Kecskemeti, "Outer Space and World Peace," in Goldsen, *op. cit.*, p. 37.

[63]See Vernon Van Dyke, *Pride and Power, The Rationale of the Space Program* (Urbana: University of Illinois, 1964), pp. 189-190, 223.

[64]*Ibid.*, pp. 192-193.

[65]Erlend A. Kennan and Edmund H. Harvey, *Mission to the Moon, A Critical Examination of NASA and the Space Program* (New York: William Morrow and Co., 1969), pp. 77-78.

[66]Van Dyke, *op. cit.*, p. 193.

[67]See *ibid.*, p. 236.

[68]John M. Logsdon, *The Decision To Go to the Moon: Project Apollo and the National Interest* (Cambridge, Mass.: MIT Press, 1970), pp. 70-71.

[69]See Shelton, *Soviet Space Exploration, op. cit.*, p. 44.

[70]Vladimirov, *op. cit.*, pp. 15, 24-25, 53-54, 57-58.

[71]Logsdon, *op. cit.*, pp. 87, 88. The report is reprinted in US Congress, Senate, Committee on Aeronautics and Space Sciences, *National Space Goals for the Post-Apollo Period*, 89th Congress, 1st Session, 1965, pp. 242-243.

[72]*Missiles and Rockets, 7,* October 10, 1960, pp. 12-13.

[73]Logsdon, *op. cit.*, pp. 109-120.

[74]General Bernard A. Schriever, "Does the Military Have a Role in Space?" in

Levy, *op. cit.*, p. 67.

[75]Etzioni, *op. cit.*, p. 136.

[76]See Arnold W. Frutkin, *International Cooperation in Space* (Englewood Cliffs, N.J.: Prentice-Hall, 1965), pp. 37-38; and Don E. Kash, *The Politics of Space Cooperation* (Purdue University Studies, Purdue Research Foundation, 1967), p. 29.

[77]Eugene M. Emme, *Aeronautics and Astronautics* (Washington, D.C.: US Government Printing Office, 1961), p. 67.

[78]Kash, *op. cit.*, p. 30.

[79]Holmfeld, "Organization of the Soviet Space Program," *op. cit.*, pp. 71-73.

[80]Kennan and Harvey, *op. cit.*, xii (Introduction by Ralph Lapp), pp. 98-105, 266; and James R. Kerr, *Congressmen as Overseers: Surveillance of the Space Program* (Ph.D. Dissertation, Stanford University, 1963), p. 173 as cited by Van Dyke, *op. cit.*, p. 194.

[81]See von Braun and Krieger in Grey, *op. cit.*, pp. 185-186.

[82]See Kennan and Harvey, *op. cit.*, p. 238; and Fortune editors, *The Space Industry, America's Newest Giant* (Englewood Cliffs, N.J.: Prentice-Hall, Inc., 1962), pp. 126-127.

[83]Public Law 87-624; 76 Stat. 419 (August 31, 1962).

[84]Jones, *op. cit.*, p. 106; and Nicholas deB. Katzenbach, "The Law in Outer Space," in Levy, *op. cit.*, pp. 70-72.

[85]John A. Johnson, "The International Activities of the Communications Satellite Corporation," in *International Cooperation in Outer Space, Symposium*, US Senate, Prepared for Committee on Aeronautics and Space Sciences, 92 Congress, 1st Session, Senate Document No. 59, December 9, 1971 (hereinafter known as *International Cooperation in Outer Space, Symposium*), pp. 198-204.

[86]L. V. Berkner, *The Scientific Age, The Impact of Science on Society*, Based on the Trumbull Lectures delivered at Yale University (New Haven: Yale University Press, 1965), pp. 54-55, 73.

[87]See Mose L. Harvey, "The Lunar Landing and the US-Soviet Equation," in Rabinowitch and Lewis, *op. cit.*, pp. 83-84; and *Public Papers of the President: Dwight D. Eisenhower, 1960-1961* (Washington, D.C.: Government Printing Office, 1961), pp. 1038-1039.

3 Military Operations in Space

As was noted in Chapter 1, the space programs of both the superpowers developed out of explicitly military programs designed to create ballistic missiles for the delivery of nuclear weapons. Moreover, the launching of Sputnik took place during a period characterized as one of cold war and an arms race. It should not be considered surprising on this account that close attention has been given by both sides to the military implications of earth satellites and space weapons systems.

This chapter will examine Soviet views on the military use of space, the role of the military in the organization of the Soviet space program, military operations in space, the impact of space on military strategy and the implications of space for future military activities. The Soviet aspects will be compared with American views and activities to highlight differences and similarities and to give some indication of the possibilities in areas where the pertinent facts on the Soviet side are not available.

DOCTRINES OF THE MILITARY USE OF SPACE

It has long been the position of Soviet leaders and publicists to emphasize the peaceful intentions of the USSR in all phases of its activities and to contrast them to the belligerent aims of its capitalist rivals. Space is no exception. In January, 1952, *Krasnaia Zvezda* attacked suggestions by von Braun and former Defense Secretary Forrestal that satellites be employed for military reconnaissance.[1] With a few exceptions[2] Soviet policy has been to overlook the military implications of its own space activities and to react negatively to suggestions along these lines by spokesmen in the West.[3] A good early example of this policy appeared in the December 25, 1958, issue of *Sovetskii Flot*, the Soviet naval newspaper. Entitled "Why the USA Is Straining To Get into Space," the article followed a typical pattern of combing American newspapers and periodicals for statements by US space

personnel, congressmen, military figures and others on the military applications of space technology. Of course, these people, many of whom have
little or no responsibility for American policy and who enjoy relative freedom to speak their minds, frequently make statements which can be characterized as expressing aggressive intentions. These can in turn be embel-
lished, as in this case where the author has an American general "foaming
at the mouth" as he demands an American military base on the moon.
Another tactic is to cite American criticisms of US programs openly published in the press. Inevitably a contrast is made between the peaceful,
scientific nature of Soviet programs and the aggressive, military nature of
American programs.[4]

The early Soviet position was to consider the military use of space for
whatever purpose as illegal and to indicate that the Soviets wanted nothing
to do with introducing military elements into their programs. This extended,
for example, to reconnaissance satellites against which one Soviet international jurist indicated that the USSR would make diplomatic and other
"reprisals and retaliation of a nonmilitary nature."[5] A major general and
doctor of sciences, Georgii I. Pokrovskii (a frequent contributor of articles
on space topics), wrote to *International Affairs* in July, 1959, explaining his
opposition to military use of space and his alarm at discussion of military
space systems in the US.[6]

These sentiments of Soviet publicists were also the official position of the
Soviet Government, which in its disarmament proposals of June, 1960,
sought to ban all military devices from space.[7] The Soviets were particularly
sensitive to the prospect of reconnaissance satellites which would have the
effect of offsetting much of the security advantage they enjoy by virtue of
the closed nature of their society. Basically, their argument was that spying
is illegal, therefore spying from space is illegal as well.[8] The Soviet Government persisted in holding to this position in international negotiations and
demanded American agreement even after it had begun to launch its own
reconnaissance satellites.[9] Even before these efforts, their view that espionage was illegal certainly did not inhibit them from more mundane efforts
in spying, an art in which the Russian reputation is second to none.

Although the Soviet opposition to reconnaissance satellites has continued,
if abated, in published sources the government's official position was modified in 1963, probably in light of its own activities in this field and as a
concession to obtain agreement on the 1963 UN Declaration on the Peaceful Uses of Outer Space.[10] In the months following this change Khrushchev
himself privately acknowledged the existence of officially secret Soviet
reconnaissance satellites to Belgian Foreign Minister Spaak and to former
US Senator William Benton, through whom he facetiously proposed to

President Lyndon B. Johnson that they trade pictures taken by each other's satellites.[11]

As the possibilities for exploiting space for military purposes grew, and as the proficiency of the United States in this capacity developed, it would have been amazing if the USSR did not seek to embark on its own programs. Indeed, uncorroborated Soviet informants have indicated that the Sputniks did infrared target-mapping of the USA and that Gagarin took pictures of military interest from *Vostok* I.[12] Certainly Khrushchev was not shy about using the Soviet ICBM capability for rocket-rattling when it suited him.[13]

An early indication of the intention to use space for military purposes was Khrushchev's 1960 threat that American espionage satellites would be "paralyzed and rebuffed."[14] More extensive justifications of military efforts did not come until 1962. In two articles in *Krasnaia Zvezda* (March 18, 21) Lt. Colonel V. Larionov noted the American intention to develop space weapons and mentioned the strategic benefits that could be gained in space. In the second article he contended that the USSR could not ignore "these preparations of the American imperialists and is forced to adopt corresponding measures to safeguard its security . . ."[15]

The same line was more extensively developed that year in the revolutionary treatise on the "new look" in Soviet military thinking entitled *Voennaia Strategiia,* edited by Marshal V. D. Sokolovskii, the former chief of the General Staff. The book reviewed the American military space program and the published plans and statements by American officials. It concluded that the American intention was to launch a preemptive attack on the USSR. This led to the conclusion that the Soviet Union must maintain superiority in space as the only way to prevent a devastating war.[16] The paradox of taking this position while continuing to claim that the Soviet space program was peaceful and nonmilitary was evident in an editorial in *International Affairs,* which stated that while the flights of *Vostoks* 3 and 4 were "not military, they would give a 'cold shower' to Western plans for a preemptive strike."[17] The legal writers E. A. Korovin and G. P. Zhukov helped to obfuscate the issue by arguing at the same time for the demilitarization of space and tying this in turn to general and complete disarmament.[18]

The position developed in *Voennaia Strategiia* has remained about the same in the years since 1962. It is certainly open-ended enough to justify operating any sort of military space systems so long as the rivals of the USSR are doing so and while no agreements have been reached prohibiting them. The Soviet Government has backed away from the insistence on general disarmament to the extent that it has agreed in subsequent treaties to bans on certain specific military activities in space. Meanwhile, an unacknowledged but extensive military space program has been developed.

In the United States the Eisenhower Administration was generally skeptical of the uses of space, including military uses. After Sputnik Eisenhower justified the relatively leisurely pace of the American satellite program by characterizing it as purely scientific and "not for security." Eisenhower's reaction to Sputnik was that as far as American security was concerned it "does not raise my apprehensions, not one iota."[19] When he did give in to considerable public and congressional pressure to accelerate the American program and submitted legislation establishing NASA in 1958, Eisenhower successfully urged that the program be exclusively under civilian management.[20] In general it was the policy of the Eisenhower Administration, including officials with responsibility for space and defense, to justify space programs in terms of either scientific or military benefits. They were inclined to think that the scientific benefits envisioned were not commensurate with the costs involved and that the military benefits were few, if any.[21] Later administrations more anxious to justify ambitious programs and more sensitive to "prestige" in addition to science and security were considerably more generous to both NASA and DoD programs.

The Eisenhower approach was not without its critics in Congress and the Armed Services. Before the formation of NASA in 1957 an ad hoc Air Force committee and a committee chaired by the "father of the H-bomb" and strategic-weapons enthusiast, Edward Teller, both recommended that the Air Force lead an American space program aimed at manned lunar landing.[22] Later the Air Force attempted to get responsibility for the entire manned space program.[23]

Some members of Congress also perceived greater military significance in space than did the Eisenhower administration. The House Select Committee on Astronautics and Space Exploration (later the standing Committee on Science and Astronautics) justified a national space program on urgent grounds of national defense and as insurance that space be "utilized for peaceful purposes."[24] Shortly afterward, the committee viewed outer space as the "heart and soul of advanced military science."[25] Senator Stennis thought space technology would become the "dominant factor in determining our national military strength" and quipped in Mackinderesque style that "whoever rules space controls the world."[26] Likewise, Lt. General Bernard A. Schriever, Chief of the Air Force Ballistic Missiles Division, and later a colonel general and Chief of the USAF Systems Command, saw national defense as the "compelling motive for the development of space technology."[27]

There was probably some degree to which the Air Force, which before Sputnik had viewed its major role largely as the arm of strategic aircraft, afterward felt that its future, if it would have an important one, would be in space.[28] Even before Sputnik the Air Force made plans for a manned orbital bomber (later Dyna-Soar) on the consideration that ICBMs would be inaccurate.[29] The very day that President Eisenhower proposed formation of a civilian space agency, the Air Force Chief of Staff, General Thomas White, secured the approval of the Joint Chiefs for the placement of manned military flights under Air Force responsibility. With some uncertainty as to NASA's role, and considering the fact that a new administration would take over in 1961, the Air Force campaigned throughout 1959 and 1960 for the leadership of the manned spaceflight program, intensifying its efforts when the change of administrations took place.[30]

When NASA was formed in 1958, it was suggested that von Braun's ABMA group in Huntsville be transferred to the new agency. The Secretary of the Army opposed the transfer, threatening to resign if it took place.[31] At the same time the Air Force wanted the Huntsville people for themselves. Eisenhower intervened in 1959 and the group, in accordance with its own wishes, went to NASA. Further Air Force efforts in the first months of the Kennedy Administration to secure direction of the operational aspects of the American space program while leaving the research work to NASA were similarly unsuccessful.[32]

It would be indeed surprising if similar rivalries did not occur between the Soviet military services and "civilian" agencies and among those services themselves. Certainly they are not allowed open expression in public media. However, certain factors would probably limit such rivalries to less than American proportions. First, the political leadership undoubtedly exercises much tighter control over all the various agencies involved in the Soviet space effort. Second, no Soviet agency corresponding to NASA exists to present an active competitor to the military services. Last, the veil of secrecy over the Soviet space program makes it possible to unify military and nonmilitary activities while simultaneously claiming to pursue exclusively peaceful and nonmilitary scientific ends.

In light of these factors, it may make little difference in practice that certain Soviet activities such as unmanned planetary flights are conducted by civilians, and, as the late Hugh Dryden of NASA suggested, manned Soviet operations are "probably" conducted by the military.[33] Manned flights by both countries have involved secret activities.[34] Ostensible characteristics of the manned flights themselves do not indicate that one or the other side is conducting military operations. The locations of Cape Kennedy and the Soviet cosmodromes are such that all Soviet orbital flights including manned flights pass over the United States, while no US *manned* flights pass over the Soviet Union. If the Soviet flights did not overfly the United States,

neither would they fly over the Soviet Union where all the manned craft
have landed. American flights of high orbital inclination, which trace orbits
further into the extremes of latitude, are usually launched from Vandenberg
Air Force Base, but all manned flights are launched at Cape Kennedy.
Almost all the American astronauts who have made space flights have been
members of the various Armed Services (28 of 34 through 1972). Seven-
teen of the 25 Soviet cosmonauts have been members of the Red Air Force
(including the cosmonette); seven have been civilians.[35] NASA trains its
astronauts. The Red Air Force trains the cosmonauts under the direction
of Colonel General Nikolai P. Kamanin.* While NASA launches its own
flights, most Soviet launches are conducted by the Strategic Rocket Forces,
a separate military service.[36]

Although American military flights had already taken place, Dwight D.
Eisenhower was the first and last president to suggest that *all* military flights
from space be prohibited.[37] Eisenhower had taken this line from the start,
and in this case was holding to it even after the U-2 incident deprived the
US of the possibility of using aircraft for reconnaissance over the USSR. It
has been the US position since, as is discussed in a subsequent chapter, to
insist on the "peaceful" use of space, whereby peaceful is meant *nonaggressive* as opposed to the Soviet definition of peaceful as *nonmilitary*. Although
the Space Act, which formed NASA in 1958, committed the agency to
peaceful exploration, it recognized this dichotomy by allotting to the Department of Defense concurrent responsibility for military space activities.[38]

President Kennedy generally did not give particular emphasis to the
military aspect of space activities.[39] As was the case with his predecessor,
this stance drew critical fire from Congress and the military, reaching significant proportion in 1962—about the same time that Soviet military doctrine
in space emerged and took shape.

Trevor Gardner, board chairman and president of Hycon (an aerospace
firm) and the former Air Force Assistant Secretary for Research and Development, thought the American space program was too scientific and over-
looked military missions such as support functions, reconnaissance, bombing, boost-glide vehicles (Dyna-Soar), satellite interceptors, moon-based
bombs and "other retaliatory vehicles as yet to be defined." General
Schriever regretted that the American public did not appreciate the military
import of space and echoed the Soviet argument that military work must
go on in the absence of an international agreement prohibiting it.[40] Senator
Dodd saw our survival at stake in space, and Senator Cannon felt civilian
space pursuits were emphasized to the exclusion of vitally needed military
capabilities.[41] Their counterparts in the House space committee, Congressmen Teague and Fulton, were concerned for the lack of a DoD man-in-

*Former Cosmonaut Shatalov, a major general, has evidently replaced Kamanin
in this job.

space program, Fulton believing that the US must be the first nation to send manned and unmanned vehicles into space with nuclear and conventional weapons.[42]

By 1963 *Air Force* magazine took a more strident line, criticizing Kennedy and Defense Secretary McNamara for "rigid opposition to military space" and "abandoning space near earth to the Soviets."[43] Purportedly the Air Force at this time was making a greater and more successful effort than NASA in presenting its views and needs to the Congress, with the effect that many congressmen demanded more emphasis on military space efforts.[44]

Doctrine supporting US military space efforts was more complex than the Soviet doctrines reviewed above, and the difference in social system and information policy assured it a more open hearing than the Soviet arguments, which were largely confined to military books and periodicals. One theme which contrasted sharply with Soviet doctrine held that military and peaceful uses of space are compatible in purpose and technologically inseparable, a contention that is dealt with below. Early expressions of this view came from Army Generals Medaris and Gavin.[45] Later this argument was made by Arthur R. Kantrowitz, vice-president of Avco Industries,[46] and Vice-President Lyndon B. Johnson, who said it was not national policy (?) to have separate space programs for NASA and Defense and that we have "space missions to help keep the peace and space missions to improve our ability to live well in peace."[47]

Another argument, already noted, was that national security by its very nature takes precedence over other justifications for a space program, so that if the US has a space program the program should reflect this principle.[48] Still another argument held that there are specific military "requirement proofs" of programs the several services must have to fulfill their missions and which are inherently military in nature. This precluded their development by NASA toward which it was felt that "the dominant bias in Congress" extended as a result of the Act's enunciation of "peaceful purposes" and "the policies of the Eisenhower Administration."[49] It is likely that the Soviet military has to justify its policies to the Politburo, the Council of Ministers and *Gosplan* in this fashion, but the issue of competition with another agency does not apply.

Closely allied to justification by "requirement proofs" is the thesis that military space activities are "merely extensions of regular service roles and missions."[50] These two justifications probably carry the most weight in the proposals of military leaders in both the Soviet Union and the United States.[51] Basically, this doctrine contends that what a particular service does on the ground, sea or in the air either requires some space system to do

well or can itself be better done in space. Such activities of the three services
as mapping, meteorology, communications, surveillance and early warning
are pertinent here. The Navy and Air Force can use satellites for navigation
and to do mapping operations to improve the targeting of Polaris and
Minuteman missiles.

The Air Force, as the service with the greatest interest in space, has
carried this justification *considerably* further with its coinage of the term
"aerospace." Aerospace means that space *and* the air constitute *one*
medium.[52] Here the argument seeks to extend the jealousy with which states
guard their airspace to proximate concern with the vacuum of space into
which the atmosphere subtly blends. This idea, not dissimilar to Senator
Dodd's quote above, holds that if the US does not have "aerospace supremacy," including superior destructive capacity in space, its freedom to use
the aerospace medium will be denied it by the USSR.[53] There are hints of
this same argument in Lt. Colonel Larionov's article, although nothing so
forthright. Moreover, since the Soviet Union has developed satellite interceptors and (presumably earthbound) fractional orbital bombs while the
US has not,* it can be assumed that the Soviet leaders take this position
seriously if it comes from Soviet military spokesmen, and too seriously if it
comes from assuming that American military spokesmen would get their
way.

The most far-reaching doctrine in support of the military role in space
is the "building block" theory. This theory holds that the military should
develop basic multipurpose capabilities even if it cannot define at present
to what use these capabilities will be put when they are achieved. The
theory is based in part on the assumption that the Soviet Union seeks to
dominate space and prevent American access to it.[54] This gives rise to a
space race of technological competition which necessitates rapid development of American military space capabilities as "insurance" against possible
Soviet efforts in the future.[55] The building-blocks theory has been championed by various defense spokesmen such as Dr. Harold Brown, then
director of Defense Research and Engineering,[56] General Thomas S. Power,
then commander of the Air Force Strategic Air Command,[57] John H. Rubel,
Assistant Secretary of Defense, and General Curtis LeMay, former Air
Force Chief of Staff.[58] It has met with relatively little sympathy from DoD,
the Congress and the Executive.

Some other justifications offered in support of military space programs

---

*The US program for an interceptor was canceled or postponed indefinitely in the
mid-1960s. Many similar proposals such as orbital bombs have not gone past the
stages of research and preliminary development, but they have generally received early
and extensive publicity, often in military and aerospace trade publications.

are (1) that Khrushchev and the Soviet generals have made threatening remarks implying military space goals,[59] and (2) that by developing a superior military space capability the US can pressure the USSR into agreeing to reserve space for peaceful purposes.[60] The technique of looking for threatening remarks is difficult since Soviet spokesmen seldom make them except in the most general terms. It is much easier for the Soviets to use this technique since the American press has been loaded with specific military programs promoted by various spokesmen including highly placed people. Moreover, when such programs secure preliminary approval, this is openly published. Soviet military personnel in the early 1960s could easily point to various explicitly military space programs operated or planned by the US. Their American counterparts could not do this for the USSR. Thus, the Soviet military could rely largely on the doctrine that America was "forcing" them into military space activities, but the American military could not use this argument until later when Soviet efforts became evident.

A substantial restraining effect on military space in both countries comes from the Test-Ban Treaty, the UN Resolution on the Peaceful Uses of Outer Space in 1963, the Space Treaty of 1967 and the Strategic Arms Limitation Agreement of 1972. Both sides, however, have continued with extensive military efforts in space even after these argreements. NASA itself, although a civilian agency having no Soviet counterpart, maintains close ties to the American military and, especially since its budget began to shrink after 1965, has often justified its own activities from the standpoint of national security.[61] In doing so NASA has stressed that it takes DoD technology requirements and research activities into account in the formulation of "practically all of its research and technology programs."[62] NASA has developed a program to apply aerospace technology to limited war, perhaps including the Vietnam War. President Nixon's science advisor, Dr. Lee DuBridge, indicated in 1969 that the military and civilian aspects of the national space program would henceforth be "more balanced,"[63] quite possibly referring to criticisms by military spokesmen and others that there was a bias toward NASA and that Secretary McNamara and the Kennedy-Johnson Administrations had been too stingy with money for advanced military projects, especially those without "strict requirement proofs."[64]

ORGANIZATION OF MILITARY PROGRAMS

The United States maintains ostensibly separate civilian and military space programs under the respective leadership of NASA and the Department of Defense. The Soviet Union does not acknowledge a military program so that it need not maintain separate agencies. There may be instead a single commission which, under the appropriate state and party authorities, has general direction over many agencies, civilian and military that are involved in various ways in the Soviet space program.[65] US military flights are the responsibility of the service(s) involved under DoD. The Soviet pattern is such that various military services and agencies are simply more involved in some projects designed to suit their own purposes and less so in others where civilian agencies such as ICIC-CEUS likely predominate. Any individual in the Soviet Union who has done advanced technological work, such as Korolev or Glushko, is almost certain to have or have had connections with various military programs. This same tendency operates in the US, although probably to a lesser extent. It is exemplified, for example, by the career of von Braun, who has worked on his own, with Oberth, for the *Wehrmacht*, for the US Army, for NASA, and lately for a private aerospace firm. Distinctions are difficult to make. The current chairman of ICIC (now called the Commission on the Exploration and Use of Outer Space), Academician A. A. Blagonravov, is a former lieutenant general of artillery.[66] One of the prominent members of ICIC, Professor Georgii I. Pokrovskii, is also a lieutenant general and an authority on explosives.[67]

Before the formation of NASA, largely civilian agencies relating to space research, such as the Jet Propulsion Laboratory at the California Institute of Technology, the Naval Research Laboratory, and projects organized under the DoD's erstwhile Advanced Research Projects Agency (ARPA), worked closely with DoD and the various military services. After 1958 many of these agencies reverted to NASA management, and DoD dropped purely scientific projects carried out under ARPA and the various services.[68]

Lacking any motive for such a separation, the Soviet leaders have very probably continued military and civilian space projects under an organizational pattern similar to American practice before 1958,[69] without the same degree of automony of the various groups and with somewhat more central coordination. If a single coordinating agency exists, it is probably much more like NASC or ARPA than NASA, coordinating explicitly military with civilian projects and agencies and without any substantial in-house capability to perform complex operations by itself. As noted in the previous chapter, the former chairman and vice-chairman of this agency were civilians, the first a man with an administrative background in the defense industry (Rudnev), and the second the aerospace scientist (Korolev). Kirillin, possibly the current head of such an agency, has made his career as a civilian scientist and state administrator.

A "substantial" proportion of Soviet defense expenditures goes into the Soviet space program. This amount would include both funding for projects of a military nature and support for projects of no particular military significance. Throughout the fifties and sixties the amount of military expenditures devoted to advanced technology including space rose as manpower levels declined and total defense expenditures increased.[70] A similar pattern emerged in the United States. American expenditures for military space started at modest levels, growing rapidly after 1958, as Table III indicates.

TABLE III
US Military Space Budgets (in $ billions)

FY	$	FY	$
1958	.21	1964	1.56
1959	.49	1965	1.59
1960	.52	1966	1.64
1961	.71	1967	1.67
1962	1.03	1968	1.89
1963	1.37	1969	2.10 (est.)

Sources: B. W. Augenstein, *Policy Analysis in the National Space Program*, p. 16; and Etzioni, *The Moon-Doggle, Domestic and International Implications of the Space Race*, pp. 128-129 for 1958 and 1959. The figures in the two sources differ for 1960, 1961 and 1964. Augenstein's have been adopted here.

US military space budgets have climbed to a level of over $2.0 billion dollars. The DoD estimates the combined Soviet military/civilian space effort at about $5.0 billion dollars annually with no breakdown as to categories.[71] With a military space effort similar to that of the US, and without an expensive manned lunar program, the USSR probably spends a greater proportion of its total space budget on military missions than the US, which has spent about $6.0 to $7.5 billion dollars a year since 1965 with one-quarter to one-third of this figure being spent by DoD.[72]

SEPARATION AND SEPARABILITY OF MILITARY AND
CIVILIAN SPACE PROGRAMS

As has been described earlier, the space programs of the USA and the USSR emerged in large measure out of military efforts to develop ballistic missiles. Most of the men in space on both sides have been military personnel. While the Strategic Rocket Troops launch Soviet vehicles, the US Navy recovers NASA spaceships.[73] The Soviet military cooperates with the Academy of Sciences in the tracking and recovery of spaceships.[74]

This raises the question of whether one can separate military and civilian space activities and technologies as the Eisenhower Administration endeavored to do in the formation of NASA.[75] All of the Soviet launch vehicles with the exception of the putative "very heavy launch vehicle," which has not yet been successfully launched, have been based on ballistic missiles developed to deliver nuclear weapons.[76] The American case is similar with space boosters like Viking, Redstone, Jupiter, Thor, Atlas and Titan developed as ballistic missiles and the larger Saturns I and V as primarily nonmilitary boosters.[77]

Ostensibly civilian efforts in weather observation, communications, geodesy, and observation of earth from space can usually be put to military uses.[78] Moreover, a satellite which is launched for scientific purposes may also carry in "piggyback" fashion instruments for military use, and astronauts may find the time to engage in activities of military interest.[79] According to Colonel Penkovsky, this was the case with Sputnik and *Vostok*.[80] On the American side the Gemini program was more or less a joint NASA-military effort.[81] Apart from other activities of military interest, the Gemini astronauts carried infrared equipment to measure the radiations given off by rocket boosters and did visual sightings of rocket takeoffs and a warhead reentry.[82] It seems probable that the cosmonauts in *Vostok*, *Voskhod* and *Soiuz* have engaged in similar activities.

There certainly is a great overlap in the interests of military and civilian planners in space. Technologies are comparable, and many of the same problems are faced. Some indication is seen in the fact that 85 per cent of NASA's contracting with other government agencies is with the three armed services.[83] The recent US Skylab mission began as the Air Force Dyna-Soar that developed into the Air Force Manned Orbiting Laboratory (MOL) before becoming Skylab, which was NASA-conducted but involved experiments of military interest.[84] The program is roughly similar to the Soviet one involving the *Soiuz* and *Saliut* spacecraft, which got underway in 1971.[85]

Despite all this overlap, there are areas in which civilian and military programs diverge, although the separation cannot be made so neatly as the Eisenhower Administration and Communist spokesmen[86] would have it. To begin with, unclassified conclusions about Soviet military missions in space are based on deductions made from the orbital characteristics and telemetry of Soviet satellites and on the fact that there are hundreds of satellites from which the Soviets have published no scientific findings.[87] Thus, military flights, announced or not, will usually have characteristics distinguishing them from other flights. Even Gemini, which was vigorously supported by DoD, was not the kind of flight DoD would have put up on its own.[88]

As has been shown in the case of the Soviet *Vostok* booster, certain launch vehicles may be advantageous for nonmilitary purposes but lacking in many respects as military boosters.[89] The very large boosters now employed by NASA and the USSR are too big and expensive for most presently defined military missions but necessary for advanced manned and planetary programs. That the hardware requirements of NASA and the military diverge significantly has been pointed out by General Schriever.[90] The proposed space shuttle would be an exception, but even given reliance on the same vehicles, orbital characteristics for most military missions are distinct enough so that one side can have a good idea of what the other is up to at least as soon as the first satellites start to fly. Beyond that, general capabilities suitable for civilian or military use can be assessed by current levels of achievements and patterns observable from flight tests of new vehicles. For example, in the Titan III and "F-1," SS-9 or Scarp, both sides now have reliable military boosters large enough and flexible enough for a great variety of military missions. The Soviets have used the SS-9 booster for tests of a fractional orbital bombardment system (FOBS) and a satellite interceptor.[91]

SECRECY

Both the US and USSR have characterized their space programs as peaceful, and the USSR has been at pains to prove that its activities are scientific and nonmilitary. The United States has boasted of the "openness" of its programs, whereas the Soviet Union has maintained a cover of secrecy over most of its space activities. This section examines secrecy as it applies only to the military launches of both countries.

The United States acknowledges its military programs, the USSR does not.[92] By maintaining and acknowledging a separate military program, the US has opened up NASA to broader participation domestically and internationally. The Soviet policy, by contrast, reaps a propaganda advantage in supporting the claim to a purely peaceful program.[93] The Soviets do make vague threatening references to the strategic value of their "scientific" accomplishments, while the US claims its military flights are nonaggressive and nondestructive.[94] The paradox of the secrecy surrounding *all* Soviet flights and the claim that their program is wholly scientific has been justified on the grounds that military vehicles are used as space boosters.[95]

Descriptions and conjectures about American military space programs appear in the aerospace and popular press as well as in congressional testimony in which specific details are often censored. Nothing of the like appears in the Soviet Union, and it is precisely those flights and missions which are *not* discussed in scientific publications and the popular press that are presumed to be military.[96]

Beginning with the first Discoverer flights in 1959, the United States adopted a relatively open policy concerning its military space programs according to which the missions and hardware were openly discussed although results were kept secret.[97] SAMOS and MIDAS were early-warning and reconnaissance satellites developed as follow-ons to Discoverer. Beginning in the fall of 1961, these projects came under strict secrecy and were no longer officially mentioned. No domestic announcement was made of two launches of DoD satellites in November and December, and they were also omitted from the President's annual space report to Congress in January, 1962. As a part of voluntary reporting of satellite launchings and orbital parameters to the UN, the United States was later reporting satellites to the UN (but not the names or nature of them) that were domestically secret. Moreover, some military flights were domestically secret, and some were not.[98]

This policy was changed in March, 1962, so that *all* military flights would be secret (joint military-civilian projects excepted), although their orbital elements would appear in the President's report to Congress and in NASA's Goddard Satellite Situation Report as well as in semimonthly reports to the UN.* The same month the Soviets adopted a different expedient for avoiding the onus of announcing their military flights.[99] This was the *Kosmos* label, first used in the case of *Kosmos* 1 launched March 16, 1962. An extremely general announcement was made describing *Kosmos* as a series of "scientific" satellites and giving the orbital elements of *Kosmos* 1. The first probable use of the *Kosmos* label as a cover for a military flight was the launch of *Kosmos* 4 on April 29, 1962, which was likely a recoverable observation satellite used for military reconnaissance.[100] As has been the case with all *Kosmos* launches, reference was made back to the March 16 announcement. Through 1973, 627 *Kosmos* satellites have been launched, and more than half of them have probably served some military purpose.[101]

By way of comparison with American policy, the Soviet military satellites are given a name (*Kosmos*) and their orbital elements announced immediately; the launch vehicles are not specified. The announcements of American military flights do not include the name or mission of the spacecraft, although the launch vehicles are specified. Announcement of orbital elements of American military satellites is deferred until the publication of the Goddard Satellite Situation Report and the semimonthly listings with the UN Secretary-General.[102] In at least one case the American policy led NASA to withdraw from a joint project with DoD (the geodetic satellite "Anna") when the project was classified.[103] Odd as it seems, the best un-

---

*In subsequent years DoD has made selective exceptions to this policy. See Sheldon, *Review of the Soviet Space Program*, p. 123.

classified source of information on the details of the Soviet military launches is a group of space hobbyists at a grammar school in England.[104] By examining the orbital elements and telemetry modes of earth satellites and by studying recurrent patterns, this group has been able to identify with some assurance the missions of secret military flights.

Opening up the military programs of both countries is probably avoided because it goes against the instinct of military planners for secrecy, would seem an affront and a challenge to the country over which such missions are flown, might aid attempts at detection and countermeasures and would saddle each side with the onus of conducting admittedly military activities in space. No one, save the general public at home and abroad, is fooled.[105]

In any case, both sides have maintained the policies of secrecy they adopted in 1962, with a few exceptions which deserve brief mention. The first two tests of the Soviet FOBS in 1966 were not announced at all.[106] Subsequent FOBS tests took place under the *Kosmos* label and blanket-mission announcement. The US announced no orbital data for a new Air Force reconnaissance satellite launched on August 6, 1968, although these were detectable and subsequently published in Britain.[107]

MILITARY OPERATIONS

As mentioned above, American military use of space reached operational status in 1959 and Soviet use in 1962. The apparent delay on the part of the Soviets can be accounted for by their desire to discourage the US from military activities, especially reconnaissance, and perhaps by their earlier use of scientific flights for military and paramilitary purposes.

Although DoD launched vehicles as part of the Explorer program in 1958, these were not intended for military purposes but resulted from the American effort to orbit something and from the difficulties of the civilian Vanguard Project. Five satellites in the Discoverer series were launched in 1959 designed to test systems required for military missions. The Soviet effort began in 1962 with the launch of *Kosmos* satellites designed to test and/or carry out reconnaissance by recording imagery and returning to earth.[108] A rough estimate of the extent of Soviet and American military efforts is provided in Table IV.

The American military space program involved from 100,000 to 200,000 persons in 1966 and has probably risen somewhat since then.[109] As noted earlier, American military space expenditures in most recent years have averaged somewhat over $2.0 billion annually, and the Soviet figure, assuming a somewhat larger proportion of the total space budget goes for military programs, would be about the same.[110] Most American military launches,

TABLE IV
PRESUMPTIVELY SPECIALIZED MILITARY SPACE LAUNCHINGS*

YEAR	DoD	USSR
1957	0	0
1958	0	0
1959	5	0
1960	10	0
1961	19	0
1962	33	5
1963	26	7
1964	33	16
1965	33	29
1966	34	28
1967	26	45
1968	22	52
1969	16	53
1970	16	55
1971	13	62
1972	13	55
1973	10	59
Totals:	309	466

*Vehicles orbiting multiple payloads are counted here as only one launching.
Source: Sheldon, *United States and Soviet Progress in Space* (Library of Congress; 1973), p. 37.

beginning in 1959, have come from Vandenberg AFB in California. In 1966 the Soviets began launches from an unacknowledged base near Plesetsk, almost all of which have been of a military nature. It is from Plesetsk that most military flights have come in recent years.[111]

Military satellites currently in use can be summarized under these headings: observation, communications, navigation, weather, geodesy, mapping, ferret, bombardment, and inspection/destruction. Proposed military programs not yet implemented by the US or USSR generally involve bombs stationed in orbit, manned operations and use of "boost-glide" vehicles, which are craft that can perform as airplanes in the atmosphere and space vehicles outside it. Some indication of the extent of effort put forth by the USA and the USSR in these areas can be taken from Table V. Certain categories such as communications, geodesy and weather include military as well as "civilian" flights which may serve military ends as well.

As is evident from this table, the greatest military effort by both sides has been in the observation category. This category includes reconnaissance for targeting and general intelligence purposes, early warning, and monitoring for nuclear tests. The United States has acknowledged reconnaissance systems (Discoverer, SAMOS and, later, more advanced projects), an early-warning system (MIDAS) and a satellite that monitors nuclear tests (Vela Hotel).[112]

The United States made an earlier top priority commitment to these kinds of projects, as was revealed by General Schriever in testimony before the House select space committee in April, 1958.[113] The Soviet programs that very probably duplicate the American ones began only in 1962 and have been carried out under the *Kosmos* program, which does not delineate separate military missions and, in fact, does not acknowledge them at all. Most of the Soviet recoverable observation flights have used the *Vostok* booster which launched the Sputniks and the Soviet manned ships. The *Vostok* program itself may have played a large part in the development of reconnaissance satellites.[114] Since 1962 these observation satellites have been improved by extending their orbital life from four to eight and subsequently twelve days and adding the capability to maneuver. Also, in some cases, scientific payloads have been carried pick-a-back fashion. In recent years most observation satellites have been launched from Plesetsk.[115] At least one recoverable observation satellite was blown up or shot down, possibly to prevent it from falling into non-Soviet hands after recovery failure.[116]

The US has operated various military support systems such as communications, command and control (West Ford, Advent, IDSCP), weather (Tiros), navigation (Transit), geodesy (Anna) and secret "ferret" missions which probe the radar and communications systems of other countries.[117] The Soviets have surely developed similar programs under the *Kosmos* label. They probably also use their announced communications (*Molniia* I & II) and weather (*Meteora*) systems for military purposes.[118] They may also use communications satellites for foreign espionage activities.[119] The Russians have claimed but not named a navigation system which must be included in the *Kosmos* launches, including small vehicle launches from Tiuratam, Kapustin Iar and Plesetsk employing the "B-1" (SS-4 or Sandal) MRBM and "C-1" (SS-5 or Skean) IRBM vehicles. B-1 vehicles for ferret purposes were probably first launched from Kapustin Iar in 1964, and other B-1 launches, mostly from there and a few from Plesetsk, may have served other military missions such as navigation, components testing and radar calibration.[120] Many of the C-1 flights from Tiuratam and Plesetsk have probably been for use in navigation, ferreting and possibly military command and control. Certainly mapping and geodesy must be carried on in a similar fashion.[121]

A separate class of weapons employed for strategic space use rather than as support of terrestrial missions are vehicles for inspection and/or destruction of hostile satellites and orbital and fractional orbital bombs

TABLE V

MILITARY AND RELATED SPACE PAYLOADS BY MISSION CATEGORY

	1966		1967		1968		1969		1970		1971		1972		1973		CUMULATIVE**	
	US	USSR	US	USSR	US	USSR	US	USSR	US	USSR	US	USSR	US	USSR	US	USSR	US	USSR
Communications	11	2	19	4	11	4	6	2	6	5	6	3	4	56*	4	32	91	111
Weather	6	2	6	4	4	2	3	2	5	5	4	4	4	3	2	2	55	27
Navigation/Ferret	4		3	4	1	6	1	6	1	16		27	1	−23*	1	9	27	70
Geodesy	4		1		1		1		1		0		0		0		17	
Military Observation (total)	(38)	(25)	(28)	(30)	(24)	(39)	(26)	(44)	(18)	(41)	(14)	(40)	(12)	(40)	(11)	(46)	(326)	(359)
Low Orbit Recoverable	23	21	19	22	16	29	12	32	9	29	7	28	7	29	6	35	211	266
Low Orbit Nonrecoverable	12	4	7	8	7	10	11	12	4	12	6	12	3	9	3	10	83	86
Intermediate Orbit	3													2		1	10	7
Synchronous or Higher Orbit	0		2		1		3		5		1		2		2		22	
Fractional Orbital Bomb Tests		1		9		2		1		2		1						17
Military Inspector/Destructors and Targets for Intercept		2		2		5		2		4		8		2		1		28
TOTALS	63	32	57	53	41	58	36	57	31	73	24	83	21	78	18	90	516	612

*The Soviet figure for navigation/ferret satellites for 1972 is either an error or the result of reclassifying previous satellites into other categories including communications.

**Includes figures from 1957-1965.

Sources: Sheldon, "Overview, Supporting Facilities and Launch Vehicles of the Soviet Space Program" and "Postscript" in *Soviet Space Programs, 1966-1970*, pp. 119-121, 508; *Soviet Space Programs, 1971*, p. 5; and Sheldon, *United States and Soviet Progress in Space*, 1972, p. 42 and *United States and Soviet Progress in Space*, 1973, p. 42.

(FOBS). Another military use of space is as an environment for nuclear weapons tests. Before the Test-Ban Treaty, both the USA and the USSR detonated nuclear weapons in space,[122] probably as part of the development of anti-ballistic missiles (ABMs).

Almost as soon as Sputnik was up, one prominent American general considered it "inconceivable" that the United States could tolerate reconnaissance of its territory from space and that it was "urgently necessary" that the US develop a satellite interceptor to destroy such hostile satellites.[123] In the early 1960s projects were under way in the US to develop such weapons under the names of Projects Saint and INSATRAC.[124] The United States also had in planning systems for interception of ICBMs by satellites in orbit such as Projects SPAD and BAMBI, the latter calling for up to 100,000 satellites in random orbits.[125] Although the US did modify a few Thor IRBMs and Nike-Zeus ABMs in the mid-1960s to serve as satellite interceptors, no *orbital* interceptor of the type planned in these projects has been developed, and the projects themselves have been dropped or cut back to mere investigations.[126] In the meantime, and especially before DoD adopted strict secrecy on its space systems in 1962, these projects were widely discussed in the aerospace trade press and the popular press as well. The American point and area defense ABM systems, now in the preliminary stages of deployment, do not have orbital capabilities.

In spite of the announced American intention to build an inspector/interceptor, it was the Soviet Union which very probably developed one without ever announcing its plans to do so or the existence of the completed spacecraft. This craft employs as a first stage boosters the large combat launch vehicle "F-1," SS-9 or Scarp, which is also used to launch the Soviet FOBS. Its first test flights, completely unannounced, probably occurred in September and November, 1966. Some 28 flights connected with the testing of this weapon or weapons have taken place through 1973. The pattern of the flights involves substantial maneuvering in orbit usually leading to rendezvous with another (Soviet) satellite. Several of the craft have exploded after conducting rendezvous, either as a part of their combat role or to prevent their being inspected by other nations. The latter seems more probable since the most recent flights have maneuvered into orbits from which they decay rapidly and have been destroyed by reentry rather than by explosion.[127]

There have been substantial variations in the orbital characteristics of the satellites orbited in this general group. Some have maneuvered in the absence of any target. Both these factors seem to indicate a variety of possible missions, extending beyond that of interception. Acknowledgement that the Soviets had acquired a satellite-based interception capability was made by Dr. Foster, director of research and engineering at DoD, in February, 1972. The system has not yet been used to destroy any satellite, Soviet or otherwise.[128]

In contrast to the satellite interceptor, the USA has never made firm plans for or announced the intention to build a nuclear-armed bombardment satellite (NABS) or fractional orbital bombardment system (FOBS), although the Air Force has claimed "systems requirements" for these weapons.[129] Nevertheless, the Soviets have frequently accused the United States of the intent to build orbital bombs.[130] Just as was the case with the satellite interceptor, so with FOBS the Soviet press vehemently attacked the United States for plans to develop such a weapon, while the USSR went on to build one and the US did not.[131]

The USSR has been more open about FOBS than the satellite interceptor and even made threatening claims about its capabilities in this field. In early 1963 Marshal Biriuzov, then commander of the Strategic Rocket Forces, made the claim that a bombardment satellite "has now become possible."[132] Khrushchev may have been referring to an orbital bomb in 1964 when he boasted of a "monstrous new terrible weapon," which was possibly the SS-10 or Scrag first paraded in May, 1965. In July, 1965, Brezhnev boasted of a supply of orbital rockets, and in November the SS-10 was described as an orbital rocket by TASS.[133] One analyst has suggested that Soviet boasts about an orbital bomb were "a blunder" from a propaganda viewpoint since they violated "the wording, [doubtful] as well as the spirit" of the 1963 UN Resolution on the Peaceful Uses of Outer Space. He concludes that the Soviet intention was to commit the US to the "tremendous expense" of an expanded Nike-X ABM system.[134] This contention gains some support from the fact that the SS-10 was never tested as an orbital bomb, and its use as such may have been a dead-end project reserved for parade use.[135] The SS-9 Scarp was in fact the missile used in the development of the Soviet FOBS. Its first testing with upper stages was in 1965, and its first fractional orbital test came in early 1967 as *Kosmos* 139.[136] In November, 1967, the modified SS-9 was identified by TASS and the US Secretary of Defense as a fractional orbital weapon. It was also so identified in 1969 in a publication of the Institute of Strategic Studies. The American Defense Secretary did not consider it a violation of the 1963 UN Resolution or 1967 Space Treaty since it flew partial orbits only and probably was not armed with nuclear weapons in the tests.[137] Orbital tests of SS-9 with added stages, known as "F-1-r" in its orbital configuration, have not passed over the USA, although some of the debris has landed here. After extensive testing in 1967 and afterward, the FOBS is now probably an operational weapon. Occasional flights in the past few years have probably been for troop training or testing minor improvements.[138]

It is worth noting some of the characteristics of orbital and fractional orbital weapons, since one of the superpowers has chosen to develop them and the other has not. First, compared to conventional ICBMs, FOBS cut radar warning time from about fifteen to five or six minutes, although the difference in warning time obtained from satellites like MIDAS would be less. FOBS may be fired in trajectories that do not pass through the early-warning radars, which both countries deploy extensively only on their northern frontiers. Because an FOBS might be indistinguishable from an ordinary satellite until it deorbited and because of its low trajectory and warning-time advantage, countermeasures such as dispersal, alert and interception would all be too late or more difficult. On the other hand, FOBS is probably less accurate than conventional ICBMs, and the same booster, when used for FOBS, must carry a smaller warhead.[139] If launched *in sufficient numbers* to be of assured strategic impact, an FOBS attack could be detected and countermeasures taken.

It is not publicly known how many SS-9s the USSR has in FOBS configuration, and it may be that the Soviet motive for developing FOBS was more for use as an element of threat and/or blackmail for diplomatic purposes rather than as a strictly military weapon. This would be a repetition of Soviet policies based around the original Soviet ICBM and the claims made for the orbital (?) SS-10 or Scrag calculated to give an exaggerated impression of Soviet strategic confidence and strength vis-a-vis the United States.

THE IMPACT OF THE SPACE AGE ON MILITARY AND STRATEGIC POLICY

Even before the Sputnik the USSR was using the same preeminence in ballistic missiles that led to its space successes to press for advantage in the arena of international politics. The same TASS announcement of August 27, 1957, which heralded the successful flights of the Soviet ICBM, announced the completion of a series of high altitude tests of nuclear and thermonuclear weapons. Joining the two announcements together was obviously intended to underscore the strategic significance of the new ICBM. Immediate pressure was focused on the disarmament talks, which had just gotten under way in London, by the concluding statement of the announcement, which blamed the West for lack of progress in arms control and disarmament.[140]

Khrushchev himself was quick to claim that the success of the first two Sputniks was proof of Soviet military might. In an interview with the Hearst papers in November, 1957, he made it clear that the Sputnik launch vehicle was an ICBM "with a different warhead" and that "we can launch ten, even twenty . . . tomorrow."[141] In January, 1958, Khrushchev took the Sputniks as a sign that "a change in favor of the socialist states" had taken place in the world balance of power.[142] In 1959 Khrushchev quipped that in the event of an attack the Soviet ICBM would make it possible to "wipe our . . . enemies off the face of the earth."[143] That same year, after the launching of the first deep-space rocket, Khrushchev said, "If the Soviet Union knows how to send a rocket over hundreds of thousands of kilometers into the cosmos, it can send powerful missiles to any spot in the world without fail."[144] In August, 1961, at a reception for the cosmonaut Gherman Titov, Khrushchev insisted that the USSR could not be threatened from a "position of strength" because of its own might. He noted that the Soviet Union had bombs of 50 and 100 megatons and more, and that the USA did not. He then suggested that the cosmonauts could be replaced with "other loads" that could "be directed to any place on earth."[145]

These quotations above evince a comparison of Khrushchev's use of rockets and spacecraft for strategic bluster with the Soviet nuclear testing policy. During 1958 the USSR carried out 25 nuclear tests, more than doubling the cumulative total since 1949.[146] It will be recalled from those years that the USSR in 1961 developed and tested nuclear weapons of enormous yield, including 50 and 100 megaton monsters for which little military justification was seen in the USA. As with military space systems like the FOBS, and like Khrushchev's attempts to equate Soviet space successes with military superiority, these weapons probably had a political purpose which overrode their military disadvantages. That is, they were meant to give substance to threats and implied threats aimed at the United States and the Western system of alliances.

The roots of this policy of strategic bluster, besides Khrushchev's personal identification with it, were very likely the reverses suffered by the socialist camp in 1956, differences among the Western powers dating from the same year and the quarrel between the USSR and China, which developed and grew in wake of the 20th Party Congress of the Soviet Communist Party. Presumably the Soviet leaders wished to make up for past reverses, split the Western allies and sound fierce enough to lay to rest China's claims that they had gone soft and become "revisionists."[147] One aspect of their difference was that the Chinese put more faith in Soviet claims of strategic superiority than the Russian leaders did themselves, and were unhappy about their unwillingness to share it.[148]

Soviet space successes helped achieve these ends by (1) lending credibility to the Soviet ICBM, its range, warhead capacity, accuracy and reliability (the USSR does not acknowledge space failures), and (2) giving the impression that the Soviet Union had sufficient ICBMs to make war with her undesirable, especially for those who might be "dragged into" a conflict by their ties with the Western alliance system.[149]

It should be recalled that at this time (1958–1962) the USSR generally took a hard line in its relations with the West, marked by ultimatums on Berlin in 1958–1960 and 1961, the cancellation of the summit meeting in 1960, and Khrushchev's famous appearance at the UN that year. At the same time, as indicated by the publication of Marshal Sokolovskii's book in 1960 and 1962, the Soviet military was changing its doctrine so as to consider strategic nuclear missiles the decisive weapons of modern warfare.[150] The book acclaimed the Strategic Rocket Forces as the "leading branch" of the armed forces, and the development of the Rocket Troops is presumed to have owed a great deal to Khrushchev's own efforts.[151] Khrushchev's championing of the "new look" in military strategy is also not unconnected to his ouster of Marshal Zhukov in 1957 and probably helped him to gain an upper hand against the "old hands" in the Red Army who resented Zhukov's eclipse. The support of the theorists of missile war, the Rocket Troops and the Air Defense Command (PVO), was probably sought in Khrushchev's attempted reductions in manpower levels in the Armed Forces and the substitution of more modern weapons.[152]

The Soviet policy of strategic bluster was also applied to the development of anti-ballistic missiles (ABMs). As was the case with ICBMs and FOBS, the claims made for the ABM were either exaggerated or premature. The first such premature claim was made by Marshal Malinovskii at the 22nd Party Congress in October, 1961.[153] More notable and picturesque was Khrushchev's claim in July, 1962, that the USSR had missiles which "can hit a fly in outer space." Subsequent claims for an ABM were made in the military parade of November 7, 1963, and by Marshal Biriuzov, the chief of staff, and Air Marshal Sudets, commander of the National PVO, the man responsible for antiaircraft and ABM batteries.[154] In fact, it was not until 1965–1966 that the limited "Galosh" ABM system was first deployed, as acknowledged by Secretary McNamara in November, 1966.[155] Since deployment was completed in 1967, the USSR has not added to its limited ABM capability, which includes 64 launchers deployed to provide an exospheric defense from four sites around Moscow.[156]

Exaggerated claims about Soviet weaponry helped to set off a strategic arms race which has shown signs of abating only in the past few months. The Gaither report delivered to President Eisenhower in late 1957 warned that Soviet ICBMs could destroy the Strategic Air Command bombers on the ground by 1960 and urged a step-up of defense spending. Some details of the report leaked out and were used by Democratic critics against the

Administration and in the 1960 political campaign. In spite of the boasts of Khrushchev and others, American U-2 flights revealed that the balky Soviet ICBMs were not being mass-produced or deployed in large numbers.[157] Administration policy precluded mention of these flights or information obtained by them, so that outside the very highest councils of the government it was Khrushchev and not the apparent calm of the Eisenhower Administration that was taken seriously.

The effect of these policies was to give the Soviet Union added prestige and to add *temporary* credibility to Khrushchev's threats. The long-run effect was a stimulus to the arms race and space race.[158] This happened because the "missile gap" became a strong campaign issue in the Democrats' victory in 1960. Kennedy himself went even further than Khrushchev in linking space success with military power in a campaign statement of October, 1960, to the effect that we were in a space race with the Soviets which was being lost and that Soviet control of space would mean Soviet control of earth and a threat to peace and freedom.[159]

In the fall of the year shortly after the Kennedy Administration took office, it learned that it was faced not with a missile gap but with a missile glut. The information came from photos obtained by Discoverer and SAMOS satellites. While it had been estimated earlier that the Soviets could have 400 operational ICBMs by mid-1961, they had only fourteen, and the US had some 50 ICBMs, 80 Polaris IRBMs and a tremendous superiority in strategic bombers.[160] For 1961 the Institute for Strategic Studies pegged the Soviet ICBM total at 50, the American at 63 (see Table VI).[161] Perhaps it was not a coincidence that in October Khrushchev lifted his ultimatum in Berlin.

For whatever reasons the Kennedy Administration went ahead with its plan to develop 200 Minuteman ICBMs in hardened silos, evidently responding to Khrushchev's claims rather than Soviet capabilities.[162] Later the total was increased by steps to 1,000. The Soviet response was a mirror image of the Americans. The USSR began developing smaller ICBMs with storable fuels capable like the Minutemen of being placed in hardened launch sites.[163] Military influence in the Kremlin increased as signified by the election of Marshals Grechko and Gorshkov to the Central Committee of the Communist Party.* The Soviet defense budget for 1962 was increased 25 per cent, cuts in the armed forces were canceled along with a promised reduction in income taxes, and prices for meat and butter increased. Nuclear testing was resumed with the detonation of the huge bombs mentioned

---

*The former is now the first professional soldier since Zhukov, and the only other, to sit on the Politburo as a full member.

TABLE VI
STRATEGIC MISSILE FORCES: US AND USSR*

	ICBM		SLBM	
YEAR	US	USSR	US	USSR
1959	0	some	0	0
1960	18	35	32	0
1961	63	50	144	some
1962	294	75	144	some
1963	424	100	224	100
1964	834	200	416	120
1965	854	270	496	120
1966	904	300	592	125
1967	1054	460	656	130
1968	1054	800	656	130
1969	1054	1050	656	160
1970	1054	1300	656	280
1971	1054	1510	656	440
1972	1054	1527	656	560
1973	1054	1527	656	628

*Figures are for the middle of the years listed. In recent years the US has begun to equip its ICBMs and submarine-launched ballistic missiles (SLBM) with multiple, independently targeted warheads (MIRV). This has resulted in retention of a slight numerical American superiority in targetable missile warheads, while the Soviets have a numerical superiority in targetable missile megatonnage.
Sources: International Institute of Strategic Studies, *The Military Balance, 1969-1970*, p. 55; *The Military Balance, 1973-1974*, p. 70.
The Interim Strategic Arms Limitation Agreement on Offense Arms limits the number of ICBMs of both sides to those deployed or under construction as of July 1, 1972. The agreement runs through 1977.

previously. The boomerang effect of Khrushchev's space and missile boasts and the shock to the West of the Soviet Sputnik had come back to haunt the Soviet Union.

It was this position of strategic inferiority after the initial Soviet boasts that probably led to the Soviet attempt to smuggle MRBMs into Cuba, rather than any great concern about the fate of Castro, who had no control over their use. Although unsuccessful, the Soviet move was much like the American dispatch of Jupiter and Thor missiles to Britain, Italy and Turkey after the development of the Soviet ICBM.[164]

By 1963 the American missile glut as a result of Minuteman deployment had reached an imbalance of 500 ICBMs to 100.[165] The Soviet Union continued to lag into the late 1960s and largely gave up the policy of strategic bluster, quietly catching up with and surpassing the United States in its total number of intercontinental missiles by the end of the decade.

In anti-ballistic missiles earlier Soviet claims were also exaggerated, but here they went on to develop a limited system which evidently includes the capabilities of intercepting both ICBMs and satellites.[166] At the time of this writing US efforts in the ABM field are just beginning to enter the deployment phase and are likely to remain, as with Soviet efforts, at a level lower than that permitted by the terms of the bilateral Strategic Arms Limitation Agreement (SALT) with the USSR (i.e., 100 ABM launchers at each of two sites).

That the long period of missile competition is now abating somewhat has much to do with intervening developments in observation satellites. The US and the USSR have the capability of unilaterally discovering the numerical missile strength of the other side.[167] Qualitative improvements such as increasing accuracy or adding additional warheads are difficult to monitor by satellite but may in some cases be detectable during testing and deployment. In 1967 President Johnson claimed that this capability alone was worth ten times the $35-45 billion that the US had spent on space up to that time.[168] According to Colonel General Earl Wheeler, then chairman of the Joint Chiefs, observation satellites also make it possible for the US to know whether the Soviets are living up to the terms of the 1967 Space Treaty which bans nuclear tests and weapons of mass destruction in space.[169] This also applies to the 1972 Strategic Arms Limitation Agreements. These capabilities may enable the United States to save $10-15 billion annually in defense expenditures. Perhaps this stabilizing influence has affected the Soviet decision to tolerate American reconnaissance satellites in spite of stated opposition to them and the development of ABM and satellite interceptors that could shoot them down. More likely the Soviets were anxious to avoid the kind of "war in space" and space weapons race that this would probably have led to. In the Strategic Arms Limitation Agreements signed May 26, 1972, in Moscow, both the US and the USSR pledged not to interfere with the other's "national technical means of verification" of the agreements or to attempt measures of concealment to impede such verification.[170]

Although smaller weapons can be hidden from observation satellites, as can large weapons in the development stages, larger missile weapons like the ABM and ICBM probably cannot be tested or deployed without detection. This enabled the United States during the SALT talks to observe both a pause and a new start in Soviet missile deployment. An added bonus has been the ability to monitor Soviet strategic attention to China, evidenced by their aiming of ABM radars at China in 1969, as probably revealed by American ferret satellites.[171]

Satellites have also found applications to limited wars. The US was able to detect Egyptian and Soviet violations of the cease-fire agreement on the Suez Canal in August, 1970. Similarly, the USSR has used observation satellites for reconnaissance of conventional wars during the India-Pakistan-Bangladesh War of 1971 and the Mideast War of October, 1973. Observation satellites can also be used to penetrate camouflage, forest and jungle without giving any warning to those below. NASA has even proposed to DoD (unsuccessfully) that it place an artificial moonlet in synchronous orbit over areas such as Vietnam to aid military operations at night.[172]

THE FUTURE OF THE MILITARY IN SPACE

Any predictions about possible futures, including the one that they will be like the present, are full of risks. Nonetheless this section attempts to set forth some indication of the possibilities that have been discussed, including new military applications of developing space technologies and the possibilities that space may be an arena for a new kind of war or is the threshold for a new era of peace.

Since the years before Sputnik, the Air Force has been convinced of the value of military man-in-space programs, even if it cannot say just what that value is.[173] It is claimed that manned missions would be more adaptable to circumstances, would provide quicker reaction time and would introduce the possibility of dealing successfully with unanticipated problems.[174] The first big building block for manned military missions was the X-20 or Dyna-Soar project, successor to the X-15 rocket plane of ARPA and NASA. Dyna-Soar was to be a winged manned "boost-glide" vehicle launched by a Titan ICBM into space where it could maneuver by rocket power, and in the atmosphere by aerodynamic lift, landing like an airplane at any location on the globe. Dyna-Soar could be a precursor to a satellite interceptor or orbital bomber; it was said that weapons systems "may evolve more or less directly from it."[175] Other uses could be for ferrying spacemen, rescue, maintenance and salvage.[176] Dyna-Soar suffered the fate of its prehistoric namesake when it was canceled by President Johnson in December, 1963.[177]

Soviet descriptions of a Dyna-Soar-like vehicle as the vehicle of the future appeared in 1957, including one by the prominent space spokesman Professor-General G. I. Pokrovskii.[178] These were general enough to also be interpreted as descriptions of a space shuttle vehicle using aerodynamic lift. In 1962 the aircraft designer Mikoian described a "space-plane" combining ballistic missile and aircraft techniques.[179] Later mention of such vehicles ceased, perhaps owing to military plans to build them and/or because the Dyna-Soar was being attacked in the Soviet press as an aggressive weapon, typical of the "militaristic" American space program.[180] There has been no indication that such a vehicle has been built or flown by the USSR.

The successor to Dyna-Soar was MOL (Manned Orbiting Laboratory), earlier known as MODS (Military Orbital Development System).[181] MOL was born in 1964 and designed for equipment testing, inspection, maintenance and repair as described by General Schriever.[182] MOL may also have had some application to reconnaissance, bombing, and space-borne command and control.[183] It was largely in this light that it was seen in Soviet publications that condemned it.[184] MOL was to remain in orbit for a long time ejecting reconnaissance and other data by recoverable capsules. Every month a crew from the Air Force's team of astronauts (now transferred to NASA) would be ferried to and from MOL in Gemini spacecraft.[185] Rising costs ($3 billion for five MOLs) prompted the Nixon Administration to "reluctantly" cancel MOL in June, 1969. MOL probably also suffered to some extent from the 1967 Space Treaty, although MOL would not have violated its terms. MOL was succeeded by the MORL project (Manned Orbital Research Laboratory), later known as Skylab and conducted in 1973–1974.[186] The technical equipment and missions of MOL and Skylab are similar to those used in 1971 in the Soviet space station *Saliut* for which no military mission has been acknowledged.

MOL's reconnaissance function has been assumed by a new generation of unmanned observation satellites, the so-called "Big Bird" launched by the large military booster Titan III-D/Agena, the biggest existing military booster. Big Bird combines area and close-up reconnaissance and may have "real time" reconnaissance capability, that is, the ability for earth observers to monitor its imagery immediately and continuously without the delays of periodic read-outs of data or physical reentry and recovery of the craft or capsules from it. An even more advanced reconnaissance craft coded by the Air Force simply as 1010 is under development.[187] The Soviets have improved their own reconnaissance satellites by extending their time in orbit and adding the capacity to maneuver. They rely exclusively for launch on the old *Vostok* (SS-6) booster which cannot put up anything so big as Big Bird.[188]

Soviet interest in a space-shuttle vehicle in an earlier period has already been mentioned. The American military (as well as NASA and the aerospace industry) is also interested in this prospect. To suit military needs such a vehicle would be cheap, reliable and reusable.[189] The military could use the shuttle for logistics, rescue and many other missions, including perhaps construction of a permanent space station which could serve as a kind of military base in space. According to NASA, "perhaps 30 per cent" of the shuttle's missions will be "defense-oriented."[190]

The possible future space weapon which has been most discussed is the orbital bomb, a device which remains in a parking orbit until such time as it is retrofired by command to land on a designated earth target. Such a weapon is illegal under the terms of the Space Treaty, except that one could be built and tested (without nuclear warheads) without violating the letter of the Treaty. Orbital bombs would be a relatively invulnerable deterrent, have the advantages (and disadvantages) of FOBS, and could be used as a terror weapon against populations who would know that these things were orbiting over their heads and could come down any time. On the negative side, they are expensive and their relative inaccuracy could be disastrous, even to the launching state.[191]

The US Air Force had plans for orbital bombs,[192] but these never passed beyond that stage. Although the USSR has the FOBS, which is the intermediate step between the ICBM and the orbital bomb, the Soviets have not tested such a weapon, and Marshal Sokolovskii indicated in 1965 that they did not need it.[193] It is highly unlikely that either side would develop such a weapon without tests, which would probably give its intentions away. Large enough numbers to have a major strategic impact could probably not be placed in orbit without alerting the opponent, and the effect of beginning such a program would probably be very destabilizing to international relations generally and very expensive if a competition to create such weapons and countermeasures against them began. The unreliability of these weapons and these other factors would seem to make their development at this stage unlikely.[194]

One American author has earned the accolades of General Schriever for a book which suggests that a future war can be fought in space; the belligerents will hide their superbombs and then attempt to destroy each others' space-based counterforce until one side succeeds and the other submits without anyone on earth getting hurt.[195] The same author sees the hardware for the Apollo mission as the "technological base" for the development of American strategic space systems that could be undertaken "well before 1980," at which time the Space Treaty, "if it is still a living instrument, will face increasingly severe tests of its viability." The Treaty, he observes, takes any of its force from "faith and trust alone." The Soviets, he notes, do not allow inspection and make no administrative distinction between their military and nonmilitary space activities.[196]

The conclusions reached by this author are similar to arguments examined above tied to the concept of aerospace, which holds that it is the intention of the USSR, if not deterred by a superior space power, to deny the use of space to its adversaries altogether.[197] Similar conclusions are reached by those in the United States who have said the US requires superior military space capabilities (building blocks) as "insurance" against the *possibility* of such Soviet actions.[198]

Perhaps the strongest motivation for both the USA and the USSR to avoid such a strategic competition in space is the monumental expense that would be involved. The USSR could not afford it and the USA, if it could, would not stand for it, as witnessed by the fate of Dyna-Soar and MOL. Seeking maximum guarantee of "security" against every imaginable danger would have to be done at the expense of other values. Moreover, the experience of the missile race has probably (and hopefully) taught the leaders of both sides that creating a deterrent against a nonexistent threat is often the quickest way to make that threat real.[199]

In the years since Sputnik the Soviet leaders have not attempted to deny the use of space to anybody despite their possession of a (limited) capability to do so. It is to be wondered why they would. They have never lodged an official *governmental* protest against American reconnaissance satellites, which constitute the main element of the American (and Soviet) military space programs. Their strategic exploitation of space has been based more on antimilitary propaganda and scientific and technical achievements (the latter being the fundamental American strategic propaganda use as well). Moreover, they have campaigned from 1958 onward for the banning of all military activities in space, dropping in 1963 the insistence that reconnaissance satellites be outlawed. These same reconnaissance satellites helped the Soviets and Americans to attain the 1972 agreements on limiting strategic arms, a goal sought by *both* sides.

Suppose the USSR were to violate the 1967 Space Treaty and the SALT agreements and to deny the US the use of space? Presuming the military value of space is as great as these spokesmen claim, is it not likely that the US would feel constrained to either (1) engage in an expensive competition to get it back, or (2) launch its ICBMs while its satellite-gleaned intelligence was not yet obsolete and before the USSR had a chance to use its newly won advantage for its own first strike? It is difficult to see how weapons in or control of space can add much to deterrence so long as the species continues to live on the planet earth, where both the USSR and the USA already have sufficient land- and sea-based forces to inflict unacceptable damage on their enemies regardless of the outcome of expensive jousts among the planets and the stars.

Soviet spokesmen have explicitly denied the possibility of confining a future war to outer space, calling it "utopian" and its authors "wild men."[200]

The opposite view from that described above is that space is the threshold of a new era of peace (not just on earth). This argument is put forth in another book which can be accused of erring on the opposite side.[201] The authors describe what they call the "Space Discovery," a process already under way by which the major powers will bury their rivalries in a cooperative effort to explore the solar system.[202] A similar stance is taken by Marion Levy, who suggests that the space program may take the place of current defense efforts as advances are made in disarmament.[203] Arguments against this possibility are made in the fictionalized but nevertheless very cogent *Report from Iron Mountain*.[204] The possibility of substituting space for defense has even drawn some credibility from those who oppose it. For example, James Killian, President Eisenhower's science advisor, expressed concern lest preoccupation with space "spectaculars" divert talent from and weaken our military effort.[205]

It is true that the USA and the USSR have gone further to achieve arms control in space than in any other area and may go further still. However, as long as the basic military competition continues, the potentiality of the military use of space, including uses not now made, even uses forbidden by international law, will remain.[206]

NASA has found that competition with the USSR rather than cooperation has promoted space spending, which still remains at levels far below those spent for defense. The very success of Project Apollo and the growing certainty after 1965 that the United States would get to the moon before the Soviet Union resulted in cuts in NASA's budget. As will be seen below, a suggestion by President Kennedy for a cooperative moon mission with the Russians prompted Congress to try to forbid use of the NASA budget for any such purpose. Nor have the Soviets been particularly cooperative, as will also be seen in a later chapter. Without any creditable enemy as in the "war system," a massive cooperative space program could not have enough political appeal to generate the economic sacrifices put up with to protect and promote "national security" and the respective "ways of life."[207] Even if the imagination and wonder of a new space age could be substituted for the tested Hobbesian motives of hostility and fear, these may prove to be passing moods. The most spectacular space achievements (Sputnik, *Vostok* and Apollo) are already over. No spectacular achievements of this kind loom in the immediate future, and, in any case, the novelty of space to the popular mind is already wearing thin.

Although efforts in Soviet-American space cooperation have been made, these have been the result rather than the cause of political detente. This detente has not seen increases in space endeavors replacing military efforts which remain on a grand scale, but rather it has had a minor limiting effect on military spending in the US and a decisively *negative* impact on space spending. Unless earthbound problems are suddenly solved or exacerbated, it appears most likely in light of events since Sputnik that peace will not come fluttering down from space, nor will war go zooming off into the cosmos.

NOTES

[1]*Soviet Space Programs, 1962, op. cit.*, p. 57.

[2]*Loc. cit.*

[3]See Raymond L. Gartoff, "Red War Sputniks in the Works?" *Missiles & Rockets, 3*, May 1958, p. 134.

[4]N. A. Alexandrov, "Why the USA Is Straining To Get Into Outer Space," *Soviet Fleet*, December 25, 1958, translated by John R. Thomas, Rand Corporation, Translation #T-112, pp. i-6.

[5]E. Korovin, "International Status of Cosmic Space," *International Affairs*, No. 1, January 1959, p. 56.

[6]G. I. Pokrovskii, "On the Problem of the Use of Cosmic Space," *International Affairs*, No. 7, July 1959, pp. 105-107.

[7]Donald G. Brennan, "Arms and Arms Control in Outer Space," in Lincoln P. Bloomfield, editor, *Outer Space: Prospects for Man and Society* (New York: Frederick A. Praeger, Inc., 1968), p. 174.

[8]G. P. Zhukov, "Space Espionage and International Law," *International Affairs*, No. 10, October 1960, p. 56; and Lincoln P. Bloomfield, "The Prospects for Law and Order," in American Assembly, Columbia University, *Outer Space: Prospects for Man and Society* (Englewood Cliffs, N.J.: Prentice-Hall, Inc., 1962), p. 173.

[9]Philip J. Klass, *Secret Sentries in Space* (New York: Random House, 1971), pp. 124, 125.

[10]Alton Frye, "Soviet Space Activities: A Decade of Pyrrhic Politics," in Bloomfield, *Outer Space, Prospects for Man and Society, op. cit.*, p. 189.

[11]Klass, *op. cit.*, pp. 127, 127n.; and Charles Sheldon II, "The Challenge of International Competition," Address to AIAA/NASA, Houston, November 6, 1964, in *International Cooperation and Organization for Outer Space*, Staff Report prepared for the Committee on Aeronautics and Space Science, US Senate, 89th Congress, 1st session, 1965, Document No. 56 (hereinafter known as *International Cooperation and Organization for Outer Space, 1965*), p. 459.

[12]*Soviet Space Programs, 1962, op. cit.*, p. 57; and Oleg Penkovsky, *The Penkovsky Papers*, translated by P. Denatin (London: Collins Clear Type Press, 1966), p. 243.

[13]See Klass, *op. cit.*, pp. 43, 58, 62.

[14]Quoted in Zhukov, "Space Espionage and International Law," *op. cit.*, p. 57.

[15]Thomas W. Wolfe, *Soviet Strategy at the Crossroads*, The Rand Corporation (Cambridge, Mass.: Harvard University Press, 1964), pp. 203-204; and *Soviet Space Programs, 1962, op. cit.*, pp. 223-224.

[16]V. D. Sokolovskii, editor, *Soviet Military Strategy*, translated by Herbert S. Dinerstein, Leon Goure and Thomas W. Wolfe, A Rand Corporation Research Study (Englewood Cliffs, N.J.: Prentice-Hall, Inc., 1963), pp. 424-427.

[17]"Leading on Earth and in Space," (anonymous), *International Affairs*, No. 9, September 1962, p. 5.

[18]See Wolfe, *supra*, p. 202; and G. P. Zhukov, "Problems of Space Law at the Present Stage," International Law Association Memorandum, Brussels, 1962, p. 49.

[19]Quoted in *Documents on International Aspects of the Exploration and Use of Outer Space, 1954-1962*, 88th Congress, 1st session, US Senate, Committee on Aeronautics and Space Sciences, 1963, Document No. 18, pp. 43, 46-47.

[20]*The National Space Project*, 85th Congress, 2nd session, US House, Select Com-

mittee on Astronautics and Space Exploration, 1958, House Report No. 1758, pp. 7, 37.

[21]John M. Logsdon, *The Decision To Go to the Moon: Project Apollo and the National Interest* (Cambridge, Mass.: MIT Press, 1970), ix-x.

[22]Thomas A. Sturm, *The USAF Scientific Advisory Board: Its First Twenty Years, 1944-1964* (Washington, D.C.: Government Printing Office, 1967), pp. 81-83; and Loyd S. Swenson, Jr., James M. Grimwood and Charles C. Alexander, *This New Ocean, A History of Project Mercury* (Washington, D.C.: NASA, 1966), p. 73.

[23]Logsdon, *op. cit.*, p. 47.

[24]*The National Space Project, op. cit.*, p. 37.

[25]*The United States and Outer Space*, 85th Congress, 2nd session, US House, Select Committee on Astronautics and Space Exploration, House Report No. 2710, 1959, p. 6.

[26]*Congressional Record, 107* (96), June 8, 1961, p. 9174.

[27]Quoted in Eugene M. Emme, editor, *The Impact of Air Power* (Princeton: Van Nostrand, 1959), p. 844.

[28]See Amitai Etzioni, *The Moon-Doggle, Domestic and International Implications of the Space Race* (Garden City, N.Y.: Doubleday & Co., Inc., 1964), p. 140.

[29]C. D. Perkins, "Man and Military Space," *Journal of the Royal Aeronautical Society, 67* (631), July 1963, pp. 407, 409; and Logsdon, *op. cit.*, p. 44.

[30]Logsdon, *op. cit.*, pp. 29, 31-32.

[31]Major General John B. Medaris, *Countdown for Decision* (New York: G. P. Putnam's Sons, 1960), pp. 245-247; Robert Rosholt, *An Administrative History of NASA, 1958-1963* (Washington, D.C.: NASA, 1966), p. 47; and *Manned Space Flight Programs of the NASA: Projects Mercury, Gemini and Apollo*, 87th Congress, 2nd session, US Senate, Committee on Aeronautics and Space Sciences, 1962, pp. 162-164.

[32]Logsdon, *op. cit.*, pp. 33, 51, 77.

[33]Dryden in Jerry and Vivian Grey, editors, *Space Flight Report to the Nation* (New York: Basic Books, 1962), p. 181.

[34]Penkovsky, *op. cit.*, p. 243; and Erlend A. Kennan and Edmund H. Harvey, *Mission to the Moon, A Critical Examination of NASA and the Space Program* (New York: William Morrow and Co., 1969), p. 228.

[35]Barbara M. DeVoe, "Appendix D—Biographies of the Soviet Cosmonauts," *Soviet Space Programs, 1966-1970, op. cit.*, pp. 603-605. See also Barbara M. DeVoe, *Astronaut Information: American and Soviet (Revised)*, Library of Congress, Congressional Reference Service, February 12, 1973, pp. 3-47.

[36]John D. Holmfeld, "Organization of the Soviet Space Program," in *Soviet Space Programs, 1966-1970, supra*, pp. 86, 87.

[37]September 22, 1960 address before the UN, in John C. Cooper, *Explorations in Aerospace Law, Selected Essays*, edited by Ivan A. Vlasic, (Montreal: McGill University Press, 1968), p. 287.

[38]*Review of the Soviet Space Program*, 90th Congress, 1st session, US House, Committee on Science and Astronautics, November 10, 1967 (hereinafter *Review of the Soviet Space Program, 1967*), p. 80.

[39]Richard Witkin, "Pros and Cons," in Walter Sullivan, editor, *America's Race for the Moon. The New York Times Story of Project Apollo* (New York: Random House, 1962), p. 150.

[40]Trevor Gardner (Chairman of the Board and President, Hycon Manufacturing Company), "Military Effects," in Greys, *op. cit.*, pp. 128-131; and General Bernard A. Schriever in the same volume, p. 193.

[41]*Congressional Record, 108* (151), August 24, 1962, pp. 16, 445.

[42]*NASA Authorization for Fiscal Year 1964*, Hearings, 88th Congress, 1st session, US Senate, Committee on Aeronautics and Space Sciences, 1963, Part 2, p. 633.

[43]Etzioni, *op. cit.*, p. 139.

[44]John R. Walsh, "NASA: Talk of Togetherness with Soviets Further Complicates Space Politics for the Agency," *Science, 142* (3588), October 4, 1963, pp. 37-38.

[45]Lillian Levy, "Conflict in the Race for Space," in Levy, editor, *Space: Its Impact on Man and Society* (New York: W. W. Norton and Co., Inc., 1965), pp. 193-194.

[46]Arthur R. Kantrowitz (Vice-President, Avco), in Greys, *op. cit.*, p. 194. See also Etzioni, *op. cit.*, p. 131.

[47]Cited in Robert H. Puckett, *The Military Role in Space—A Summary of Official Public Justifications*, The Rand Corporation, Paper #P-2681, August 1962, p. 24. This quote brings to mind how President Johnson inspired the resurrection of the expression "guns and butter."

[48]*Ibid.*, pp. 28-29; and General Bernard A. Schriever, quoted in Emme, *The Impact of Air Power, op. cit.*, p. 844.

[49]Puckett, *op. cit.*, pp. 8-11. Quoted material on p. 9.

[50]*Ibid.*, p. 3.

[51]Perkins, *op. cit.*, p. 399.

[52]General Thomas D. White, Chief of Staff, US Air Force, in *Department of Defense Appropriations for 1962*, Hearings, 87th Congress, 1st Session, US House, Committee on Appropriations, Subcommittee on Department of Defense Appropriations, Part 3, 1961, pp. 409-410.

[53]General Curtis E. LeMay, Chief of Staff, US Air Force, in *Air Force Information Policy Letter for Commanders, XVI* (9), May 1, 1962, as cited in Puckett, *op. cit.*, p. 8.

[54]White in *Department of Defense Appropriations Hearings, 1962, op. cit.*, p. 291.

[55]Roswell L. Gilpatric, Deputy Secretary of Defense, in *Air Force Information Policy Letter for Commanders*, Supplement, No. 108, July 1962, p. 16, as cited in Puckett, *op. cit.*, pp. 12-13.

[56]Dr. Harold Brown, Director of Defense Research and Engineering, in *Department of Defense Appropriations for 1962*, Hearings, *op. cit.*, pp. 335.

[57]Etzioni, *op. cit.*, pp. 128-129.

[58]*Ibid.*, p. 121; and J. H. Rubel, "The Military in Space," *Bulletin of the Atomic Scientists, 19* (5), May 1963, pp. 20-22.

[59]Etzioni, *supra*, p. 131.

[60]Alton Frye of The Rand Corporation as quoted by Edwin Diamond, *The Rise and Fall of the Space Age* (Garden City, N.Y.: Doubleday and Co., Inc., 1964), p. 14.

[61]John N. Wilford, *New York Times*, August 14, 1968, as cited in Kennan and Harvey, *op. cit.*, pp. 196, 310; and William Shelton, *Soviet Space Exploration* (New York: Washington Square Press, Inc., 1968), pp. 257, 287.

[62]1968 NASA Authorization Hearings quoted in Kennan and Harvey, *op. cit.*, p. 207.

[63]*Washington Post* and *Aviation Week* as cited in *ibid.*, p. 208, xvi.

[64]See Puckett, *op. cit.*, pp. 8-11, 16.

[65]Nicholas Daniloff, *The Kremlin and the Cosmos* (New York: Alfred A. Knopf, 1972), pp. 77-83; Foy D. Kohler and Dodd C. Harvey, "Administering and Managing the US and Soviet Space Programs," *Science, 169*, September 11, 1970, p. 1050; and John D. Holmfeld, "Organization of the Soviet Space Program," *Soviet Space Programs, 1966-1970, op. cit.*, pp. 70, 71.

[66]Holmfeld, *supra*, pp. 83-84; and Daniloff, *supra*, pp. 43, 77.

[67]Daniloff, *supra*, p. 56.

[68]Charles S. Sheldon II, *Review of the Soviet Space Program With Comparative US Data* (New York: McGraw-Hill Book Co., Inc., 1968), p. 157.

[69]Robert Salkeld, *War and Space* (Englewood Cliffs, N.J.: Prentice-Hall, Inc., 1970), p. 186; and Holmfeld, "Organization of the Soviet Space Program," *Soviet Space Programs, 1966-1970, op. cit.*, pp. 84-87.

[70]Holmfeld, *supra*, p. 112.

[71]John S. Foster, Jr., before Senate Armed Services Committee, March 18, 1971, pp. 2-2 and 2-4 as cited in *ibid.*, p. 114.

[72]B. W. Augenstein, *Policy Analysis in the National Space Program*, The Rand Corporation, Paper #P-4137, July 1969, p. 16.

[73]See Schriever in Levy, *op. cit.*, p. 67.

[74]Holmfeld, "Organization of the Soviet Space Program," *Soviet Space Programs, 1966-1970, op. cit.*, pp. 86-87.

[75]*The National Space Project, op. cit.*, pp. 7, 37.

[76]Charles S. Sheldon II, "Overview, Supporting Facilities and Launch Vehicles of the Soviet Space Program," *Soviet Space Programs, 1966-1970, op. cit.*, pp. 130-148. Evidently even the heavy *Proton* vehicle, which is not used for military purposes, was sold to Khrushchev by Korolev as an ICBM capable of delivering a 100-megaton warhead. See Jim Orberg, "Korolev," *Space World*, K-5-125, May 1974, pp. 18, 21.

[77]See Schriever in Levy, *op. cit.*, p. 66.

[78]Carl H. Amme, Jr., "The Implications of Satellite Observations for US Policy," in Frederick J. Ossenbeck and Patricia C. Kroeck, editors, *Open Space and Peace, A Symposium on Effects of Observation* (Stanford, Calif.: Stanford University, The Hoover Institute, 1964), p. 109.

[79]Augenstein, *op. cit.*, p. 49.

[80]Penkovsky, *op. cit.*, p. 243.

[81]Augenstein, *op. cit.*, pp. 24-25, 47; Kennan and Harvey, *op. cit.*, pp. 91-92, 94; and Etzioni, *op. cit.*, pp. 135-136.

[82]Klass, *op. cit.*, p. 178.

[83]Etzioni, *op. cit.*, p. 136. For a thorough discussion of NASA-DoD relationships, see Augenstein, *op. cit.*, pp. 45-58.

[84]Etzioni, *op. cit.*, pp. 135-136; Kennan and Harvey, *op. cit.*, pp. 35, 310; and William J. Normyle, "NASA Molding Post-Apollo Plans," *Aviation Week and Space Technology*, 89 (26), December 23, 1968, p. 17.

[85]Shelton, *Soviet Space Exploration, op. cit.*, p. 257. See also Charles S. Sheldon II, "Postscript," *Soviet Space Programs, 1966-1970, op. cit.*, pp. 516-519, 526-529.

[86]See Gyula Gal, *Space Law* (Leyden: A. W. Sijthoff and Dobbs Ferry, N.Y.: Oceana Publications, Inc., 1969), p. 168.

[87]Charles S. Sheldon II, "Program Details of Unmanned Flights," *Soviet Space Programs, 1966-1970, op. cit.*, pp. 174-176.

[88]Puckett, *op. cit.*, p. 26.

[89]See also Perkins, *op. cit.*, pp. 406-407.

[90]Schriever in Levy, *op. cit.*, pp. 60-61.

[91]Sheldon, "Overview . . . ," *Soviet Space Programs, 1966-1970, op. cit.*, pp. 143-146.

[92]Paul Kecskemeti, "Outer Space and World Peace," in Joseph M. Goldsen, editor, *Outer Space in World Politics* (New York: Frederick A. Praeger, 1963), p. 33.

[93]See Sheldon, *Review of the Soviet Space Program . . . , op. cit.*, p. 81; and Joseph G. Whelan, "Political Goals and Purposes of the USSR in Space," *Soviet Space Programs, 1966-1970, op. cit.*, pp. 43-44.

[94]Kecskemeti, *op. cit.*, pp. 31-32.

[95]Sheldon, *Review of Soviet Space Program . . . , op. cit.*, p. 121.

[96]Sheldon, "Program Details of Unmanned Flights," *Soviet Space Programs, 1966-1970, op. cit.*, p. 172.

[97]Charles S. Sheldon II, "Soviet Military Space Activities," *Soviet Space Programs, 1966-1970, op. cit.*, pp. 324, 334.

[98]Sheldon, *Review of Soviet Space Program . . . , op. cit.*, pp. 122-123; and Klass, *op. cit.*, pp. 109-110.

[99]Sheldon, *supra*, p. 123.

[100]The announcement appears in Sheldon, "Program Details of Unmanned Flights," *Soviet Space Programs, 1966-1970, op. cit.*, p. 173. It mentions no military uses whatsoever. See also *ibid.*, p. 180.

[101]*Ibid.*, p. 174; Appendix A in *Soviet Space Programs, 1966-1970, op. cit.*, p. 558; Sheldon, *Soviet Space Programs, 1971, op. cit.*, pp. 64-70; and "Satellite Report," in *Space World*, K-6-126, June 1974, p. 30.

[102]Sheldon, *Review of the Soviet Space Program, op. cit.*, p. 123.

[103]D. S. Greenberg, "Space Accord: NASA's Enthusiasm for East-West Cooperation Is Not Shared by Pentagon," *Science, 136*, April 13, 1962, p. 138; and Don E. Kash, *The Politics of Space Cooperation* (Purdue University Studies, Purdue Research Foundation, 1967), p. 35.

[104]Sheldon, "Program Details of Unmanned Flights," *Soviet Space Programs, 1966-1970, op. cit.*, pp. 175-176.

[105]Klass, *op. cit.*, pp. 109, 187; Kash, *op. cit.*, p. 35; Diamond, *op. cit.*, p. 106; and Sheldon, *Review of the Soviet Space Program . . . , op. cit.*, pp. 125-126.

[106]Sheldon, "Program Details of Unmanned Flights," *Soviet Space Programs, 1966-1970, op. cit.*, p. 188; and Appendix A in *ibid.*, p. 541. These may have been interceptor tests.

[107]Klass, *op. cit.*, pp. 180-181.

[108]Sheldon, "Program Details of Unmanned Flights," *Soviet Space Programs, 1966-1970, op. cit.*, pp. 181-183; and Appendix A in *ibid.*, pp. 533, 534.

[109]Seth T. Payne and Leonard S. Silk, "The Impact on the American Economy," in Bloomfield, editor, *Outer Space, Prospects for Man and Society, op. cit.*, p. 98.

[110]Holmfeld, "Organization of the Soviet Space Program," *op. cit.*, pp. 112-114.

[111]Sheldon, "Overview . . . ," *op. cit.*, pp. 127-128.

[112]Sheldon, *Review of the Soviet Space Program, op. cit.*, pp. 129, 132; Fortune editors, *The Space Industry, America's Newest Giant* (Englewood Cliffs, N.J.: Prentice-Hall, Inc., 1962), pp. 21-22; and Robert W. Buchheim *et al., New Space Handbook: Astronautics and Its Applications* (New York: Vintage Books, 1963), pp. 183-184.

[113]Salkeld, *op. cit.*, p. 138.

[114]Shelton, *Soviet Space Exploration, op. cit.*, p. 244.

[115]Sheldon, "Program Details of Unmanned Flights," *op. cit.*, pp. 180-186.

[116]*Ibid.*, pp. 180-181; Appendix A in *Soviet Space Programs, 1966-1970*, pp. 536, 544; and Shelton, *supra*, p. 244.

[117]See Klass, *op. cit.*, pp. 183, 190-192; Payne and Silk, *op. cit.*, p. 96; *The Space Industry . . . , op. cit.*, p. 23; and Erin B. Jones, *Earth Satellite Telecommunications Systems and International Law* (Austin: University of Texas, 1970), p. 30.

[118]Sheldon, "Soviet Military Space Activities," *op. cit.*, p. 327; and Klass, *supra*, pp. 182-183.

[119]Penkovsky, *op. cit.*, p. 320.

[120]Sheldon, "Overview . . . ," *op. cit.*, pp. 122-124; Sheldon, "Program Details of Unmanned Flights," *op. cit.*, pp. 177-178; and Sheldon, "Soviet Military Space Activities," *op. cit.*, pp. 333-334.

[121]Sheldon, "Program Details of Unmanned Flights," *op. cit.*, pp. 178-180; and Sheldon, "Soviet Military Space Activities," *op. cit.*, pp. 327-331.

[122]Arnold W. Frutkin, *International Cooperation in Space* (Englewood Cliffs, N.J.: Prentice-Hall, 1965), p. 148.

[123]Lt. General James M. Gavin, *War and Peace in the Space Age* (New York: Harper and Bros., 1958), p. 224.

[124]Michael N. Golovine, *Conflict in Space, A Pattern of War in a New Dimension* (London: Temple Press, Ltd., 1962), pp. 90-91; and Etzioni, *op. cit.*, p. 131.

[125]Golovine, *loc. cit.*; and Gal, *op. cit.*, p. 162.

[126]See Klass, *op. cit.*, p. 216; and Sheldon, "Soviet Military Space Activities," *op. cit.*, p. 339.

[127]Sheldon, *supra*, pp. 339-343; and "Postscript" in *Soviet Space Programs, 1966-1970*, *op. cit.*, pp. 523-525. See also *Soviet Space Programs, 1971, A Supplement to the Corresponding Report Covering the Period 1966-1970*, Staff Report Prepared for the Use of the Committee on Aeronautics and Space Sciences, US Senate, by the Science Policy Research Division, Congressional Research Service, Library of Congress, April 1972, 92nd Congress, 2nd session (hereinafter *Soviet Space Programs, 1971*), pp. 49-53; and Sheldon, *United States and Soviet Progress in Space: Summary Data Through 1972 and a Forward Look*, Congressional Research Service, Library of Congress, January 29, 1973 (hereinafter *US and Soviet Progress in Space*, 1972), p. 36.

[128]*Soviet Space Programs, 1971*, *op. cit.*, pp. 50-51.

[129]Etzioni, *op. cit.*, p. 132; and Golovine, *op. cit.*, pp. 89, 93.

[130]For example, see V. Glasov (probably Colonel V. Glazov), "Cannibals in Space," *New Times*, No. 24, June 11, 1963, pp. 12-13; B. Teplinskii, "The Pentagon, The Mad Men, and the Moon," *Red Star*, January 10, 1965; and Colonel V. Glazov, "Cosmic Weapons," *Red Star*, January 26/27, 1965, as cited in *Soviet Space Programs, 1966-1970*, *op. cit.*, pp. 324, 325, 334.

[131]Sheldon, "Soviet Military Space Activities," *op. cit.*, pp. 326, 334.

[132]Moscow Domestic Service, February 21, 1963, as cited in Wolfe, *Soviet Strategy at the Crossroads*, *op. cit.*, p. 208.

[133]Salkeld, *op. cit.*, p. 127. See also Sheldon, "Soviet Military Space Activities," *op. cit.*, pp. 334-335.

[134]Shelton, *Soviet Space Exploration*, *op. cit.*, pp. 252-263.

[135]Sheldon, "Soviet Military Space Activities," *op. cit.*, pp. 335-336, 337. Two SS-9 flights in late 1966 were probably either interceptors or FOBS according to Sheldon (see note #129 above).

[136]Sheldon, "Soviet Military Space Activities," *supra*, pp. 337-338; and International Institute for Strategic Studies, *The Military Balance, 1969-1970* (London: 1969), p. 6.

[137]Sheldon, *ibid.*, pp. 335, 336; and Ian Smart, *Advanced Strategic Missiles: A Short Guide* (International Institute for Strategic Studies, Adelphi Paper No. 63, December 1969, London), p. 24.

[138]Sheldon, *ibid.*, p. 338; and Sheldon, "Postscript," *Soviet Space Programs, 1966-1970*, *op. cit.*, p. 523. See also *Soviet Space Programs, 1971*, *op. cit.*, p. 49.

[139]Sheldon, "Soviet Military Space Activities," *op. cit.*, pp. 337, 338; and Salkeld, *op. cit.*, pp. 129, 137.

[140]See Klass, *op. cit.*, p. 20; and Fermin J. Krieger, *Behind the Sputniks, A Survey of Soviet Space Science* (Washington, D.C.: The Rand Corporation, Public Affairs

Press, 1958), p. 233, which contains the text of the announcement.

[141]*Journal-American*, November 25, 1957, pp. 1, ff, as quoted in Daniloff, *op. cit.*, p. 127.

[142]Interview with the editor of *Dansk Kolkestyre* in *For Victory in Peaceful Competition with Capitalism* (New York: E. P. Dutton and Co., 1960), p. 23.

[143]November 14, 1959 speech to All-Union Congress of Soviet Journalists as quoted in Klass, *op. cit.*, p. 43.

[144]*Pravda*, January 28, 1959.

[145]Quoted in Klass, *op. cit.*, p. 62.

[146]*Ibid.*, p. 36.

[147]See Adam B. Ulam, *Expansion and Coexistence, A History of Soviet Foreign Policy, 1917-1967* (New York: Praeger, 1971), p. 636.

[148]Frye, "Soviet Space Activities: A Decade of Pyrrhic Politics," *op. cit.*, pp. 202-203.

[149]See Arnold L. Horelick, "The Soviet Union and the Political Uses of Outer Space," in Goldsen, *op. cit.*, pp. 51-52; and Joseph M. Goldsen and Leon Lipson, *Some Political Implications of the Space Age*, Rand Corporation, Paper #P-1435, February 24, 1958, pp. 5-6, 7; and Joseph M. Goldsen, *Outer Space and the International Scene*, Rand Corporation, Paper #P-1688, May 6, 1959, pp. 7-8.

[150]Sokolovskii, *op. cit.*, pp. 306, 313.

[151]Shelton, *Soviet Space Exploration, op. cit.*, p. 41.

[152]See Merle Fainsod, *How Russia Is Ruled* (Cambridge: Harvard University Press, 1963), pp. 487-489.

[153]Klass, *op. cit.*, p. 70. See also *Pravda*, October 25, 1961 and *Pravda*, February 23, 1963.

[154]Thomas W. Wolfe, *Soviet Strategy at the Crossroads*, Rand Corporation (Cambridge, Mass.: Harvard University Press, 1964), pp. 190-192. See also *Izvestiia*, July 19, 1962 and November 8, 1963; and *Krasnaia Zvezda*, November 13, 16, 1963 and March 28, 1964. ABM claims are also summarized in *Sokolovskii*, *op. cit.*, pp. 315-316, note B.

[155]Klass, *op. cit.*, pp. 196-197. See also *The Military Balance, 1968-1969, op. cit.*, p. 5.

[156]International Institute for Strategic Studies, *The Military Balance, 1970-1971* (London, 1970), p. 7; and IISS, *The Military Balance, 1972-1973* (London, 1973), pp. 7, 83.

[157]Klass, *op. cit.*, pp. 28-29, 38.

[158]Diamond, *op. cit.*, p. 20; and Vernon Van Dyke, *Pride and Power, The Rationale of the Space Program* (Urbana: University of Illinois, 1964), p. 140.

[159]*Missiles and Rockets*, 7, October 10, 1960, pp. 12-13.

[160]Klass, *op. cit.*, pp. 67, 106-107. See also Joseph Alsop, "Facts About the Missile Balance," *Washington Post*, September 25, 1961, p. A-13.

[161]*The Military Balance, 1969-1970, op. cit.*, p. 55. The author cannot account for the differences between the figures cited by this source and those of Klass and Alsop referred to above.

[162]Klass, *op. cit.*, p. 117.

[163]Diamond, *op. cit.*, p. 27.

[164]Klass, *op. cit.*, p. 120; Etzioni, *op. cit.*, pp. 123-124; and Diamond, *supra*, pp. 25-27.

[165]Etzioni, *supra*, p. 123.

[166]Shelton, *Soviet Space Exploration, op. cit.*, pp. 255-256.

[167]Donald G. Brennan, "Arms and Arms Control in Outer Space," in Bloomfield, *Outer Space, Prospects for Man and Society, op. cit.,* pp. 163-164.

[168]Shelton, *supra,* pp. 242-243; and Klass, *op. cit.,* xv.

[169]Bloomfield, "The Quest for Law and Order," in Bloomfield, *supra,* p. 125.

[170]Klass, *op. cit.,* pp. 215, 217-218. The texts of the Strategic Arms Limitation Agreements of 1972 are available in *Arms Control, A Selection of Readings from Scientific American* (San Francisco: W. H. Freeman & Co., 1973), pp. 260-273.

[171]Klass, *op. cit.,* pp. 198, 200, 209.

[172]Charles S. Sheldon II, "Peaceful Applications," in Bloomfield, *supra,* pp. 55, 63; and *Aviation Week and Space Technology,* 99 (17), October 22, 1973, p. 16.

[173]Lillian Levy, "Conflict in the Race for Space," *op. cit.,* p. 197.

[174]Puckett, *op. cit.,* p. 20; and Salkeld, *op. cit.,* pp. 116-117.

[175]Buchheim *et al., op. cit.,* p. 183.

[176]Shelton, *Soviet Space Exploration, op. cit.,* p. 260.

[177]Etzioni, *op. cit.,* p. 133.

[178]See *Soviet Writings on Earth Satellites and Space Travel* (New York: Citadel Press, 1958), pp. 196, 199-200.

[179]*Krasnaia Zvezda,* January 9, 1962, pp. 2-3.

[180]Shelton, *Soviet Space Exploration, op. cit.,* p. 260.

[181]Etzioni, *op. cit.,* p. 133.

[182]Levy, "Conflict in the Race for Space," *op. cit.,* p. 197.

[183]Etzioni, *op. cit.,* p. 133.

[184]See Brennan, *op. cit.,* p. 160.

[185]Klass, *op. cit.,* p. 169.

[186]*Loc. cit.,* and Salkeld, *op. cit.,* pp. 149-150.

[187]Klass, *supra,* pp. 170-172, 181; and Ted Greenwood, "Reconnaissance and Arms Control," *Arms Control . . . , op. cit.,* pp. 223-234 (from the February 1973 *Scientific American*).

[188]See Sheldon, "Program Details of Unmanned Flights," *op. cit.,* pp. 180-186; and *Soviet Space Programs, 1971, op. cit.,* pp. 15-16.

[189]Perkins, *op. cit.,* p. 407.

[190]Salkeld, *op. cit.,* p. 104; and *Space World,* J-9-117, September 1973, p. 30.

[191]Etzioni, *op. cit.,* pp. 118-119.

[192]Golovine, *op. cit.,* pp. 89, 93.

[193]Frye, "Soviet Space Activities: A Decade of Pyrrhic Politics," *op. cit.,* p. 191.

[194]See Sheldon, "Soviet Military Space Activities," *op. cit.,* pp. 348-349.

[195]See Salkeld, *op. cit.,* entire book.

[196]*Ibid.,* p. 168.

[197]White, *Department of Defense Appropriations for 1962,* Hearings, *op. cit.,* p. 291; and General C. H. Mitchell, Vice Commander, US Air Force Systems Command, in *Missiles and Rockets, X,* January 15, 1962, p. 10.

[198]Roswell Gilpatric as cited in Puckett, *op. cit.,* p. 13.

[199]Van Dyke, *op. cit.,* p. 71.

[200]See V. Pechorkin, "The Pentagon Theoreticians and the Cosmos," *International Affairs,* No. 3, March 1961, pp. 34, 36; and Professor Georgii Pokrovskii, "Crime in Space," *New Times,* No. 25, June 20, 1962, pp. 9-11.

[201]See Frank Gibney and George J. Feldman, *The Reluctant Spacefarers. A Study in the Politics of Discovery* (New York: New American Library, 1965), entire work.

[202]*Ibid.,* see pp. 150-174.

[203]Levy, "Conflict in the Race for Space," *op. cit.,* p. 210.

[204]*Report from Iron Mountain on the Possibility and Desirability of Peace* (New York: Dial Press, 1967), pp. 62-63, 66, 77-78, 86.

[205]James R. Killian, Jr., "Shaping a Public Policy for the Space Age," in American Assembly, *Outer Space, Prospects for Man and Society, op. cit.,* pp. 187-188.

[206]Brennan, *op. cit.,* pp. 146-147.

[207]*Report from Iron Mountain . . . , op. cit.,* p. 66.

4 Motivations and Goals

Speaking generally, the motives and goals behind the Soviet space program probably do not differ markedly from those of the United States. Both countries seek to expand their stores of scientific knowledge; to explore new frontiers; to associate themselves with progress and a role in the human future; to enhance the pride and prestige of each nation, its leading institutions and political personalities; to exploit space for earthbound benefits; to promote the employment of skilled persons and stimulate technical progress; to promote education; to strengthen national security, and to win new friends, keep old friends and demoralize enemies at home and throughout the world.[1] Leaders in both the US and the USSR have also expressed the hope that world peace may result from cooperation in the exploration of space.

Perhaps the most important general area of difference in space goals is that in the time frame set by the space programs of each country. In comparison with the United States, the USSR has done more in the way of setting long-range space goals just as it has in setting other kinds of goals. Working within the framework of five- and seven-year plans, the Soviet space program may not face the restrictions inherent in the yearly budgets of the United States. The continuing policy of the USSR has also been to impose restrictions on the immediate demands of consumers, collective farmers and wage workers in order to maintain high rates of investment (as in space) with the promise of payoffs in the more or less indefinite future. To the extent that scientists through the mechanism of the Soviet Academy of Sciences play a more active role in the planning of and as a constituency of support for their space program than is the case in the US, one may expect greater support for long-range planning than would come from the dominant constituencies in America such as the aerospace industry, areas of economic impact of space policy, and the public at large. In the US such long-term support is likely to come only from NASA, from a small segment of the scientific community and from a small section of the public which looks into space with the perspective of galaxies and the time frame of centuries.

The Soviet Union has consistently emphasized its long-range goal in space as being the manned exploration and colonization of the solar system.[2] Soviet spokesmen, particularly the scientific community, have revealed their plans for the exploration of space several years before programs reached the operational stages. The considerable and early emphasis that the Soviet Union placed on developing programs of manned orbital flight is an indicator of its long-term goal of manned exploration of the planets.

One author contends that the greater emphasis in the Soviet program on bioastronautics, or the study of the problems associated with sustaining human life in space, is indicated by the Soviets placing a woman in orbit and orbiting a physician to make medical observations.[3] The same may be said of the Soviet choice of an earthlike atmosphere and air-lock system for the *Voskhod* and *Soiuz* spacecraft.* These considerations may point to requirements for sustaining life in space for periods considerably beyond the time frames involved in all manned Soviet flights to date. That such policies have often involved ostensibly inefficient use of booster thrust, spacecraft load and weight, and orbit time may in part be explained by the prospects for future use when what is now inefficient will be vital. For example, the considerable weight added to *Voskhod* by its air-lock and time delays occasioned by depressurization required in atmospheres other than the pure oxygen used in American flights may indicate plans for much larger Soviet vehicles with larger crews and longer stays in space. Air locks will permit the main body of a crew to work in shirtsleeves while particular members egress in space suits.[4] Similarly, breathing a pure oxygen atmosphere over very long periods produces adverse physical reactions which do not occur in the "natural" nitrogen-oxygen atmosphere at sea-level pressure employed in all Soviet manned flights. It should be noted that American flights after Sputnik operated generally under constraints of weight and time owing to relatively smaller booster size for comparable missions and the psychology of a "space race" with the Soviets which required more "efficient" utilization of parameters than Soviet flights. With-

*Beginning with the space shuttle, the US plans to shift to use of spacecraft cabin atmosphere like that used by the Soviets from the start.

out these pressures the Americans may have made different plans, but the fact remains that they did not.*

Looking at many of the same Soviet missions, another author has come to dramatically different conclusions. He contends that the *Vostok* and *Voskhod* missions, which included such "firsts" as putting a woman, a scientist and a physician in space, use of an air lock, and a space walk, were the result of hasty efforts to best the Americans in a race for prestige that was forced on Soviet space scientists by political leaders, most especially Khrushchev.⁵ The test of time appears to have confirmed this analysis. *Vostok* with six missions and *Voskhod* with only two were largely dead-end projects, as was the American project Mercury. The Gemini program by contrast, and also the Soviet *Soiuz* program, developed techniques such as orbital maneuver, long duration in space and orbital docking, which were technological prerequisites of more ambitious missions. This applies especially in the way in which Gemini developed techniques later used in the Apollo and Skylab missions.

Beyond the questions of the time frame involved in the planning of the two space programs, it is difficult to say whether one program or the other exhibits a greater degree of coordination of the various goals that both countries share in their planning and operation. Chapter 3 indicated that both countries coordinate military and nonmilitary goals at least to some extent. It was the proclaimed policy of the Eisenhower Administration, as revealed by its relative lack of commitment to a broad and massive effort in space and by the disparagement of a military-industrial complex and scientific-technological elite, to be wary of the political and social consequences associated with the planning and organization of an extensive unified space program.⁶ This is, of course, in keeping with the traditional American philosophies of limited government, competition, and political pluralism. The political leaders of the USSR take generally opposite points of view and are themselves to a large extent members of the very groups stigmatized by the American president.

Although the Congress and subsequent administrations had fewer qualms about the consequences associated with the organization and implementation of an ambitious space program, criticisms persist that American space efforts suffer from a lack of coordination in planning and a policy which puts the emphasis on major missions (man in space, man on the moon, meteorology and communications) rather than on the more distant and

---

*Weight constraints led in part to the choice of a pure oxygen atmosphere with tragic results for the crew of Apollo 4. The nitrogen in the atmosphere of *Soiuz* 11 caused fatal pulmonary embolisms for its crew when a valve failed.

general goals envisioned in the Soviet program. In particular the lack of coordination between NASA and DoD has been criticized along with an exclusive emphasis on the moon mission and the absence of consensus on American space goals in the post-Apollo period.[7]

As they compare with Soviet programs, major US missions like Gemini, Apollo, Skylab and the space shuttle exhibit *technical*, if not philosophic or scientific, continuity. It is NASA's style, indeed America's style, to move from one technical milestone to the next without specifying some penultimate goal that might emerge from the whole process. Indeed, NASA is afraid to associate the space shuttle with the manned planetary missions that the shuttle will make possible, because these missions are more likely to be the targets of congressional and public attack than the technological building blocks that make them possible. The Soviet style of proclaiming goals in space, like the Soviet style of proclaiming goals of Communism, overtaking the West, etc., seems to focus on the ultimate, while often ignoring the technical, means of arriving there. Thus, the Soviets proclaim their intention to colonize the planets, while NASA, shying away from such proclamations, seeks to build the equipment necessary.

Differences in political processes between the US and the USSR unquestionably contribute to the styles of planning and formulation of space goals in the two countries. The question of time frame has already been discussed. In addition, American policies and plans are shaped by and must ultimately be approved by a much wider and more diverse constituency than Soviet plans. In particular the congressional committee structure, the public media and the constituencies of administrative agencies like NASA serve to put greater emphasis on the question "*Cui bono?*" and to lay greater stress on the details of hardware, personnel and mission activities. The requirements of assuring broad and diverse public and interest group support has tended to focus American attempts on particular missions to the exclusion of questions about where the successive missions ultimately lead and why the US is in space in the first place. As will be further detailed in this chapter, the planning of American space policies and their justification to the public and special elites has focused on immediate and short-range returns, whether they be in terms of material payoffs in applications, employment and technological spinoff or in less tangible terms of political pride, prestige and international cooperation.

SCIENTIFIC GOALS

The Soviet Union has generally emphasized scientific discovery and exploration in the service of man as the *exclusive* goal of its space program. As one cosmonaut put it, "Soviet space explorers are moved by one aim: to gain maximum scientific information about space in general, and the planets of the solar system in particular, with the least expenditure of effort and resources."[8] Soviet scientists probably agree to a large extent with some American scientists who see the opportunity to visit the moon and the planets as the "most exciting scientific prospect of our time."[9] It should be emphasized that in fact the Soviet program's scientific motives in space are not associated with the advancement of science generally or the advance of science for the sake of science. Conditioned by the long-range goals of colonizing and exploiting the planets, the Soviet emphasis is on science for the sake of man's needs, albeit indistinct needs, which will be met in the distant future and at the expense of very pressing, distinct and immediate needs.[10]

Is the Soviet claim to an exclusive interest in space science merely put forth as a claim for propaganda purposes as a cover for other motives with less popular appeal and more difficult to reconcile with the proclaimed benevolence and progressiveness of all Soviet policies? The obvious answer is yes. This is evident from the previous chapter, which observed that most Soviet and American launches in recent years are of a clearly military nature. The Soviets maintain their claim to a wholly scientific program simply by refusing to acknowledge the existence of their military flights. If one overlooks this obviously false aspect of the Soviet claim and excludes from consideration the roughly equivalent military space efforts of the two powers, the Soviet claim does acquire a certain validity.

Certainly the justifications publicly offered and the goals specified for the Soviet space program have been expressed in terms of broad scientific goals. As already indicated, certain characteristics of the missions themselves, such as the orbiting of a woman and a physician (and a doctor of science), the use of air locks and earthlike environments, appear to be in keeping with the broadly scientific goal of Communist conquest of the planets.* However, the accomplishment of these feats at a time when the USSR is so far from the technical capability of manned flight to the moon or to the planets appears to expose them as intended, at least in part, to accomplish propaganda purposes.

---

*A US scientist, a Ph.D. in Geology, Dr. Harrison H. Schmitt, was landed on the moon as a part of the mission of Apollo 17. More recently a medical doctor, Naval Commander Joseph P. Kerwin, was part of the crew of Skylab 2 in May-June, 1973. Owen Garriot aboard Skylab 3 was a Ph.D. engineer, as was Edward Gibson on Skylab 4. There are no women in the American astronaut corps or any plans to use women astronauts, although NASA tests recently indicated that women are physically qualified, and NASA has said women will be passengers on the space shuttle.

The Communist and Russian emphasis on eschatology and the high place accorded science in both Communist ideology and the Soviet system augur well for Soviet goals of scientific exploration of space. The interest in science in the later writings of Engels and Lenin, along with the identification of Marxism or *scientific* socialism as *the* science of society, has contributed to the prestige of Soviet science and the practice of the regime of wrapping its political decisions in the mantle of scientific certainty. Projections of life under Communism find their only Western counterparts in the tales of science fiction. One may even take note of the Communist notion of freedom as expressed by Engels and Lenin that "freedom is the perception of necessity" or the ability of men to acquire and put to use the laws of science for their own ends. This is certainly a far cry from Western notions of freedom, which ignore scientific criteria and emphasize the autonomy of individuals and groups from any central control whether or not that control is founded on "scientific" certainty.

The comparative Soviet emphasis on long-range planning already described is witnessed above all in Soviet scientific plans elaborated well in advance which seem to culminate in and are a logical progression toward manned conquest of the solar system.[11] The discussions of Soviet scientists and political leaders generally extend beyond those missions already in the stages of implementation and execution to the most distant goals. American statements by contrast have tended to focus on particular missions already underway.[12] The main reasons which account for this difference are the asymmetry of the Soviet and American philosophies already described and the fact that American programs and goals are subject to public and elite political criticism, while Soviet goals are not. Thus, the proclaimed Soviet goals have the flavor of idealism and propaganda, while the American goals focus on what can be achieved technically and supported politically.

It was noted earlier that the emphasis on science in the Soviet space program does not take as its basic goal the expansion of scientific knowledge generally or the advancement of science for the sake of science. In fact, there are some indications that the space program has quite possibly siphoned off funds and talent from other fields of science and been subject to criticisms from the Soviet scientific community not wholly dissimilar to criticisms by American scientists of the scientific impact of the US space program. It is the view of James R. Killian, formerly science advisor to President Eisenhower, that the general development of Soviet science and technology has been inhibited by overinvestment in the space field motivated by the goals of political prestige.[13] Some credence accrues to Killian's argument from the unusually frank confession of the leading Soviet space scientist, Leonid Sedov, that the competition of the US and the USSR in space has increased levels of efforts in Soviet space science.[14] Robert E. Marshak relates that Soviet scientists outside the space field have complained that the Soviet space program drains off personnel and funds which could have been otherwise allotted to their own programs.[15]

American leaders have not shared the Soviet propensity to justify the costs of space programs by scientific benefits, particularly where these are of a long-range nature. Indeed where scientific benefits are promised, it is very often evident that they serve as a rationale for policies determined by criteria of national pride, prestige and the desire to achieve technical accomplishments. Taking a *narrower* view of science than that evident in Soviet justifications of space activities, Vernon Van Dyke holds that no one in the United States, including the vigorous champions of American space efforts, claims that the American manned and lunar programs can be justified in terms of scientific goals.[16] He singles out competition with the USSR rather than the conquest of nature as being the principal motivating factor behind the American space program.[17] It is probably wrong to dichotomize these two categories. In space technology the US competes with the USSR *through* the conquest of nature. The discovery and perfection of techniques as a method of subduing nature and enriching the life of man seems peculiarly characteristic of Americans, who excelled in this field long before the advent of cold-war competition with the USSR.

Some American scientists have been particularly critical of the expenses involved in the crash program to put a man on the moon as slighting the prospects for scientific discovery available from a more broadly based effort in space, in particular one which puts less emphasis on manned missions. The critics have come to include nearly one-half of America's living Nobelists.[18] One might indeed expect much more of such criticisms from Soviet scientists were they equally free to make them.

There has been considerable change in the attention paid to science as compared to other goals in the United States. The most salient shift in goals took place between the Eisenhower and Kennedy-Johnson Administrations. President Eisenhower generally saw the justifications for any space effort as being either scientific or military.[19] Reacting to the launch of Sputnik, Eisenhower saw the importance of earth satellites as being largely scientific rather than military.[20] In matters concerning space Eisenhower preferred to accept the advice of the scientific community.[21] His scientific advisor, James R. Killian, advocated a balance between scientific and prestige efforts in space and other areas of scientific and technological progress, fearing that too great an emphasis on space would have deleterious effects in these other areas.[22]

In general the reverse was true of the views of the Kennedy and Johnson

Administrations. Whereas the Space Act of 1958 put the advancement of science at the top of the list of goals to be served by NASA, an ad hoc committee chaired by President-elect Kennedy's science advisor, Jerome B. Wiesner, put it third behind national prestige and national security. Wiesner explicitly repudiated Killian's contention that a massive emphasis on space could weaken scientific and technological development in general, saying that an ambitious space program would create a favorable atmosphere for general support of science. Kennedy himself, emphasizing goals other than science, tended to overlook the advice of the scientific community in his space planning.[23] The concentration of money and manpower associated with Kennedy's (and later Johnson's) goal of putting a man on the moon during the 1960s relied largely on known technologies and very probably did not contribute as much to scientific and technological progress as a more broad expenditure of effort would have done.[24]

In closing this section, one qualification should be made. While it is the case that, especially after 1961, public statements by leading space personnel in the USA and the USSR seemed to betoken a greater emphasis on scientific motives and goals in the latter nation, this difference is in large part ascribable to a peculiar view of what constitutes "scientific" goals among Soviet spokesmen. While a difference in the priority of various motives and goals appears evident, the policies of the two nations which emerged from those differing goals were remarkably similar. If science is defined as a process of applying certain methods whereby knowledge is acquired and tested, then it is certainly not the foremost goal of either the Soviet or the American space program. What the Soviets call science fades into the realm of what ought to be called science fiction or futurology. Similarly, American leaders seem far more interested in *doing* through technical accomplishments than *knowing* as these priorities are reflected in their space program.

GOALS OF EARTHBOUND APPLICATIONS AND IMPACTS

Earthbound applications can generally be divided into two classes—those space activities that provide direct benefits to society by virtue of the usefulness of the space activities themselves, and those that have a more general secondary impact on national economies and societies.

A number of space activities can be directly applied to human benefit. Some mention has already been made of military missions served by use of space vehicles. In the nonmilitary sector direct benefits accrue in the fields of communications, weather forecasting, locating earth resources, navigation, mapping and traffic control. In general, space missions with direct benefits are largely earth-orbital flights, while the payoffs from deep-space flights are mostly confined to scientific discoveries with no immediate earth benefits. In terms of the number of missions mounted by both countries, one may conclude that the US has put substantially more emphasis on achieving direct and immediate earth benefits from its space program than has the USSR. The US began satellite programs in communications, meteorology and navigation before the USSR and has placed a much greater relative emphasis on these programs in its total space effort. Although as Table V in Chapter 3 reveals, the USSR has launched more satellites in the navigation/ferret and communications categories, it is likely that most of these are intended for military uses. Thus far neither country has mounted extensive efforts in the fields of earth resources or traffic control, although both have indicated their intentions to move ahead in this field[25], the Americans and Soviets through manned orbital stations[26] and the US in its Earth Resources Technology Satellite program (ERTS). Ongoing and future improvements in space activities of direct benefit will increase in cost-effectiveness through prolongation of the useful life of applications satellites and employment of cheaper and reusable boosters.[27]

Although it appears that the US puts more emphasis than the USSR on direct earth benefits in its space planning, it should be emphasized that both nations rate this factor relatively far down in their lists of priorities in space. The Soviet emphasis on long-term scientific and socio-economic goals like colonization of the planets has already been mentioned, along with the fact that the lion's share of the launchings of both countries are connected with military missions. In listing five major goals of the American program, the ad hoc committee of President-elect Kennedy put nonmilitary applications fourth after prestige, national security and science.[28] The number of American launchings indicates that it is, however, given a higher priority than science, which the committee may have chosen to put high on the list for propaganda purposes.

Unlike Soviet space spokesmen, NASA maintains a public distinction between its applications and man-in-space programs. As previously noted, man-in-space has taken some two-thirds of NASA's budget, with applications and all other programs dividing the remaining third. Although Soviet applications flights did not begin until 1966, or 1964 at the earliest, they have generally increased in the years since then, and it is probably fair to assume that the Soviet leaders consider the economic returns on such flights sufficient to justify them aside from any other motives.[29]

Applications programs have the further advantage in that they can be used to gain public support by assuring domestic taxpayers that they are getting a fair return for their taxes, and thus to justify the overall space program to a public which is often skeptical.[30] This is especially important in the United States. Although public approval is more of a necessity for NASA's programs than for Soviet space activities, no less a Soviet spokesman than Leonid Brezhnev has had occasion to claim for Soviet space efforts "direct practical use" as well as long-term benefits to Soviet citizens.[31] To peoples of the less developed countries who might feel that space rubles and dollars come out of funds available for economic aid, Chairman Kosygin has seen fit to declare that they benefit from Soviet applications flights as well.[32] Regardless of these additional motives and a recent emphasis on short-term benefits,[33] the major difference between proclaimed American and Soviet goals continues to be the long time frame and indistinct focus in which the Soviets foresee the greatest benefits to themselves and mankind.

The fact that the United States has developed a larger program of applications satellites may be accounted for in part by the attempts of the Kennedy Administration to connect these programs to its most basic space goal, prestige and/or pride. The special committee led by Vice-President Lyndon B. Johnson and Secretary of Defense McNamara, which conducted a hasty review of the American space program in May, 1961, decided to emphasize applications programs because this was one of the areas in which America could score "firsts" against the Russians[34] (as the US proceeded to do). Kennedy's famous remarks of May 25, 1961, which in effect challenged the Soviets to a moon race, also called for accelerated efforts in space communications and meteorology.[35] In addition, the American philosophical style of pragmatism has the effect of favoring applications programs more than the visionary Soviet style stemming from the millenarian aspects of Communism.

The second way in which space activities find application in the economy is more indirect. In the main this consists of stimuli to the economy by increased employment and general growth as well as by the application of technological spinoffs from space programs made in nonspace industries.

To the extent that the Soviet economy is completely under the direction of the state to begin with and that there is presumably no unemployment in the USSR, these motives and goals would seem to be irrelevent to Soviet political leaders and economic planners. Although this situation undoubtedly makes the goals of full employment and economic growth more important in the USA than in the USSR, there are some indications that they may enter into Soviet decision-making as well. For example, a lead editorial in the newspaper of the Soviet Communist Party justified high investments in science with the claim that a ruble investment in science pays four times the dividend in national income of a ruble investment in production capital in the absence of technological progress.[36] If true, this is a strong argument for space spending in a country such as the USSR, where immense annual investments have in recent years ceased to produce a rapid rate of growth.

According to Raymond A. Bauer, the American space program is supported by American businessmen in the hopes of creating economic opportunities derived from technical developments which will find application in the private sector. Because the space program, as military spending, can be justified on grounds of national prestige and security, businessmen (a vital political constituency) can support it as a kind of "pump-priming" or "Keynesianism," which they might find ideologically distasteful in a more obvious form.[37]

One should recall that President Kennedy had promised in his campaign to "get America moving again"[38] and that the aerospace industry was suffering the effects of a deflationary trend in 1960–1961.[39] Administration spokesmen argued that space expenditures were not wasted but would be repaid by stimulating economic growth. The Deputy Administrator of NASA even argued that space investments (in the South) could effect desirable social changes.[40] Not only the 1960–1961 recession but the general resistance of the Congress to the President's attempts to cure economic and social ills may have led to frustrations which the social, economic and emotional impact of an expended space program might overcome.[41]

Unquestionably both the US and the USSR have benefited from technological spinoffs from space programs in the form of new materials, techniques, processes and quality conrtol.[42] It is doubtful, however, that similar benefits could not have been won at a cheaper price through more direct attempts to achieve them. In the United States private firms are perhaps unwilling to undertake the necessary work, or it may be politically impossible for the government to subsidize such work *directly*. In both the USA and the USSR it may be that the best scientific minds and the biggest and most powerful organizations will work more willingly in the space field than in more mundane areas of industry. In the case of the biggest industrial enterprises, space and defense may provide the only areas of challenges matching them in scale. In such cases the challenge and opportunities for growth and employment may overcome considerations of profit, even in the United States. Even so, the billions spent on space could hardly be justified wholly or even largely by spinoff. As such, spinoff is more likely to be involved as an element of post hoc justifications and rationalizations rather than as an important motivating factor.[43]

Beyond direct and indirect benefits to technology, space success also creates an "image" of technological competence which strengthens foreign markets for American products[44] and provides an even greater improvement of the Soviet "image," which has until recently been one of technological backwardness. Once again, however, technological progress must be discounted as a motive in that improved goods will probably do more to expand trade than the "image" of technological sophistication. Moreover, as a motive, technological "image" belongs more in the category of prestige, which is considered separately in this chapter.

Neither the Soviet nor the American space program has advanced technology sufficiently outside the space and defense fields to indicate that general technological progress is a goal which is served by space research. Most techniques discovered in these fields are either completely inapplicable to the civilian sector or so expensive as to be impractical.

Just as supporters of the Soviet and American space programs have used technological development as a post hoc justification and rationalization of expensive space activities, opponents (in the US) have argued that space expenditures on the major programs have in fact inhibited or tended to inhibit broadly based technological development.[45] By 1964 NASA absorbed some five per cent of America's total supply of engineers and scientists on its own and through its contractors.[46] NASA Administrator James Webb himself admitted that the Agency "[does] not seek to justify our program on the basis of industrial applications that flow from it."[47] Moreover, the Rand Corporation's Alton Frye has argued that space investments have hurt the Soviet economy (and by implication the American economy?).[48]

MOTIVES OF PRESTIGE AND PRIDE

Most writers in the West have seen prestige and power through prestige as perhaps the most important goal of the American and Soviet space programs.[49] Prestige may be defined as a reputation for four qualities: (1) the pursuit of goals that are creditable and which respond to the challenges of the time; (2) the capacity to achieve those goals; (3) the determination to achieve those goals without sacrificing other desirable goals, and (4) an assured future in which these other qualities will be preserved or enhanced.[50] As applied in particular to national space programs, prestige means a reputation for extremely high technological competence and the reputation of a commitment to conquering problems thought to be vital to the human future, even if this involves giving up a measure of present comforts.

Vernon Van Dyke has correctly pointed out that prestige needs to be distinguished from pride. Prestige is an aspect of the perception and evaluation of a subject or subjects such as a person, a leader or leaders, a political party, a nation, movement or social system *by others*; pride is the perception or evaluation of a subject or subjects *by themselves*. It is often difficult to distinguish between the desire for prestige and the desire for pride as motives behind the national space programs. It is the view of Van Dyke, for example, that although prestige *appears* to be the most important goal of the American space program, from ostensible actions and often from the statements of leaders, pride is in fact the dominant goal.[51]

Clearly there is a general correspondence of the kinds of policies aimed at enhancing both pride and prestige. Above all these include being *first* to achieve important milestones in space and *spectacular* space activities, especially those involving the participation of men-in-space. One reason Van Dyke advances for the preeminence of pride over prestige is that it is tolerable for Americans and especially American leaders to bear up under the weight of a lack of prestige abroad, but intolerable for these people to retain an opinion of themselves as being second behind others or second-rate in general.[52]

Some have questioned whether an ambitious space program pays off with benefits of prestige (and pride) any better than efforts to improve housing, education, medicine, military power, foreign aid, economic growth or consumer goods.[53] In assessing pride and prestige as motives, particularly in the case of the United States, one must consider the fact that leaders and citizens are apt to view the attainment of pride and prestige as emerging from a general *competition*, whether of individuals, organizations or nations. To the extent that this is true, each country would be most sensitive about those areas in which it feels lacking. Unlike space, the other areas suggested by DuBridge and Eisenhower, while they may contribute to American prestige, may do less for American pride in that the United States is so far ahead in most of those fields that no real competition exists. Moreover, space is a new field, one many people think will play an increasingly important role in the future. More drama attaches itself to space than to more mundane activities, and space offers the opportunity for accomplishment of unequaled scale. The United States was particularly likely to feel damage to its pride and prestige because the Soviet breakthrough with Sputnik came in advanced technology, a field in which Americans had felt themselves to be the unchallenged leaders of the world—"the people with know-how, the people who are bold, innovative pioneers—first in the world in technological achievement."[54]

Did Sputnik markedly increase the prestige of the USSR at the expense of the lessened prestige of the United States? The indicators, such as they are, yield a mixed answer. The prestige impact of Sputnik and subsequent Soviet space firsts seems to have been most marked outside the United States. Referring to unspecified post-Sputnik polls of opinion in Western

Europe, Alton Frye indicates that whatever the impact on US prestige, a sense of solidarity between the United States and Europe was in fact promoted.[55] Other, later surveys, some taken after the Gagarin flight in April, 1961, revealed that pluralities in France, Britain, India, Holland, Uruguay, Switzerland, Norway, Italy and West Germany saw Russia as "leading in science" in the next ten years. Only Greece gave the nod to the United States.[56] Other polls indicated that those who viewed the United States as leading in science in the present and future were much more inclined to favor close ties between their own countries and the US,[57] but it is questionable here whether it is general anti-American or pro-American sentiment or the impact of Soviet space successes which is the dominant independent variable.

The impact of Sputnik on American public opinion seems to reveal that opinion leaders in politics, science and communications were much more disturbed about Soviet success than the population in general. Sputnik was judged to have improved Soviet credibility abroad, especially in the Near East, but not to have resulted in a loss of face or prestige either abroad or among the American public.[58] All this seems to reinforce the notion that it was American pride and not prestige that was hurt.

Soviet leaders were probably surprised by the extent to which many people around the world viewed Sputnik as a blow to American pride and prestige and a great victory for the Soviet Union. As such, pride and prestige probably played little role in the decision to build and launch the first Soviet satellites and space probes but were, rather, goals to be sought with later satellites and generally in the *postlaunch* phase through vigorous propaganda efforts, especially during Khrushchev's tenure.[59] To the extent that pride and prestige became important motivating factors, this did not occur for either country until after the first stages of the exploration of space by both. Unlike American leaders and public spokesmen, the Soviets have consistently denied their interest in taking part in any space race with the USA.[60] The Khrushchev years, which more or less coincided with the period of the Soviet lead in space, were marked by a propaganda policy of "rubbing it in," or making the most of Soviet preeminence at US expense.[61] The most prominent role in this campaign was played by political leaders, most notably Khrushchev himself, and by the popular press including publications aimed at both domestic and foreign audiences. This policy was probably reinforcing and had the effect of increasing the importance of prestige as a motive in space decision-making in the USSR, which played its space program for maximum propaganda benefit right up to the time preeminence in space technology clearly belonged to the United States and Khrushchev, the USSR's leading space propagandist, left the scene. Moreover, Khrushchev's policy of maximizing prestige in fact goaded the United States into pressing ahead with vigorous programs of space research and strategic-weapons development that turned the Sputnik into "Khrushchev's boomerang."

As was the case with the Soviets, American policymakers did not expect space activities to have the impact on national prestige that they did.[62] Even after the first two Sputniks, when Project Vanguard was given priority status, the Eisenhower Administration refused to view the space program in the context of a race with the USSR.[63] Indeed, the measures undertaken to develop the American space program such as the formation of NASA and increases in space spending were largely the results of congressional efforts which had to overcome the resistance of the Eisenhower Administration.[64] The Eisenhower Administration generally chose policies which maximized security rather than prestige or other "political goals." It could only view the space program as justifiable on the grounds of national security and/or science. Eisenhower did not believe that space activities would greatly affect American prestige one way or the other.[65] Moreover, as previously noted, he did not see space activities as particularly important to national security. This left science as the most important goal in the space program, and Eisenhower chose to rely upon the scientific community much more than his successors in formulating plans for America in space.[66] Typical of the views of most American scientists were those of Eisenhower's science advisor, James R. Killian. Killian contended that a prestige race in space with the USSR would actually weaken American science and technology.

As seen in Chapter 1, the Congress and especially its Democratic members reacted differently to Sputnik and subsequent Soviet space successes. Majority Leader Lyndon B. Johnson viewed the Sputnik as a "technological and propaganda defeat," led a subcommittee investigation into its causes, and became interim chairman of the Senate space committee.[67] In the election campaign of 1960 candidate Kennedy chose to emphasize the importance of national prestige and security in space to the exclusion of the largely scientific goals envisioned by Eisenhower.[68]

As President-elect, Kennedy appointed an ad hoc committee chaired by his science advisor, Jerome Wiesner. As noted, the committee put national prestige at the top of its five principal motivations for an expanded space program. In his inaugural address Kennedy may have been trying to recoup what he saw as a loss of American pride and prestige through his claim that despite the Soviet lead in booster capability, the US was "ahead in the science and technology of space."[69]

The importance of pride and prestige to the Kennedy-Johnson Administrations is most evident in the circumstances surrounding and included in the commitment of Kennedy to send an American to the moon and back during the decade of the sixties, made in May of 1961. April, 1961, had seen the failure of the American-sponsored Bay of Pigs invasion of Cuba and the first manned orbital flight of Soviet cosmonaut Iurii Gagarin. Leaders of the Administration viewed these two events as crippling blows to American prestige (and very probably to their own and the nation's pride).[70]* It was in the wake of these events that a high-level ad hoc group met over a weekend in May to consider the future of the American space program. The group consisted of Johnson, NASA Administrator Webb, McNamara and other highly placed officials. It acted within the limits of what was obviously enormous time pressure, having its conclusions prepared in two days' time. The committee's emphasis on prestige (and pride) is evident from the fact that it was most interested in identifying those areas of space exploration in which the United States stood a good chance of achieving "firsts," i.e., "beating" the USSR.[71] It was from the conclusions of this group that emerged Kennedy's commitment to the moon mission on May 25, 1961.

Later in the year NASA Administrator Webb indicated the high priority he accorded prestige in the space program, saying that "in the minds of millions space achievements have become today's symbol of tomorrow's scientific and technological supremacy." According to Webb, "The Soviet Union has recognized this . . . [w]e cannot afford to yield to them by default the next great prize in this competition . . . When we can win, we must win."[72] Here is a clear indication of the American conception of a space race, something in which the Soviets have always denied being contestants. Webb stated that not only was prestige at stake in a USA-USSR competition, but that the US was "the symbol of democratic government." If the US yielded to "the leading advocate of the Communist ideology" in the race to the moon, American pride and prestige would suffer to the extent that "we could no longer stand tall in our own image or in the image that other nations have of us and of the free society we represent." The chief of Project Apollo claimed that if not first to the moon, the US would not be first in space "and one day soon we will not be the first on Earth"—a kind of celestial domino theory.[73]

Similar views were expressed by other American space officials. Edward C. Welsh, Kennedy appointee as executive secretary of NASC, thought it "absurd" to pretend that America was not in a "space race" with its survival

*There is no way of knowing at present whether Soviet space plans may have been similarly affected by reversals in Poland and Hungary in late 1956.

and the "spiritual advantages of a free society" at stake. "[T]he Communist threat," he believed, left the US "no honest choice."[74]

Although a public-opinion poll indicated that the population in general did not back the race to the moon,[75] the press and the Congress were anxious for this sort of race, their reaction to the Gagarin flight extending in some cases to the point of "panic" and "hysteria." Logsdon notes that one congressman even suggested painting the moon red, white and blue.[76] It is certain that Soviet political leaders do not have to face this kind of pressure.

The indications are that Kennedy himself shared this sort of attitude so uncharacteristic of his predecessor. In April, 1961, he had asked his space administrators to tell him in which space feats the US could catch or "leapfrog" the USSR.[77] In his memorandum to the Vice-President setting forth the tasks of the May ad hoc committee, he asked this question about the moon program in particular. NASA informed the Vice-President that a crash program could put a man on the moon by 1967, which information was expressed by Kennedy as "before the end of this decade." Evidently it was NASA's intention to be on the moon before the fiftieth anniversary of the Bolshevik Revolution, at which time they expected the Soviets to make their moon shot. NASA also indicated to Johnson that there was prestige value in communications and meteorological satellites as well.[78]

Because the deliberations were viewed as political rather than scientific, neither the President's science advisor nor his Scientific Advisory Committee played any important role in the ad hoc group. The science advisor himself believed that the expense of the moonshot would never be accepted on scientific grounds.[79] A memorandum sent from Webb to McNamara at this time, and never officially made public, explicitly rejected "the scientific, commercial or military" justifications of the moon project in favor of "national prestige" as an element of "international competition" with the Soviet Union in the "cold war."[80] Logsdon contends that Kennedy reached his decision about the moon program and an expanded space program on the basis of the philosophy expressed in this secret memorandum from Webb to McNamara. That is, he was determined that the United States win what he saw as a space race by winning power through prestige as a strategy of the cold war.[81]

That Kennedy's view was shared by the Congress (which probably put some greater emphasis on military benefits) is evident from the ease with which his proposals, including a 61 per cent boost in space spending, gained legislative approval.[82] Under Kennedy and Johnson the civilian space budget grew from the Eisenhower Administration's figure of around a half billion for fiscal 1961 to over $5 billion in 1966. American leaders have continued to view the US space program as above all else a competition with the Soviet Union and perhaps exhibit a psychological tendency toward projection in attributing the same motives to their Soviet counterparts. The Soviets themselves deny that this is the case.[83] It has been argued that the American view of space as a competition has resulted in a policy of undue haste which contributed to the fatal Apollo fire at Cape Kennedy in January, 1967.[84] Four Soviet cosmonauts have perished during reentry in two accidents involving a parachute failure and a failed valve. (The cause of the second accident was revealed by Soviet spokesmen to NASA officials who were concerned about the safety of *Soiuz* for the upcoming Apollo-*Soiuz* mission.*) Whether these fatalities are attributable to undue haste on the part of Soviet space planners and designers is extremely difficult to judge. Stories of multiple cosmonaut deaths before the Gagarin flight have never been substantiated. Both nations have on the whole exhibited great caution in their manned flights in terms of safeguarding the lives of spacemen, but haste undoubtedly contributed to the failure of many early unmanned orbital attempts by the USA and to unmanned lunar and planetary failures by both countries, especially the USSR. In the case of other Soviet missions, including many manned ones, haste has probably reduced the scientific and technical payoffs from missions that should have been conducted with greater deliberation and planning.

At least partly because decision-making in the Soviet space program is largely cloaked in secrecy, it is much harder to find the kind of emphasis on prestige that is so obviously a decisive motive for American leaders. Certainly the Soviets have actively exploited the prestige value of their space achievements after the fact, but it is difficult to say what role this motive played in the planning stages. Post facto exploitation of prestige value was particularly characteristic of the Khrushchev period and has been moderated with the change of leadership in 1964 and the relative improvement in the level of American space achievements.[85] It will be recalled that Khrushchev himself made much of the "peaceful competition" through which the USSR would catch up with and surpass the USA in the fields of industrial production and technology, and that this would contribute to the achievement of the "Communist" stage of social development in the USSR in the near future.[86] It is noteworthy that his successors have had much less to say about attaining Communism or surpassing the US and do not, as Khrushchev did, attach any dates or time limits to these developments.

One may presume that prestige (and pride) is a basic motive in the Soviet space program to the extent that Soviet space decisionmakers have gone to great lengths, including high expenditures, multiple launchings and

Aviation Week and Space Technology 99 (19), November 5, 1973, pp. 20-21.

hasty attempts, to achieve firsts in space when scientific and other long-range purposes could be developed through more broadly based or patiently executed programs. Prestige may also be assessed as an important motive to the extent that the USSR considers the space plans and programs of the United States and other political factors in making its own plans and scheduling its own space launchings. As detailed above, the American moon program and other programs were planned with the aim of accomplishing their objectives before Soviet programs could do so. The same seems to have been true of many Soviet missions.[87]

In the early years it seems that the USSR was aiming at firsts rather than maximum scientific information in that there were no repeats in the missions of flights between 1957 and 1962 except where the flights failed to achieve their planned objectives. Something new was always added.[88] Prestige was also very likely a motive behind early and repeated Soviet attempts to hit the moon and the planets in the late fifties and early sixties. After launching the first biological payload, Sputnik II with the dog, Laika, in 1958, the Soviets made three attempts to hit the moon in 1959. The one which succeeded in impacting the moon, *Luna* 2, carried more than 150 hammer and sickle emblems of the USSR, which were scattered over the moon's surface for obvious propaganda purposes.[89] Whatever space and weight these took up in the craft could as easily have been reserved for scientific instruments.

At every opportunity between 1960 and 1967 the USSR made repeated attempts to hit or fly by the planets Venus and Mars. Two unannounced Mars attempts in 1960 failed even to attain earth orbit, suggesting the possibility of undue haste (and waste). Probably coincidentally* these flights corresponded in timing with Khrushchev's memorable visit to the United Nations General Assembly session in October. A defecting Soviet sailor from the ship that brought Khrushchev to New York revealed that the leader had brought with him a replica of the spacecraft to be displayed at the UN if the flight was successful.[90] Although the launch was probably not scheduled to provide Khrushchev with a toy to exhibit at the UN, the possibility of making such gestures is probably not wholly out of the minds of Soviet leaders when they make space policy decisions.

The *extent* of the Soviet effort to hit the planets at an early date despite repeated failures also probably indicates some interest in achieving firsts despite the costs of failures made more probable by haste. Venus "windows" occurred in 1961, 1962, 1964, 1965, 1967, 1969, 1970, 1972 and 1973. On every one of these occasions except 1972 and 1973, at least two attempts were made to fly by, or more probably impact, the planet Venus. It was

---

*Mars "windows" occur only for a couple of weeks about every two years.

only in 1965, after eight partial or total Venus failures, that *Venera* 3 succeeded in hitting the planet, and it returned no data. Nonetheless, it was a first, as was *Venera* 4 in 1967, which did return data from within the planet's atmosphere.[91]

The Soviets also went to great lengths to be the first to hit the moon, to stage a soft landing there, to accomplish lunar orbit and to make an unmanned landing and return ahead of the Apollo program. In 1958 both the United States and the Soviet Union were competing to achieve lunar orbit or impact. The US registered four failures (Pioneers 1-4) and the Soviets one failure before suceeding in impacting with *Luna* 2 and returning photos with *Luna* 3. Six attempts at a soft landing failed in 1963 and 1965, with *Luna* 9 the first to achieve this feat in 1966, some four months before the American Surveyor 1. After several failures by both countries *Luna* 10 achieved lunar orbit also in 1966, some four and one-half months ahead of the US Lunar Orbiter 1. By 1967 the US, plagued by a succession of failures more dismal than that of the Soviets, finally forged ahead. In that year the US mounted eight lunar missions including only one failure, while the Soviets flew no lunar flights at all. In 1968 the unmanned *Zond* 5 achieved the first circumlunar flight three months ahead of the manned Apollo 8. *Luna* 15, which was possibly intended in part to achieve an unmanned landing and return a few days ahead of Apollo 11's manned mission, crashed on the moon instead.[92] The *Zonds* were very likely precursors to manned circumlunar flights, and had they been man-rated earlier, it may have been possible for the USSR to carry out manned missions in lunar orbit before or during the Apollo missions of manned landing and return. Soviet missions on the order of Apollo were and remain impossible in that they have not, at the time of this writing, successfully tested a booster of sufficient thrust, and as such they either opted out of any race with Apollo at an early stage or encountered difficulties which forced them to delay or cancel such an effort.

Just as the Soviets demonstrated a strong commitment to be first on Venus, their efforts to probe Mars have also been rather more ambitious than those of the United States. In all, there have been 14 Soviet attempts and five American to reach the red planet. The Soviets either skipped the Mars window of 1966–67 or failed so completely that they left no evidence of any attempts. The US skipped the Mars windows of 1960, 1962 and 1973. Mariner 4 returned the first pictures from Mars to the United States in 1964. Mariner 7 repeated this task during the 1969 Mars window, when the Soviets again evidently carried out no Mars launch.[93]

In sum, the USSR has been particularly active, and perhaps "hasty" in its attempts to achieve a podhold on the moon and planets. To a large extent this probably reflects the Soviet emphasis on its particular definition of "science," that is, discovery compounded with exploration and aimed at benefits in the distant future. However, the haste involved indicates the presence of other motives; and this author is convinced that the Soviet lunar and planetary efforts reveal at least some of the concern with prestige that was so evident in American plans for manned exploration of the moon. However, the fact that the Soviets have continued very extensive unmanned planetary programs in recent years when the prestige of such missions is minimal indicates the salience of factors *other than prestige* in their space planning. It is noteworthy that the only field of civilian space research in which the USSR has sustained a significantly greater effort than the US is in sending unmanned probes to the planets. This is precisely the area in which American space scientists have been most critical of NASA for slighting scientific benefits and substituting political ones. Although Soviet spokesmen have generally denied that prestige, pride and competition with the USA enter into their space planning, there have been at least two occasions on which Soviet scientists, expressing unwonted frankness, have conceded as much as this author claims here. In a private exchange with NASA's Hugh Dryden, Academician Sedov admitted that scientists in both the USA and the USSR should be thankful for Soviet-American competition in space, for without it neither would have a manned space program.[94] Academy President Keldysh, in a more guarded public statement, allowed that competition was inevitable when scientists are working in similar fields, but insisted that "this must not be the determining factor."[95] On the American side, Dr. James Van Allen was probably being considerably more forthright than Academician Keldysh in suggesting that without the competitive aspect the space program would be funded at 20 per cent of the current level.[96] The same could probably be said of the Soviet space program, especially in the Khrushchev period.

Policies aimed at the goals of keeping and increasing prestige may seek to create an aura of prestige around different people and institutions and may be aimed at differing target groups, including various strata of opinion at home and abroad. In general, space achievements may increase the prestige of the launching country, the leaders of that country and the social system or ideology which reigns there.

Joseph G. Whelan, Soviet and East European Affairs specialist with the Library of Congress, considers power through prestige as the "main goal" of Soviet space programs. Since this is more clearly true of American programs in 1961–1969, Dr. Whelan may be projecting. Whelan himself admits that the Soviets have denied this to be the case.[97] Certainly American space officials felt pressure to claim that the USSR was racing to beat

America in space in view of criticism of the expensive programs they had mounted.[98] At least until the successes of the Apollo program the Soviets denied being in a space race but generally insisted that they were ahead of the US in space, anyway.[99]

Perhaps the earliest indication of a Soviet concern with prestige was in the choice of frequencies for the "beep-beep" of Sputnik. These were not the frequencies agreed to by the IGY committee on the initiative of the United States but rather frequencies readily receivable by standard household shortwave sets throughout the world.[100] Soviet leaders and public spokesmen have consistently held that Soviet space achievements reflect glory on, and are legitimate grounds for, pride of the Soviet nation and people. The proclamation of April 12, the anniversary of Iurii Gagarin's orbital flight, as Cosmonautics Day is a good indication of such attempts to build pride and prestige domestically. Although Soviet claims about being ahead of the USA in space have moderated and declined in recent years, different groups still emphasize the relative preeminence of Soviet space achievements in particular fields. The scientists are somewhat more apt to claim preeminence than the political leaders, the press commentators are more boastful still, and the cosmonauts have been the most boastful of all.[101]

Even if Soviet preeminence in space has of late come into question, the record established so far must certainly have overcome much of whatever inferiority the leaders and people of the USSR have suffered through decades of economic and technological backwardness vis a vis the West. Space achievements have also lent support to Soviet claims to "superpower" status evident in Khrushchev's plan to reorganize the UN Secretariat and in Chinese complaints about the Soviets trying to divide the world with the US. The enhancement of national prestige in technology has aided Soviet foreign policy especially in relation to the countries of the Third World by attracting students to the USSR and boosting markets for Soviet equipment.[102] As a component of national power, prestige aids Soviet foreign policy to attain such established goals as splitting the Western alliance system, solidifying the Communist system of states and promoting respect and admiration for the USSR around the world.[103]

There is one aspect in which the USSR is clearly not in a space race with the USA. That, as noted in Chapter 2, is space spending. By relying so far on a relatively modest stable of boosters and opting out of the moon race, the Soviets avoided matching US expenses ruble for dollar. Nonetheless, the relative cost to their less wealthy economy must have been and continues to be comparable, if not greater, and in terms of achievements, "competitive" from a prestige, if not a scientific and technical, perspective. Besides this, both the USSA and the USSR have attempted to enhance the prestige-building aspects of their national space programs at no additional cost through attacks on the motives and policies of one another. The Soviets have been most active in this regard. Basically, they have argued that the American program is less deserving of prestige than their own because military and profit-making motives rather than "scientific" goals are behind it.[104] Perhaps reacting in the mirror-image or stimulus-response fashion of the cold war,[105] American officials have chosen (to a lesser extent than Soviet spokesmen) to taint Soviet space successes with the image of Soviet militarism and secrecy.[106]

It should be observed that the quest for prestige in space may have backlash and boomerang effects as well. Not only did Khrushchev's "rubbing it in" of Soviet space firsts provoke the "leapfrog" reaction of the wealthier superpower in the 1960s, but the space race may have resulted in a loss of prestige among those who saw it as a potlatch ceremony or contest to see who could waste the most treasure.[107] This was the case especially among opinion elites in the Third World as revealed at the UN Conference on Trade and Development. There, a majority of the 120 nations assembled "severely castigated the space powers" for excessive expenditures. Similar reactions to the space race have occurred among scientists and the general public in many countries.[108]

Space feats also serve to build prestige for the political leaders in the US and the USSR who have always hastened to associate themselves with space achievements. President Nixon saw fit to hobnob with American astronauts on the moon and greet them on their return. Perhaps he should have been warned that Mr. Khrushchev's last public exposure before his fall from grace was a publicly broadcast telephone conversation with the cosmonauts aboard *Voshod 1*, who landed to find the erstwhile Chairman and First Secretary enjoying the obscurity of private life. While still riding high, Khrushchev had been publicly acclaimed as the principal architect of Soviet space achievements, receiving the Order of Lenin, while the scientists and engineers responsible remained anonymous.[109] Although the new leaders have not developed the cult of personality which surrounded the late Khrushchev, the impression is still given that space successes evidence the "correctness" of decisions made by the highest party and state officials.[110]

The beneficiaries of today's space cults of personality are not so much the political leaders as the cosmonauts and astronauts. In both countries they are put on public display, sent on good-will junkets around the world and fawned on by reporters whether from *Pravda* or CBS.[111] Soviet cosmonauts have been honored by election to the Supreme Soviet and in two cases by burial in the Kremlin Wall. Cosmonauts have also been employed to lend their prestige to justifications of Soviet foreign policies and condemnations of US policies.[112] The explicitly political activities of American astronauts are evidently more their own affairs. One has run for Congress, and another winged off to Vietnam on behalf of American POWs. All have been spokesmen for NASA and an ambitious space program. Both countries have also filled various international expositions with their space hardware, which often literally hangs from the ceiling.

Space-borne prestige has also been exploited by both nations to enhance the international reputation of their respective ways of life. The Soviet Union has been more active in this area probably because it has always been more vocal in its socialist propaganda than the US has been about the virtues of free enterprise, and more specifically because some might question whether American space feats are the product of free enterprise or creeping socialism. Although American successes have had a moderating effect in recent years, the Soviets have consistently attributed their pre-eminence and achievements in space to the advantages of the socialist system.[113] They have even gone so far as to credit Marxism-Leninism and Lenin in particular for the favorable climate enjoyed by science in the USSR.[114] All the cosmonauts have been Communists, including Bykovskii, who was enrolled in the CPSU while in orbit.[115] The Soviets have also hailed the cosmonauts as exemplary of the "new Soviet man" who will presumably occupy the Communist future.[116]

Perhaps the American efforts have not been aimed so much at proving the superiority of capitalism (a word with at best mixed international connotations) but the nonsuperiority of socialism. President Eisenhower was probably correct in asserting that consumer goods dangled in the faces of the inhabitants of the socialist states and uncommitted nations would serve the former purpose better. Advertising both consumer and space technology would accord with what Stanley Hoffman has identified as a policy of "technological anti-communism" which emerges from "the basic elements of the American style." Hoffman distinguishes the American drive to master nature from the attempts of other societies to master man, but mastery over nature is perhaps even more characteristic of Communist philosophy, albeit with mastery of men included as part thereof.[117] As suggested previously, the American fascination with the conquest of nature long antidates the space age and the cold war. Outside these areas it also emerges in the "edifice complex" of American institutions which often belies the narrow profit motive ascribed by Marxist and classical capitalist ideologies.

Insofar as the US and the USSR have been motivated by the thirst for pride and prestige through space, their efforts have met with diminishing returns in recent years. Both have accomplished so much that neither can be called a clear winner. Moreover, the public the world over has probably become rather sated with space spectaculars and monumental efforts to derive the fullest propaganda impact therefrom. Sputnik, Gagarin and Apollo have left the space powers with little to do for an encore. More important than anything else, the United States has overcome with Apollo what some of its leaders viewed as a crippling blow to its pride and prestige.[118] It is mute testimony to the importance of pride and prestige as motives that as diminishing returns set in, civilian expenditures declined in the US, and the Soviets dropped or postponed their own manned lunar program.

NATIONAL SECURITY AS A MOTIVE

It will be recalled that President Eisenhower saw science and security as the only goals of space activity that could justify the cost. If the satellite program served important national security goals, Eisenhower believed it should get urgent priority. However, to his mind satellites did not affect national security to any degree, so that Sputnik did not raise his apprehensions "not one iota."[119] Thus, Eisenhower appears to have been ready to have accorded goals of national security top priority in the space program, something subsequent administrations did not do. However, feeling that security was not at stake and other goals were of lesser priority, he was content with a very modest space effort. Despite the reaction in some circles of the press and Congress, the American public reaction to Sputnik coincided with that of the President—no imminent threat to national security was perceived.[120]

As detailed above, it was on American pride and prestige rather than on national security that the Kennedy-Johnson Administrations put their major emphasis. However, because they linked prestige to power in the cold war, and because they needed authorizations and funds from a security-minded Congress, these administrations also emphasized the protection of national security in space more than the Eisenhower Administration had. In his campaign statement cited previously, Kennedy referred to a *strategic* space race in which the assurance of "peace and freedom" required that America be first. By controlling space Kennedy said the Soviets could control earth, much as powers controlling the seas have done in past centuries.[121] The ad hoc committee of the President-elect listed national security second after national prestige on its list of goals for an expanded space program.[122] Edward C. Welsh, executive secretary of NASC under Kennedy and Johnson, saw our survival at stake in the space race, as mentioned previously. Similar and even more extreme views were characteristic of the Congress. There, congressmen attacked what they saw as a slighting of military programs by the Executive. Among the "major policies of our government which interact . . . to shape our space program," a high State Department official mentioned the insuring of "our national security from hostile threats."[123] Spokesmen for the military have often taken a similar view, among them General Schriever, who considers national defense "the compelling motive" for the development of space technology.[124]

Despite these pressures, motives of national prestige and pride remained dominant. Otherwise, Van Dyke contends, there would have been no Project Apollo. As a result, the Air Force was unhappy with the exclusive emphasis on the moon mission and DoD rejection of proposals for manned military spaceflight.[125] As noted previously, the Air Force may have taken some consolation from the tailoring of the Gemini missions to fit its interests.

National security is also involved in US-USSR cooperation in space. American scholars and statesmen have often subscribed to the view that there is a vital connection between Soviet Communism, the closed nature of Soviet society, and the expansionist and anti-Western policies of the USSR. Thus, by pushing ahead American space efforts the US will increase the temptations for and benefits to Soviets from space cooperation with the US. In doing so, moreover, the US will be opening up Soviet society to the influence of Western ideas and speeding the transformation of their society into a form more acceptable to the West.[126] In a not dissimilar fashion, American cooperative space programs generally increase the interdependence of the Western powers, attract the uncommitted and exert some centrifugal force on the socialist allies of the USSR. Needless to say, the more modest Soviet cooperative programs, although hardly aimed at opening up the West, may reflect motives of cementing ties with allies, attracting neutrals and dividing opponents.[127]

Perhaps the most important goal of Soviet space programs and policies vis a vis national security has been the effect of making Soviet military might and deterrent capability *creditable* to the world at large.[128] It was Chairman Khrushchev who went to great lengths to achieve this end. This was evident in the coordination of missile and space announcements with those of nuclear tests. Moreover, some analysts have concluded that Soviet space successes were behind the confidence that led Khrushchev in 1958–1959 and again in 1961 to make ultimatums aimed at forcing the Western powers out of Berlin.[129] As mentioned above, whether by chance or by design, Khrushchev's visit to the UN in 1960 coincided with a Soviet shot at Mars. A similar coincidence took place when *Luna* 2 was launched on the eve of Khrushchev's visit to the United States in 1959[130] and in August, 1961, when Gherman Titov was launched into orbit at the same time the Berlin Wall was being erected.[131]

Between the Sputnik and the Cuban missile crisis, Khrushchev attempted to squeeze the greatest possible advantage from Soviet space successes, and the author is led to assume that this must have influenced him to be more favorable to the proposals of Soviet space scientists. One must recall that before the Sputnik, American allies reposed under the comforting certainty of the continued strategic and technological preeminence of the United States. Whatever problems they may have faced, they did not have to go to great lengths at great expense to guarantee their external security, which was already protected by American nuclear weapons ("massive retaliation") and by American strategic air superiority based on massive fleets of existing and planned strategic bombers. The assumption of the military leaders around President Kennedy, most notably Generals Gavin and Taylor, was that the Soviets had secured a strategic breakthrough with the ICBM as symbolized by the Sputnik. This meant that the strategic pose of deterrence based on the threat of massive retaliation was no longer creditable in view of the Soviet capability to reply in kind to, if not preemptively destroy, the American deterrent on the ground. As a result, a revolution occurred in American military thinking with the shift to a strategy of "flexible response," among the outcomes of which were the dispatch of half a million American troops to Vietnam and the building of what was at the time the largest army in the world (3.5 million men in the US armed forces, a number which has subsequently been reduced).

Of course, the Soviet intention could hardly have been to provoke this kind of buildup by the United States which has cost them so dearly. Rather, one may conclude that the Soviets intended at least in part to encourage a reassessment in various countries of the costs, risks and benefits of allignment with the USA and the Western alliance system. In brief, the Soviets hoped, by overcoming American strategic superiority, to encourage weaknesses and differences among the Western allies and to push the United States into withdrawing from its most advanced outposts along the front of the cold war. Space achievements could also help to promote neutralist and anti-Western sentiments among the unaligned and new states in part through the appeal of communism to African and Asian intellectuals as the wave of the future as evidenced in space.[132] It is just in this sense that many of the space accomplishments of both the USSR and the United States are aimed not so much at creating military power but rather strategic influence and international self-confidence.

In large measure the use of space in these strategies was based on the tendencies of international publics to view space as an index of scientific, technological and military strength.[133] The Soviet policy, associated especially with the statements of Khrushchev, was to deliberately create and strengthen this tendency of international publics.[134] It is in a certain measure ironic that Soviet policy in this regard strove to achieve a maximum short-term strategic benefit from space at the expense of helping to promote an even greater commitment to space in the USA and escalating the arms race to a level very expensive for the USSR.[135] One must conclude that Soviet attempts to gain maximum strategic advantage from space while at the same time attempting to ban all military activities in space (including reconnaissance) were viewed as sufficiently contradictory by American leaders as to preclude their achieving long-range success in both aims or even in either one.[136]

THE GOALS OF PEACE AND INTERNATIONAL COOPERATION

Both Soviet and American spokesmen have expressed the view that exploration of space can be a venture which is carried out in a spirit of international cooperation and can promote the achievement of a stable peace between the two great space powers. Both have argued that space research fosters the cause of peace in a number of ways. First, and probably most important, space research by its very nature encourages international cooperation which can break down hostilities and increase the interdependence of nations because (1) science is by nature a cooperative activity, international in scope, (2) the complexity and expense of exploring space encourages international pooling of resources, (3) space exploration can serve as a safer alternative to the arms race and as an outlet of national energies, and (4) satellites can be used to police arms-control agreements.

In (falsely) characterizing its space program as wholly "scientific," the USSR has from the outset claimed that it can serve only peaceful ends. With the exception of Khrushchev's linking space to strategic capabilities, this has been a constant theme of Soviet descriptions of their space program.[137] As described above, the Eisenhower Administration also emphasized that the exploration of space was above all a peaceful and scientific endeavor. This view had its impact on the creation of NASA in 1958 as a civilian agency. It was hoped by the President and others that as a civilian agency NASA would secure "the fullest cooperation of the scientific community at home and abroad."[138]

Although the Kennedy Administration viewed the national space program in a different light than its predecessor, it retained some of the emphasis on peace and international cooperation. The ad hoc committee of the President-elect listed "possibilities of international cooperation" and the replacement of the existing atmosphere of competition by one of cooperation as the fifth of the five major motivations for the American space program.[139] Some have argued that President Kennedy's commitment of the nation to a race to the moon was in fact an attempt to goad the USSR into choosing to cooperate with the US in the conquest of the moon and planets rather than run second in the space race. Indeed, Kennedy did appeal for US-USSR cooperation in space in his inaugural address and his speech to the UN in September, 1963, which specifically called for a joint rather than a competitive lunar program.[140] His doing so may have been more an attempt to score propaganda points against the USSR than the reflection of a commitment to cooperation as an overriding goal. Suffice it to say at this point that Kennedy's remarks evidently contradict the competitive goals he set forth in his memorandum to Vice-President Johnson of April 20, 1961, alluded to previously.[141] Nonetheless, the goals of promoting international cooperation in space remained high on the list of priorities in the American space program.[142] That they do so remain and have found some measure of attainment in recent years is partly accounted for by the fact that cooperation with the US may have the effect of "opening up" the USSR and driving "a wedge in the iron curtain."[143]

The thesis that space endeavors including competitive endeavors could provide an alternative to the arms race was examined in Chapter 3. For the most part, it has been American spokesmen who have put forth this thesis, but Leonid Sedov, the leading spokesman for the Soviet space program internationally and long-time confidante of NASA's late Hugh Dryden, praised peaceful space competition as promoting the transfer of war efforts to peaceful purposes.[144] More recently, Academician Sedov has held that space research "makes for broad international cooperation and understanding and thereby serves the cause of world peace."[145] His earlier thesis that competition is a good thing has been largely repudiated by other Soviet space spokesmen who have always been at pains to deny their interest in any kind of space race.[146] Likewise, General Secretary Brezhnev has indicated that Soviet space research is devoted entirely to "peaceful purposes" and Soviet policy in space to the goal of international cooperation.[147] The fact remains that the US has gone considerably further toward achieving broad international cooperation than has the USSR.

Soviet spokesmen have gone to much greater lengths than their American counterparts to use the supposedly exclusively peaceful purposes of their space program as a propaganda weapon against the other space power. To some extent, therefore, the Soviet commitment to peace and cooperation is not so much an independent goal, but rather a means of maximizing another goal, prestige, relative to the US. Not only have the Soviets time and again alluded to the peaceful nature of their program, but they have stressed the contrast to American programs which they claim are directed by the US military and the quest after profits by aerospace firms.[148]

NOTES

[1]See Charles S. Sheldon, II, *Review of the Soviet Space Program, With Comparative US Data* (New York: McGraw-Hill Book Co., Inc., 1968), p. 84.

[2]Charles S. Sheldon, II, "The Challenge of International Competition," Address to AIAA/NASA Houston, November 6, 1964, in *International Cooperation and Organization for Outer Space, 1965, op. cit.,* p. 454; and Bernard Lovell, "The Great Space Competition," *Foreign Affairs, 51* (1), October 1972, p. 138.

[3]William Shelton, *Soviet Space Exploration* (New York: Washington Square Press, Inc., 1968), pp. 167-168.

[4]See the remarks of General Kámanin in *ibid.,* pp. 172-173.

[5]See Leonid Vladimirov, *The Russian Space Bluff* (New York: Dial Press, 1973), pp. 77-83, 86, 91-93, 107, 110-119, 123-133.

[6]John M. Logsdon, *The Decision To Go to the Moon: Project Apollo and the National Interest* (Cambridge, Mass.: MIT Press, 1970), pp. 36-38.

[7]See B. W. Augenstein, *Policy Analysis in the National Space Program,* Rand Corporation, Paper #P-4137, July 1969, pp. 6, 14-16, 45, 53, 56-59, 80-82.

[8]Introduction by Gherman Titov in Shelton, *op. cit.,* xi.

[9]L. V. Berkner, *The Scientific Age, The Impact of Science on Society, Based on the Trumbull Lectures delivered at Yale University* (New Haven: Yale University Press, 1965), p. 18.

[10]See, for example, Academician Boris Petrov in *Pravda,* December 30, 1969, pp. 1-2.

[11]Krieger, *Soviet Space Experiments and Astronautics,* Rand Corporation, Paper #P-2261, March 31, 1961, pp. 35-37.

[12]See, for example, a comparison of statements on bioastronautics in Shelton, *Soviet Space Exploration, op. cit.,* pp. 167-168.

[13]James R. Killian, "Shaping a Public Policy for the Space Age," in American Assembly, Columbia, *Outer Space: Prospects for Man and Society* (New York: Frederick A. Praeger, Inc., 1968), pp. 183-184.

[14]Leonid Sedov, "Interplanetary Travel Soon a Reality," *New Times,* No. 38, September 1959, pp. 5-6.

[15]Robert E. Marshak, "Reexamining the Soviet Scientific Challenge," *Bulletin of the Atomic Scientists, 19* (4), April 1963, pp. 13-14.

[16]See Vernon Van Dyke, *Pride and Power, The Rationale of the Space Program* (Urbana: University of Illinois, 1964), pp. 90-91, 176; and Lovell, "The Great Space Competition," *op. cit.,* p. 136.

[17]Van Dyke, *ibid.,* p. 154. See also Don E. Kash, *The Politics of Space Cooperation* (Purdue University Studies, Purdue Research Foundation, 1967), p. 31.

[18]Lillian Levy, "Conflict in the Race for Space," in Lillian Levy, editor, *Space: Its Impact on Man and Society* (New York: W. W. Norton and Co., Inc., 1965), p. 202. Arguments of American scientists and others for and against the lunar mission are summarized by Berkner, *The Scientic Age . . . , op. cit.,* p. 18-20.

[19]Van Dyke, *op. cit.,* pp. 23, 120; and *Public Papers of the Presidents of the US,* Dwight D. Eisenhower, January 1 to December 31, 1957 (Washington, D.C., 1958), p. 146.

[20]US Congress, *Documents on International Aspects of the Exploration and Use of Outer Space, 1954-1962*, Senate, Committee on Aeronautics and Space Sciences, 88th Congress, 1st session, 1963, Document #18, pp. 43, 46-47.

[21]Logsdon, *op. cit.*, p. 155.

[22]See Van Dyke, *op. cit.*, p. 96; and Killian, "Shaping a Public Policy for the Space Age," in American Assembly, Columbia University, *Outer Space: Prospects for Man and Society* (Englewood Cliffs, N.J.: Prentice-Hall, Inc., 1962), p. 184.

[23]Van Dyke, *supra*, pp. 5, 96, 155.

[24]See Augenstein, *op. cit.*, pp. 80-81.

[25]See, for example, Academician A. P. Vinogradov (Vice-President of Soviet Academy of Sciences) in *Komsomolskaia Pravda*, February 11, 1970, p. 1.

[26]Vitalii Sevastianov (M.Sc., Cosmonaut), "Opening Up Outer Space," *New Times*, No. 25, June 1971, pp. 8-9.

[27]Augenstein, *op. cit.*, p. 72.

[28]Van Dyke, *op. cit.*, p. 5.

[29]Sheldon, "Soviet Military Space Activities," *Soviet Space Programs, 1966-1970*, *op. cit.*, p. 395; and Barbara DeVoe, "Soviet Applications of Space to the Economy," in *ibid.*, p. 297.

[30]Herman Pollack (Director, Bureau of International, Scientific and Technological Affairs [US Department of State]), "Impact of the Space Program on America's Foreign Relations," in *International Cooperation in Outer Space, Symposium, op. cit.*, p. 601.

[31]Moscow Domestic Service, January 22, 1969, as cited in Joseph G. Whelan, "Political Goals and Purposes of the USSR in Space," *Soviet Space Programs, 1966-1970, op. cit.*, p. 36. Whelan sees a general trend of emphasis by Soviet spokesmen on practical space applications (p. 35).

[32]TASS International Service, August 1, 1968, as cited in *ibid.*, p. 50.

[33]See the statements by Blagonravov (p. 379) and Keldysh (p. 381) in Sheldon, "Projections of Soviet Space Plans," *op. cit.*

[34]See Logsdon, *op. cit.*, p. 116.

[35]Levy in Levy, *op. cit.*, p. 199.

[36]*Pravda*, February 4, 1969, p. 2.

[37]Raymond A. Bauer, "Keynes Via the Back Door," *Journal of Social Issues, 17* (2), 1961, p. 52.

[38]Amitai Etzioni, *The Moon-Doggle, Domestic and International Implications of the Space Race* (Garden City, N.Y.: Doubleday & Co., Inc., 1964), ix.

[39]Erlend A. Kennan and Edmund H. Harvey, *Mission to the Moon, A Critical Examination of NASA and the Space Program* (New York: William Morrow and Co., 1969), pp. 83, 240.

[40]Lincoln P. Bloomfield, *The Peaceful Uses of Space*, Public Affairs Pamphlet No. 331, 1962, p. 27.

[41]Etzioni, *op. cit.*, xiii.

[42]Charles S. Sheldon II argues that the Soviets have benefited a great deal in "An American 'Sputnik' for the Russians?" in Eugene Rabinowitch and Richard S. Lewis, editors, *Man on the Moon—The Impact on Science, Technology and International Cooperation* (New York: Basic Books, Inc., 1969), pp. 61-62.

[43]As by Vitalii Sevastionov, *op. cit.*, pp. 8-9; and Berkner, *The Scientific Age, op. cit.*, pp. 18-19.

[44]Pollack, *op. cit.*, p. 601.

[45]See the arguments cited in Berkner, *supra*, p. 19 and the argument by Killian in

Van Dyke, *op. cit.*, p. 9, and in Killian, "Shaping a Public Policy for the Space Age," in American Assembly, *op. cit.*, p. 186. See also Etzioni, *op. cit.*, ix-x, xv; and Martin Summerfield, Princeton Professor of Aero-propulsion, in Jerry and Vivian Grey, eds., *Space Flight Report to the Nation* (New York: Basic Books, Inc., 1962), p. 118.

[46]US Congress, *1964 NASA Authorization, Hearings*, House Committee of Science and Astronautics, Subcommittee on Applications and Tracking and Data Acquisition, 88th Congress, 1st session, 1963, Part 4, p. 2916; and J. Herbert Holloman, "The Brain Mines of Tomorrow," *Saturday Review, 46*, May 4, 1963, p. 47.

[47]Raymond A. Bauer with Richard S. Rosenbloom, Laure Sharp *et al.*, *Second-Order Consequences, A Methodological Essay on the Impact of Technology* (Cambridge, Mass.: MIT Press, 1969), p. 161.

[48]Alton Frye, "Soviet Space Activities: A Decade of Pyrrhic Politics," in Bloomfield, *op. cit.*, pp. 178-179.

[49]For example, Van Dyke, *op. cit.*, Arnold L. Horelick, "The Soviet Union and the Political Uses of Outer Space," in Joseph M. Goldsen, editor, *Outer Space in World Politics* (New York: Frederick A. Praeger, 1963); Edwin Diamond, *The Rise and Fall of the Space Age* (Garden City, N.Y.: Doubleday & Co., Inc., 1964); and Joseph G. Whelan, *op. cit.*

[50]Van Dyke, *supra*, pp. 119-120.

[51]*Ibid.*, pp. 123, 271-272.

[52]*Ibid.*, ix, pp. 271-272.

[53]See testimony of Dr. Lee DuBridge in US Congress, *Scientists' Testimony on Space Goals*, Hearings, US Senate, Committee on Aeronautics and Space Sciences, 88th Congress, 1st session, 1963, p. 139; and Dwight D. Eisenhower, "Are We Headed in the Wrong Direction?" *Saturday Evening Post, 235*, August 11-18, 1962, p. 24.

[54]Van Dyke, *supra*, p. 274.

[55]Frye, "Soviet Space Activities: A Decade of Pyrrhic Politics," *op. cit.*, p. 196.

[56]Hazel G. Erskine, editor, "The Polls, Defense, Peace and Space," *Public Opinion Quarterly, 25*, Fall 1961, p. 486.

[57]USIA, Research and Reference Service, Survey Research Studies, "The Image of US Versus Soviet Science in West European Public Opinion. A Survey in Four West European Countries," WE-3, October 1961, pp. 23, 28-31.

[58]Joseph M. Goldsen, *Public Opinion and Social Effects of Space Activity*, Rand Corporation, Research Memorandum #RM-2417, July 20, 1959, pp. 1-2, 4-6. See also Sam Lubell, "Sputnik and American Public Opinion," *Columbia University Forum, 1* (1), Winter 1957, pp. 15-21.

[59]See Sheldon, *Review of the Soviet Space Program . . . , op. cit.*, p. 85. This was probably the case with the USSR putting a woman in space—a decision based on scientific motives but exploited subsequently for propaganda effect. See Shelton, *Soviet Space Exploration, op. cit.*, p. 147.

[60]For example, cosmonaut Gherman Titov's introduction to Shelton, *supra*, xi-xii, which mentions pride as a permanent and post hoc rather than a motivating factor.

[61]Horelick, "The Soviet Union and the Political Uses of Outer Space," *op. cit.*, pp. 46-47.

[62]Lloyd V. Berkner, "Earth Satellites and Foreign Policy," *Foreign Affairs, 36*, January 1958, p. 229.

[63]Van Dyke, *op. cit.*, p. 13.

[64]See Etzioni, *op. cit.*, xii.

[65]*Public Papers of the Presidents . . . Dwight D. Eisenhower, op. cit.*, pp. 127, 146.

[66]Logsdon, *op. cit.*, p. 155.

[67]Quoted in Philip J. Klass, *Secret Sentries in Space* (New York: Random House, 1971), pp. 26-27. See also Etzioni, *op. cit.*, xii.

[68]See *Missiles and Rockets*, 7, October 10, 1960, pp. 12-13; and Van Dyke, *op. cit.*, p. 23.

[69]Quoted in Etzioni, *op. cit.*, xiii.

[70]That these events contributed to the Kennedy moon commitment is affirmed by Logsdon, *op. cit.*, pp. 4, 102-103, 111-112, 144; and Etzioni, *loc. cit.*

[71]See Etzioni, *supra*, xiii-xiv; Logsdon, *supra*, p. 116; and Diamond, *op. cit.*, p. 34.

[72]James E. Webb, address of September 13, 1961, NASA News Release No. 61-205, pp. 9ff, as cited in Van Dyke, *op. cit.*, p. 126.

[73]Statements in *Space: The New Frontier*, NASA (US Government Printing Office, 1967, O-223-626), p. 2 quoted in Kennan and Harvey, *op. cit.*, p. 86.

[74]Statement in *Air Force and Space Digest*, November 1961, p. 73.

[75]See Etzioni, *op. cit.*, p. 160.

[76]According to Logsdon, *op. cit.*, pp. 102-10.

[77]Hugh Sidey, *John F. Kennedy, President* (New York: Atheneum Press, 1964), pp. 121-123.

[78]Memo of April 20, 1961, cited in Logsdon, *op. cit.*, pp. 109-110, 113; and Kennan and Harvey, *op. cit.*, p. 86.

[79]Interview with Jerome Wiesner by Logsdon, *supra*, p. 118.

[80]Quoted in Hugo Young, Bryan Sikock and Peter Dunn, "Why We Went to the Moon: From the Bay of Pigs to the Sea of Tranquility," *The Washington Monthly*, 2 (2), April 1970, p. 38.

[81]Logsdon, *op. cit.*, pp. 117, 126-127, 134, 137.

[82]See *ibid.*, pp. 126-127, 129.

[83]Whelan, "Political Goals and Purposes of the USSR in Space," *Soviet Space Programs, 1966-1970, op. cit.*, pp. 54-55, 57-58.

[84]Ralph Lapp in his introduction to Kennan and Harvey, *op. cit.*, x-xi.

[85]See Sheldon, *Review of the Soviet Space Program . . . , op. cit.*, p. 85.

[86]The theme of a collection of speeches by Khrushchev published under the title *For Victory in Peaceful Competition with Capitalism* (New York: E. P. Dutton & Co., Inc., 1960).

[87]Vladimirov, *op. cit.*, pp. 20, 86, 115-119, 123-133.

[88]Horelick, "The Soviet Union and the Political Uses of Outer Space," *op. cit.*, p. 55.

[89]Michael N. Golovine, *Conflict in Space, A Pattern of War in a New Dimension* (London: Temple Press Ltd., 1962), p. 58.

[90]Sheldon, *Review of the Soviet Space Program . . . , op. cit.*, p. 54.

[91]Sheldon, "Program Details of Unmanned Flights," *Soviet Space Programs, 1966-1970, op. cit.*, pp. 163-166, 212-213.

[92]*Ibid.*, pp. 162-163, 166-167, 196-209.

[93]*Ibid.*, pp. 165-166, 218; and Sheldon, *United States and Soviet Progress in Space* (1973), pp. 56-59.

[94]Van Dyke, *op. cit.*, p. 101; Leonid Sedov expressed a similar view in *New Times*, No. 38, September 1959, pp. 5-6.

[95]Moscow Domestic Services, November 5, 1968, as cited by Whelan, "Political Goals and Purposes of the USSR in Space," *Soviet Space Programs, 1966-1970, op. cit.*, p. 59.

[96]US Congress, *Panel on Science and Technology*, House, Committee on Science and Astronautics, Fifth Meeting, 88th Congress, 1st session, 1963, p. 25.

[97]See Whelan, "Political Goals and Purposes of the USSR in Space," in US Con-

gress, *Soviet Space Programs, 1962-1965, Goals and Purposes, Achievements, Plan and International Implications,* Staff Report to Committee on Aeronautics and Space Sciences, US Senate, December 30, 1966, 89th Congress, 2d session (hereinafter *Soviet Space Programs, 1962-1965*), pp. 32-70.

[98]See the *Christian Science Monitor,* May 26, 1964, p. 3.

[99]For example, Leonid Brezhnev as cited by Whelan, "Political Goals and Purposes of the USSR in Space," in *Soviet Space Programs, 1962-1965, op. cit.,* p. 81.

[100]Lt. General James M. Gavin, *War and Peace in the Space Age* (New York: Harper and Bros., 1958), p. 232. General Gavin is incorrect in implying that the Soviet choice of frequencies violated any rule of the IGY.

[101]See, for example, statements cited by Whelan, "Political Goals and Purposes of the USSR in Space," *Soviet Space Programs, 1966-1970, op. cit.,* pp. 21-26, 57-62; and the quotations collected by Sheldon on pp. 359-384 of the same volume.

[102]See Sheldon, "Projections of Soviet Space Plans," *op. cit.,* p. 394.

[103]Whelan, "Political Goals and Purposes of the USSR in Space," *Soviet Space Programs, 1966-1970, op. cit.,* pp. 45-53.

[104]For example, J. Y. Sejnin, *New Times,* No. 14, April 1960, p. 28; and G. Ivanov, *International Affairs,* No. 6, June 1964, pp. 64-68. See also Whelan, *supra,* pp. 43-45 for more recent attacks.

[105]See Jan Triska and David Finley, "Soviet-American Relations: A Multiple Symmetry Model," *Journal of Conflict Resolution,* 9 (1), March 1965, p. 38.

[106]Van Dyke, *op. cit.,* p. 132.

[107]Diamond, *op. cit.,* p. 1.

[108]See Alton Frye, "Soviet Space Activities: A Decade of Pyrrhic Politics," *op. cit.,* p. 203.

[109]See Arnold L. Horelick, "The Soviet Union and the Political Uses of Outer Space," *op. cit.,* pp. 49-50, especially the quoted praise of Khrushchev by Gagarin, which recalls the excesses of the Stalin cult.

[110]See Whelan, "Political Goals and Purposes of the USSR in Space," *Soviet Space Programs, 1966-1970, op. cit.,* pp. 26-27.

[111]The microscopic examination of the astronauts, their families, pets and diets along with the author's own probing of their (and his own) psyches are interestingly described in Norman Mailer's *Of A Fire on the Moon* (Boston: Little, Brown and Co., Inc., 1970).

[112]See Whelan, "Political Goals and Purposes of the USSR in Space," in *Soviet Space Programs, 1966-1970, op. cit.,* pp. 50-51.

[113]Sheldon, *Review of the Soviet Space Program . . . , op. cit.,* p. 85; and Sejnin, *op. cit.,* p. 27.

[114]Whelan, "Political Goals and Purposes of the USSR in Space," *Soviet Space Programs, 1962-1965, op. cit.,* p. 36; and Whelan, "Political Goals and Purposes of the USSR in Space," *Soviet Space Programs, 1966-1970, op. cit.,* pp. 20-21.

[115]Whelan, "Political Goals and Purposes of the USSR in Space," *Soviet Space Programs, 1962-1965, op. cit.,* pp. 38-39.

[116]Whelan, "Political Goals and Purposes of the USSR in Space," *Soviet Space Programs, 1966-1970, op. cit.,* pp. 29-30.

[117]Stanley Hoffman, *Gulliver's Troubles or the Setting of American Foreign Policy* (New York: McGraw-Hill, 1968), pp. 13, 147.

[118]See Frye, "Soviet Space Activities: A Decade of Pyrrhic Politics," *op. cit.,* pp. 194-196; and Van Dyke, *op. cit.,* p. 178.

[119]Eisenhower quote on October 9, 1957 in US Congress, *Documents on Interna-*

tional Aspects of the Exploration and Use of Outer Space, 1954-1962, op. cit., pp. 43, 46-47.

[120]Lubell, *op. cit.,* pp. 15-21.

[121]*Missiles and Rockets,* 7, October 10, 1960, pp. 12-13. Emphasis added.

[122]See Van Dyke, *op. cit.,* pp. 5-6, who says security was a "vitally important" goal in Kennedy's space planning (p. 271).

[123]Pollack, *op. cit.,* p. 601.

[124]Eugene Emme, *The Impact of Air Power* (Princeton: Van Nostrand, 1959), p. 844.

[125]*Air Force* as cited in Etzioni, *op. cit.,* p. 139.

[126]William R. Kintner, "The Problem of Opening the Soviet System," in Frederick J. Ossenbeck and Patricia C. Kroeck, editors, *Open Space and Peace, A Symposium on Effects of Observation* (Stanford, Calif.: Stanford University, The Hoover Institute, 1964), p. 119.

[127]J. M. Goldsen and L. Lipson, *Some Political Implications of the Space Age,* Rand Corporation, Paper #P-1435, February 24, 1958, pp. 7-8; and Whelan, "Political Goals and Purposes of the USSR in Space," *Soviet Space Programs, 1966-1970, op. cit.* pp. 45-47.

[128]See Horelick "The Soviet Union and the Political Uses of Outer Space," *op. cit.,* pp. 51-52; and Sheldon, "Projections of Soviet Space Plans," *Soviet Space Programs, 1966-1970, op. cit.,* p. 395.

[129]Testimony of James Webb in US Congress, *Independent Offices Appropriations for 1963,* Hearings Before a Subcommittee, Committee on Appropriations, US Senate, 87th Congress, 2nd Session, 1962, Part 3, p. 424.

[130]F. J. Krieger, *Soviet Space Experiments and Astronautics,* Rand Corporation, Paper #P-2261, March 31, 1961, p. 15.

[131]See Vladimirov, *op. cit.,* pp. 109-110.

[132]See Horelick, "The Soviet Union and the Political Uses of Outer Space," *op. cit.,* p. 48.

[133]See Bauer *et al., op. cit.,* p. 82.

[134]Horelick, "The Soviet Union and the Political Uses of Outer Space," *op. cit.,* pp. 46-47.

[135]See Frye, "Soviet Space Activities: A Decade of Pyrrhic Politics," *op. cit.,* p. 201.

[136]See Goldsen, "Outer Space in World Politics," *op. cit.,* p. 20.

[137]As seen in almost every article on the Soviet space program in *New Times, Pravda, International Affairs,* etc.

[138]US Congress, *The National Space Project,* Select Committee on Astronautics and Space Exploration, House, 85th Congress, 2d session, 1958, House Report No. 1758, pp. 7, 37. See also Section III of the Space Act of 1958.

[139]Van Dyke, *op. cit.,* pp. 5-6; and Kash, *op. cit.,* p. 10.

[140]Levy in Levy, *op. cit.,* p. 200; and Etzioni, *op. cit.,* xii, xiv.

[141]The memo from NASA's Historical Archives is cited by Logsdon, *op. cit.,* pp. 109-110.

[142]See State Department Bureau Chief Herman Pollack, *op. cit.,* p. 601.

[143]Kintner, *op. cit.,* p. 119.

[144]"Interplanetary Travel Soon a Reality," *New Times,* No. 38, September 1959, pp. 5-6.

[145]"New Stage in Space Research," *New Times,* No. 9, March 3, 1971, pp. 23-24.

[146]For example, cosmonaut Gherman Titov in Shelton, *Soviet Space Exploration, op. cit.,* xi.

[147]Moscow Domestic Service, November 1, 1968, as cited by Whelan, "Political Goals and Purposes of the USSR in Space," *Soviet Space Programs, 1966-1970, op. cit.,* p. 47.

[148]For example, see G. Ivanov, "Space Business and US Policy," *International Affairs,* No. 6, June 1964, pp. 64-68; and Sejnin, *op. cit.,* pp. 26-29. See also the remarks of General-Academician Blagonravov cited by Horelick, "The Soviet Union and the Political Uses of Outer Space," *op. cit.,* p. 66.

5 A Consideration of Policies Associated with the Soviet Space Program

Although the leadership of the USSR is *relatively* free of the restrictions on its freedom of action that constrain the leaders of less authoritarian regimes, one may assume that maintaining an expensive space program in a country in which the necessary resources of wealth and talent are sorely needed for other purposes requires some support or at least passive, if grudging, acceptance from general and special publics. Marxism-Leninism serves a dual role in providing support for the Soviet space program because of its concern with science and the future, and because it justifies the leading role of the Communist elite who, schooled and ordinated in the mysteries of the Communist faith, must be correct in their decisions on space as in other matters.

For centuries Russia lagged behind the countries of Western Europe. Even a tsar, Peter the Great, chided his fellow Russians for their backwardness. In the modern period most Soviet citizens are aware of the advantages enjoyed by their counterparts in the West despite the efforts of Communist leaders to hide these from them. Perhaps the rudest and most cogent reminder of Russian technological backwardness was the onslaught of the technically advanced German war machine in 1941–1942. Even after the Second World War, the Soviet Union was faced with the rivalry of the most technologically advanced and wealthy power, the United States, which for a time enjoyed a monopoly of nuclear weapons. Before the Sputnik these factors certainly combined to promote strong feelings of envy, inferiority and insecurity vis a vis the West among the Soviet population generally. In the absence of survey data of the Soviet public after Sputnik, it still seems safe to conclude that Soviet citizens must have felt a great deal of

117

pride and relief at their country's space achievements, which would induce them to lend support to the efforts of their leaders in space.

As described in Chapter 1, Russia has a longer and perhaps richer tradition in rocketry than any Western country. This factor, combined with what some take to be distinctive features of "Russian character," may dispose Soviet citizens to favorable emotional and even romantic feelings about their national space program.[1] In the absence of statistics it is difficult if not impossible to compare the measure of support of Soviet citizens for their space program with that of their counterparts in the United States. It might seem that with their manifold advantages American citizens would not be subject to Soviet feelings of inferiority and insecurity, even after Sputnik. Surveys of American public opinion appear to bear this out.

One study based on polls taken just after Sputnik and again in May, 1958, concluded that America was neck and neck with the USSR in the field of science, although one or the other power might be ahead in certain fields. The study found that "an overwhelming majority" preferred spending increased amounts on medical research or basic scientific research to putting a man on the moon.[2]

Corroborative findings were made by Sam Lubell, a public-opinion analyst who found no evidence of public hysteria or pessimism and a concern lest space spending jeopardize the current economic boom.[3] Following President Kennedy's public commitment to a manned lunar landing on the moon on May 26, 1961, a Gallup Poll found that 33 per cent of the American public supported expenditure of $40 billion to achieve this goal, while 58 per cent were opposed.[4] A 1963 poll by the Opinion Research Corporation found that Americans put landing a man on the moon twenty-fifth on a list of 26 priorities, just ahead of increasing financial support for artists. Just as public support for the lunar mission was low, the trend was down. Surveys taken in Iowa found that those who saw the moon goal as "very" or "fairly" important dropped from a high of 62 per cent in 1962 to a low of 39 per cent in 1963. Similarly, Harris Polls found that although a slim plurality of the public (42 per cent to 45 per cent) found the $40 billion price tag "worth it" in 1965, this had become a minority of 34 per cent by 1967.[5]

Soviet citizens are not asked whether they support the levels of funding involved in the Soviet space program, either by opinion pollsters or by their own political leaders. Revenues in the USSR are collected mostly through "invisible" taxes similar to the value-added tax proposed recently by the Nixon Administration for the United States and to the profits made by US corporations. Expenditures on space are not specifically published, and there are no competitive elections in which taxes might become a political issue as in the United States. For these reasons the Soviet public is not so likely to be concerned with footing the bill for space exploration as the American public, but rather, concerned that large sums are spent on space as opposed to other areas of more direct and immediate benefit to the ordinary citizen. In this respect the position of the Soviet general public probably corresponds not so much to the American general public as to special American publics such as ethnic minorities, scientists, educators, etc., who object not so much to the fact of government expenditure as to the fact that it is not directed to other programs dearer to their special interests. Unlike the American Government, the socialist government of the Soviet Union has assumed general and specific responsibility for the national economy, including the quantity, quality and prices of goods and the amount of wages paid to workers. As such, the Soviet public constitutes a general interest group *as consumers*, which probably best corresponds in scope to the American general interest group of taxpayers and in aims to more specialized American interest groups such as those mentioned. The point is not that the objections of the general public to space expenditures are likely to be greater on either side of the iron curtain, but rather that they will be expressed differently, Soviet citizens having more reason to connect space expenses to their plight as consumers.

Although this author knows of no public-opinion data to indicate the extent of consumer dissatisfaction in the USSR and its relation to any lack of public support of the Soviet space program, occasional articles have appeared in the Soviet press indicating that these sentiments exist. In 1960 a spate of letters appeared critical of the goods available to Soviet consumers, one of which pointed out the discrepancy between the quality of Soviet space technology and that of goods available to ordinary citizens.[6] More recently, one Soviet matron confronted with a miserable lot of potatoes at the central farm market in Moscow drew the applause of her fellow citizens for the suggestion that the Sputniks which "fly around beautifully in outer space" be used to "send these rotten potatoes into outer space too."[7] Similar sentiments are known to be widespread but are generally excluded from expression in the public media.[8]

Public opposition to space expenditure is now and in the foreseeable future of considerably less consequence in the USSR than in the USA. The USSR lacks the kind of public scrutiny of space and other government programs provided by the US Congress and news media. It also lacks any comparable means of expressing public objections and making them felt in the decision-making process. As such, the Soviet space program began without any great public support and can very probably continue without it. Certainly genuine public pride must have been built up by the successes of

Sputnik, *Vostok* and the other early flights, but this has probably diminished as the novelty of space has worn off and as a result of American accomplishments with the Apollo program. This pattern of "diminishing returns" of domestic public-relations value has done more to reduce space spending in the United States than in the USSR.

Whatever the importance of various groups and centers of power in the USSR, there can be little doubt that, compared to the USA, the extent of rivalry between the leading political institutions over basic policy questions is profoundly lesser. In particular the dominance of the single party prevents the kind of rivalry visible in the perennial tug of war between the Presidency and the Congress in the United States. This is most evident when neither American political party controls both Congress and the White House, but continues in other periods as well. Thus in the last three years of the Eisenhower Administration after Sputnik, the Congress, controlled by the Democratic Party, was strongly in favor of a drastically accelerated space program; the Republican President and his NASA Administrator were not. Congress, not the President, took the lead in establishing national space policy in 1958–1960. Led by Senate Majority Leader Lyndon B. Johnson, the first chairman of the Senate Space Committee, the Congress, by threatening to take measures of its own, got Eisenhower to propose the Space Act of 1958, which resulted in the creation of NASA.[9] The Democrats then proceeded to lambast the Eisenhower Administration in the 1960 election for its complacency in the areas of space and missile development. After the Democrats took the White House in 1961, the rivalry between the two branches died down, as both were committed to a very ambitious space program,[10] but further into the 1960s the Congress became more and more stingy in approving the Administration's budget requests for NASA.[11] Meanwhile, the Republican minority with the support of some southern Democrats chided the Administration for neglecting the role of the military in the conquest of space.[12] The American two-party system feeds upon and encourages these kinds of rivalries.

In spite of this major difference, the space programs of both countries have become more solidly entrenched with the growth of political agencies whose purpose is to manage and execute them. This is especially true in the United States as regards NASA and the space committees in the House and Senate, which have no Soviet counterparts. NASA has become a large and powerful agency with large funds to disburse. Although its first administrator, Keith Glennan, shared Eisenhower's doubts about the need for an ambitious space program, this was not true of his successors, or of his immediate subordinates who remained with NASA when administrations changed and Glennan resigned.[13] As is the case with other congressional committees, the space committees have in large measure become the clients and collaborators of the agency they oversee, and to which their own fates and influence in government are tied.[14]

Since the Soviet space agencies are more diverse, anonymous and ad hoc, they probably do not exercise the political influence enjoyed by NASA and the congressional committees. However, their anonymity may rebound to the advantage of the Soviet space program. Since the leading space workers themselves are generally unnamed, the credit which might go to them for their achievements goes instead to the leaders of the Soviet Party and Government. This was the case with the honors after the Gagarin flight. Of some 7,000 decorations awarded, the only recipients named were half a dozen officials of the party and state: Khrushchev, Brezhnev, Rudnev, Kalmykov, Ustinov and Keldysh.[15] More recently the role of the party in the space program was emphasized by having the "Internationale," or Communist hymn, broadcast back to the 23rd Party Congress from *Luna* 10 on the moon.[16]

Political support of the Soviet space program, as with most other aspects of Soviet policy, depends much more than in the United States on the dispositions of the highest leaders of the party and government. The manner in which the American Congress after Sputnik was able to establish an accelerated program in the face of administrative reluctance, and the way in which Congress has pared NASA requests in recent years, are outside the limits of Soviet political processes. Some comparisons of support by political institutions can be made when applied to the top political leaders. In particular this author finds some similarity between the results of American presidential elections and the Soviet political process generally referred to as the "struggle for power." It has been argued before that within certain limits, the attitude of the American president can have a dramatic impact on the goals and degree of commitment of the national space program. This was most evident in the shift of administrations in 1960–1961. In the Soviet case the corresponding events are the changes in leadership that took place in 1953–1955 and 1964.

This author was fortunate enough to hear at Illinois State University the address of a Soviet citizen formerly employed in the Soviet unmanned lunar program and now working in the office of the UN Under Secretary for Political Affairs. In his informal address and in subsequent conversations with this author, the speaker, Dr. Sergei P. Fedorenko, emphasized the role of the top leadership in shaping Soviet policy in science and technology. For his purposes of analysis he divided postwar Soviet scientific and technological development into three periods, which he characterized as the periods of Stalin, Khrushchev, and the "new leadership." Dr. Fedorenko corroborated some of this author's conclusions about the organization and development of the Soviet space program. In terms of the role of the top leaders, he summarized the main goals in science and technology in the three periods.

Fedorenko cited "unofficial but reliable figures" indicating that in 1945–1953 some 90 per cent of Soviet expenditures on research and development was devoted to the creation of nuclear and thermonuclear weapons and the means of delivering them from the USSR to the USA. This author has already shown how these efforts created a technological base for the Soviet space program. Nonetheless, Fedorenko corroborated statements of Soviet defector Tokaty-Tokaev that the use of rockets for launching space satellites was rejected by the Stalinist leadership, which put exclusive emphasis on military goals and evidently took a view similar to that of President Eisenhower that the rocket was strategically important only as an ICBM, not as a space booster.

Fedorenko indicated that in 1953–1964, under the leadership of Nikita Khrushchev, the main emphasis on Soviet science and technology was on the development of rockets *both* for military use and as space vehicles, each supposedly seen by Khrushchev as contributing to Soviet power and serving thereby the foreign policy goals of the USSR. It was Dr. Fedorenko's contention that after Khrushchev, the "new leadership" has pursued a "broader and more pragmatic" policy in the development of science and technology; and as a result, the degree of effort devoted to the exploration of space has declined as money and talent have been redirected into applying research and development to *all* areas of the Soviet economy.[17] If these contentions are correct, some comparison may be made between Khrushchev's view of space and impact on space policies and those of the Kennedy and Johnson Administrations. Perhaps the policies of current governments are also comparable in that, if Dr. Fedorenko is correct, space spending has declined under both. While the space budgets of the American military services have increased in recent years, comparable developments can also be deduced from the increasing number of Soviet military satellites and the Soviet buildup of intercontinental missiles and space weapons systems.

The identification of the new leadership with broader application of R & D throughout the Soviet economy and a more pragmatic approach to the space program is evident from a change in the style of justifying space expenditures and exploiting space for propaganda purposes. As indicated in the previous chapter, Brezhnev and Kosygin have been at pains to justify the Soviet space program in terms of its practical benefits, the latter claiming in an interview with James Reston of *The New York Times* that space expenditures were just as helpful to the total economy and to Soviet consumers as any other activity sponsored by the government.[18]

The leaders of the Soviet Union since Stalin have been in a position to outdo the United States in a single or a few areas such as certain products of heavy industry, types of machinery, certain kinds of military hardware, certain types of social services, etc. This possibility, of course, is one of the differences and, in a sense, advantages of a command economy. Nevertheless, overtaking the overall level of economic development in the United States has remained beyond the capacity of the USSR to achieve within the framework of long-range economic plans. Perhaps it was Khrushchev's biggest mistake, and the most important factor in his involuntary retirement, that he did not grasp this distinction. It seems reasonable to conclude that Khrushchev was influenced by Soviet (and his own) success in using the advantages of a command economy to briefly outstrip America in space accomplishments. Specifically, in the 1950s and early 1960s the USSR enjoyed something of a lead in the exploration of space and might reasonably have been expected to maintain that lead in the future. Khrushchev may have erroneously concluded that "the advantages of the socialist system" so often proclaimed in connection with the Soviet space program in that period could be further applied so that the USSR would surpass the United States in total and per capita industrial production around 1970. These goals were proclaimed in connection with the first seven-year plan (1959–1965) and personally identified with N. S. Khrushchev at the 21st and 22nd Party Congresses in 1959 and 1961. Khrushchev's ill-fated reorganization of the party and state in 1962–1963 was probably part of an attempt to deal with the initial failures of the seven-year plan by using the party to further political control of and invigorate the economy.

Khrushchev's reorganization was resisted and did not succeed in saving the goals of the seven-year plan or his own political neck. His successors have modified his command economy by loosening the reins of control and partially depoliticizing economic processes. Moreover, they seem to have recognized that the "advantage" of a command economy in outperforming a market economy in *selected* areas results in practice in imbalances which retard overall economic development. This conclusion is connected to criticism of Khrushchev's "hare-brained schemes" and to what Dr. Fedorenko characterized as a switch to "broad-based" and "pragmatic" policies for the development of science and technology under the current leadership.

The loose correspondence of policies between American and Soviet leaders may be in part an illustration of "Dupreel's Theorem," or the "multiple symmetry"[19] hypothesis, which postulates similarity and balance in the actions and reactions of antagonists in situations of protracted conflict. The change in space policies in the USSR after 1964, however, seems to be compounded of five factors, whether it suits this hypothesis or not: (1) the leadership changed, (2) America was forging ahead in the space race, (3) the economy was not achieving planned growth, in part owing to heavy space investments, (4) parity in missiles was put above parity in space, (5) difficulties were encountered in the development of the Soviet superbooster.[20] At least at this stage in the exploration of space, it appears that the support of the highest political leaders for their national space programs is lesser in the cases of Kosygin and Brezhnev as well as Nixon, compared to their predecessors.

Much more than Soviet leaders, American national leaders must acquire the "consent of the governed" in formulating policies on space and other matters. In both societies support and consent extend beyond the general public to include special publics whose interest and influence exceed those of their fellow citizens. When speaking of space policies, the most important special publics in both the USA and the USSR include the military, the scientific community, and those who as managers, planners, owners or employees of aerospace industries are directly affected by space policies.

The important role played by the military in the early stages of the development of rocket and space technology, and the present role of the military in American and Soviet space efforts, have already been described. Especially to the extent that the military is involved in the total space programs of each country and benefits in one way or another from civilian space activity, it becomes a natural supporter of space programs and space spending. Both countries have substantial military space programs and substantial "interface" of civilian and military space hardware, personnel and missions.

The American military, particularly the Air Force, has campaigned for an active role in the national space program, and one may assume that the Soviet military has done similarly. When it is decided in either country that certain missions should be ostensibly civilian and/or "peaceful" in nature, the fact that they were put forth as military programs probably adds to their support in that the military leaders will prefer half a loaf to none and political decisionmakers can satisfy themselves with killing two birds with one stone, i.e., accomplishing national security objectives in addition to the main or ostensible scientific goals of a certain mission. For example, the project to develop the Saturn booster was originally a military project.[21] Although NASA has used Saturn for the Apollo and Skylab programs, the American military is presumably happy to have the booster in the operational stage for possible use in accomplishing whatever its original intentions for the rocket were. The development of the F-1 engines in the Saturn project represents a quantum jump in space technology that was required for any mission, civilian or military, for which immense booster thrust is a technical requirement. The reusable space-shuttle vehicle planned as NASA's next major technological development is considerably more adaptable to military purposes than the Saturn family of boosters. Similarly, by taking part in NASA programs such as Gemini and Skylab, the military can perform those missions for which it could not get the permission of DoD, other executive agencies or the Congress to do on its own.[22] This not only broadens NASA's support by including the military but gives credence to NASA's claim that its efforts contribute to national security.

As mentioned in Chapter 3, Khrushchev garnered military support from one faction of the Soviet armed services that favored strategic rocket weapons over more conventional arms. Edward Crankshaw has suggested that Khrushchev relied on this faction for support in his ouster of Marshal Zhukov from leading government and party posts.[23] Presumably this faction, which emerged as the dominant voice in Soviet strategic thinking, would tend to favor a broad and ambitious space program and seek a role in it for themselves. In the early 1960s, when the Soviet Union was faced with a "missile gap" and after the Cuban missile crisis, the hand the military played in Soviet politics was, presumably, strengthened. Marshal Grechko and Admiral Gorshkov became members of the Central Committee of the CPSU, and the former is now the Minister of Defense[24] and, more recently, a member of the CPSU Politburo. It is questionable, however, that these developments represent any great plus for the Soviet space program. First, the two men made their careers in the Red Army and Red Fleet, not the Strategic Rocket Troops or Air Defense Command. Second, the Soviet Union does not have an agency comparable to NASA, nor does it seem that an alliance between space administrators and generals would be too congenial, because the former, even if they include a number of men with military ranks, are probably more closely tied to the Communist Party, the government and the scientific community. Moreover, in the Soviet Union technical and fiscal resources are more strained than in the USA, so that competition between military and space needs is likely to be at least as keen. The Soviet missile buildup of 1966–1970 probably contributed to cancellation of any manned lunar mission, if such ever existed. In this respect the Soviet drive for parity may have had an effect similar to that of the Vietnam War on NASA's budget in the same period. Even in the United States cooperation between NASA and the military has not prevented rivalry. Although Kennedy's commitment to the manned lunar mission was made in an atmosphere of crisis and cold war, the military evidently played no role in the decision.[25]

Behind the military in the USSR and the USA lies a complex of industrial organizations geared to the support of *both* military and nonmilitary space missions. In the United States this is generally referred to as the aerospace industry, the Soviet counterpart lying somewhere within various Soviet Government ministries and the industries associated with them, generally referred to as the machine-building industry or "heavy" industry. Although no profit motive as such impels the managers of Soviet enterprises to push for space "boondoggles," they presumably consider their services important and wish to maintain the important positions of power they hold in Soviet politics. Although the diversion of fiscal, technical and manpower resources to these industries creates a favorable basis for the manufacture of space hardware, it is probably true, as suggested above, that many planners and managers in this and other sectors of the economy suffered from overinvestment of resources in space and actually pushed for dispersion of such to nonspace production. Broader utilization of technical resources was the principal goal of the State Committee for Scientific and Technological Work created in 1966. The chairman of the Committee is Academician Vladimir A. Kirillin, who is also deputy chairman of the Council of Ministers and member of the Central Committee of the Party.[26]

The profit motive and a more hospitable climate for influence-group politics probably make the US aerospace companies a more important source of suport for the national space program than their counterparts in the USSR. Some 90 cents of the federal space dollar is spent with private industry.[27] Companies whose main business is the production of weapons systems like to emphasize their contributions to peaceful and "futuristic" space activities. Naturally enough, they also lend political support to NASA's proposals and appropriations. Advertising from aerospace firms supports special periodicals of the aerospace trade press, which also champions NASA's efforts. Further support comes from employees and unions concentrated in the areas where the aerospace industries and NASA facilities are located, including California, Texas, Alabama, Florida and other states in the South and Northeast.

The scientific communities in the USA and the USSR also take an interest in their countries' space programs. Of course, those scientists and engineers directly involved in space work are most likely to support ambitious programs. The views of other scientists are mixed and in many cases actually critical of the amount of resources devoted to space generally, and in particular to space activities for which the scientific benefits are disproportionally small compared to the expenses involved.

Both the Soviet and American space programs grew out of the work of small corps of space scientists. The role of these men was described in

Chapters 1 and 2. In addition, Soviet support comes from the Academy of Sciences, which enjoys an important position in the administration of Soviet space work and in international execution of Soviet policies of international space cooperation. Although it enjoys no such position of influence, the Space Sciences Board of the American National Academy of Sciences has provided substantial support to American space activities. The Board was behind the emphasis on the Apollo program and manned flight generally in the early years of the Kennedy Administration, at a time when the President's own Science Advisory Committee was critical.[28]

With this important exception the American scientific community has generally been unhappy with the spending of huge sums on manned missions, and especially Project Apollo. Other scientists, including several Nobelists and former members of the Presidents' Science Advisory Committees, have argued that the pattern of space spending hampers overall scientific and technological progress and overlooks pressing social problems on the domestic front.[29] Despite pressures to the contrary, Soviet scientists have expressed similar criticisms of their own space programs.[30] A number of American scientists have criticized NASA for sending pilots rather than scientists into space. In fact, the Soviets, who have placed greater reliance on automated spacecraft systems, have nonetheless put scientists, including a doctor of science (1964) and a physician (1964), into space at a relatively early date in their manned program. It was only in 1972 that the first American Ph.D., a geologist, was a part of the crew of Apollo 17, the last in the series. Many American scientists have also criticized NASA for neglecting to include sufficient of their colleagues among the crews of Skylab in 1973. The crew of Skylab 2 included the first American astronaut-physician. The six men who crewed the remaining two Skylab missions included four pilots and two engineers with doctorates. It is probable that the presence of trained scientists added a great deal more to missions like Apollo, *Soiuz* and Skylab than they did to earlier missions like *Voskhod*.

Those responsible for the American and Soviet national space programs have adopted various policies which are apparently intended to increase the support of space activities by the public at large and by some special groups. The United States is relatively more open about its space program, and its policy is to disseminate and encourage prelaunch publicity. This makes the total public-relations effort of NASA (in conjuction with the communications media) greater than comparable Soviet efforts, although certainly the Western press has helped to publicize Soviet space achievements. Both countries, however, permit and encourage active publicity of their space programs generally and in the postlaunch phase, although the Soviets have not opened their launch facilities to the public or identified leading space personnel.

In general the same kinds of techniques are used in both nations to broaden public support: the public is exposed to space exhibits, museums, articles in the popular press, astronaut tours, films, television, radio, books and space bric-a-brac. In order to focus public attention on Soviet space achievements, the government has set aside the anniversary of the flight of *Vostok* 1 as Cosmonautics Day (April 12). This is often the occasion for the presenting of various awards to those "collectives" responsible for the design and launching of spacecraft.

In choosing to devote a major and probably the greatest share of their space budgets to various *manned* missions, Soviet and American space planners must have been partly motivated by the consideration that manned operations generally have a greater appeal to the public than the most complex unmanned missions. It is also probably to build and maintain public suport that Soviet and American spokesmen have in recent years given particular emphasis to the practical results to be gained from space exploration. Both countries have commemorated space achievements on postage stamps, but the Soviets have probably gone further in this direction by having space flights dedicated to the 23rd Party Congress and to the fiftieth and fifty-first anniversaries of Soviet rule. The success of the flights then creates public support of the efforts of the leaders in space, as does the treatment of cosmonauts and astronauts as heroes by the political leaders and the press.[31]

NASA is a large agency with a lot of money to spend and has a strong impact on various areas throughout the country involved in meeting NASA's needs and carrying out its programs. Combined with the "pluralist" nature of the American political systems, these considerations increase the possibilities and advantages of NASA efforts to secure the support of special publics. As already suggested, NASA also builds its support among the scientific community by direct employment of scientists and engineers, by sponsoring space-connected research through grants, and by providing considerable support to advanced education in science and engineering through fellowships, grants and the financing of academic centers for space-connected work. NASA also cultivates the support of industry through its attempts to promote the widest dissemination of new materials, techniques, and hardware discovered in the work of the Agency and its contractors.[32] NASA may also consider criteria of political support in awarding its contracts.[33] Other special publics in the USA and USSR may be appealed to using the same techniques employed to attract the support of the general public.

It might seem that the unchecked authority enjoyed by the top leadership in the USSR would maximize the political consequences of any changes in leadership and make policy reversals more dramatic even when the leadership remains relatively stable. Although Khrushchev's de-Stalinization campaign and the rapid dismantling of Khrushchev's political and economic reforms by the current leadership provide examples of such dramatic changes, in space policy the United States has exhibited a more dramatic shift with the change of leaders than has the Soviet Union.

American space policy was, in effect, diametrically reversed between the Eisenhower and Kennedy and Johnson Administrations. The leadership changes in the USSR in 1964 and in the USA in 1968–1969 did not produce such dramatic reversals. President Eisenhower minimized the political significance of space. A meeting of his Science Advisory Committee in 1959 ridiculed the idea of a manned lunar landing, and Eisenhower and his budget bureau were unwilling to fund development of any manned space projects beyond the relatively modest limits of Project Mercury, the small capsule, single-man, orbital program. Unlike his successors and probably unlike Khrushchev, Eisenhower and his advisors did not view space as a field for Soviet-American competition.[34] Eisenhower's NASA administrator, Keith Glennan, shared his views.[35] Eisenhower's view also reflected those dominant in the American scientific community, to which he deferred much more than his successors in formulating policies for space.[36]

In contrast, President Kennedy did at least "at times" view America as engaged in a space race.[37] The Kennedy Administration turned to the aerospace industry, the Air Force and NASA for its support rather than to the scientific community. One of the first changes in space policy by the new administration in 1961 was to reinstate very substantial cuts in NASA's budget request for 1962 by the previous administration.[38] From a high of $.5 billion under Eisenhower, the NASA budget grew under Kennedy and Johnson to more than ten times that amount.[39]

Although President Kennedy was certainly the man responsible for space policy in 1961–1963, Lyndon B. Johnson unquestionably helped to shape space policies even before his tenures as president and vice-president. Johnson had led the Senate inquiry in 1957–1958 into America's alleged "failure" in space, and he became under Kennedy the chairman of NASC (formerly the job of the President himself). Johnson had close ties to leading members of NASA and NASC and an indirect tie to James Webb, administrator of NASA in the Kennedy-Johnson years, through his Senate crony Robert Kerr, for whom Webb had worked as an executive of Kerr-

McGee Oil. Johnson's home state of Texas is, along with California, Florida and Alabama, among the states benefiting most from space spending.[40] Johnson viewed the importance of Sputnik as "essentially political" and credited its political successes to "misjudgments of our political leadership," compounded of inadequate relationships between the scientific and political communities, the "unfortunate anti-intellectualism of the early 1950s" and the failure of our elected representatives to exercise their "responsibility to lead."[41] The facts that relative to their predecessor, Kennedy and Johnson ignored the advice of the scientific community and that Johnson had his own subsequent problems with "intellectuals" reveal a certain irony in his comments.

As with Kennedy and Johnson, Nikita Khrushchev can be identified with championing an ambitious space program and with putting great stress on the political benefits to be derived in space.[42] Although Khrushchev was not so frank as his American counterparts in acknowledging political motives for space exploration, he did exploit the political benefits to the hilt. As previously described, Khrushchev used his championship of space and strategic rockets to gain allies against his domestic political opponents and, internationally, to make exaggerated claims about Soviet strategic might. His policy of strategic bluster, compounded of space achievements and multimegaton bombs, was moderated after his bluff was called during the Cuban missile crisis. Khrushchev's successors have probably reduced space spending somewhat. They have also been reluctant to construe the exploration of space as competitive and have not put so much emphasis on space as a propaganda weapon against domestic or foreign opponents. A reduction on the priority afforded space exploration is probably indicated by what can be described as *defensive* justifications of space programs by the current leaders.[43]

BUDGETING OF VARIOUS PROGRAMS

Percentage breakdowns of Soviet space budgets are no more available to Western scholars than are the total budget figures for space. The Soviets are continuing their *Saliut-Soiuz* manned program as indicated by five launchings in 1973.[44] They are also committed to a cooperative manned project with the US in 1975. As such, manned programs probably take up a larger share of the Soviet civilian space budget than any other single category. A majority (around 60 per cent) of NASA's budget has gone into various manned programs, mostly Apollo, until the most recent years. About 14 per cent has gone into various unmanned scientific missions, 14 per cent for general space technology, six per cent for support operations, four per cent for applications of space technology and two per cent for aeronautics. By distributing the figures for space technology and support operations into the appropriate categories, manned operations would increase to some 70-75 per cent of the total NASA budget with unmanned science 15 per cent, applications and aeronautics both about five per cent.[45] Although comparisons are uncertain, the USSR probably spends a smaller proportion on manned space flight and somewhat more on unmanned science, especially unmanned missions to the moon and planets.[46] Recently NASA has canceled or cut back on many of its scientific and other unmanned programs in an attempt to insure adequate funding of the space shuttle.

Because it is a large and semiautonomous agency and seeks to build support in the context of the American political system, NASA probably sponsors a more diversified range of activities than are explicitly supported for space purposes in the USSR. In fact, NASA funding of research by educational and nonprofit organizations has increased.[47] Diversity of spending, however, has been inhibited by the concentration on a single major mission (Apollo), and it appears that this style of funding may be retained in part with a subsequent concentration on the space shuttle, which may absorb some $5-10 billion compared to approximately $30 billion for Apollo.[48]

MANNED VS. UNMANNED PROGRAMS

In both the Soviet Union and the United States there have been differences expressed over the emphasis that should be placed on manned space operations in the current stage of space exploration. The debate over manned missions has been particularly keen in the USA. As discussed previously, the American space program has exhibited a tendency to focus on a succession of major missions. Each of the successive major missions, Mercury, Gemini and Apollo, have been *manned missions*. The next major mission, Skylab, was also manned, and the space shuttle set for around 1980 is to be a piloted spacecraft which will incorporate a variety of missions, manned and unmanned, military and civilian, many yet to be defined.

Cutbacks in funding and two fatal accidents in the *Soiuz* series, combined with apparent difficulties in developing and man-rating larger boosters, have apparently held back the progress of manned Soviet flights in recent years, but the *Soiuz-Saliut* combination of vehicles still promises a variety of manned activities in earth orbit. With *Soiuz* and *Saliut*, with the possible man-rating of the *Proton* booster used to launch *Saliut*, and with the development of an even larger launch vehicle, the Soviets can continue to mount ambitious manned programs. However, it appears that budget cutbacks have curtailed Soviet manned flights, leaving unmanned orbital flights, including scientific and military payloads, unaffected. The success of Apollo probably helped curtail any Soviet plans for manned flights outside earth's orbit on the grounds that such flights would only emphasize the fact that the USSR was taking second place to Apollo. Manned operations in orbit, for which the USSR has already demonstrated the technology, could go ahead without the stigma of duplicating or falling short of past American achievements. One may also presume that the problems which resulted in the death of the crew of *Soiuz* 11 and the difficulties with the *Saliut* station in 1973 have been reviewed and that measures are being taken to prevent future recurrence. The next step in manned Soviet operations, then, may hinge on some technical problem, perhaps associated with use of a larger booster which is possibly being worked on now.

Public criticisms of the effort and expense devoted to manned missions in the US began as soon as such missions were planned. The Eisenhower Administration, including the President himself, was opposed to any manned missions beyond the limits of Project Mercury.[49] James Killian, Eisenhower's science advisor, believed that large booster and man-in-space projects would get out of hand if not deliberately "contained," resulting in a commitment to an undesirable "multi-billion-dollar space program."[50]

It was mostly around the Apollo program with its enormous price tag originally estimated at $40 billion that the debate over manned vs. unmanned space systems developed. The most outspoken critics of the emphasis on manned flight were in the American scientific community. In general their criticisms boiled down to the concentration that the scientific return from Apollo and other manned missions was not worth the price and that greater scientific knowledge could be obtained from unmanned flights at a fraction of the cost.[51] Even scientists who support NASA's Apollo program have admitted that the expense cannot be justified in terms of scientific payoff.[52] Reflecting the dominant attitude of the scientific community, even President Kennedy's science advisor did not support the overwhelming emphasis on manned missions.[53] In NASA itself scientific opinion was divided between those favoring manned vs. unmanned flights, with the preponderant opinion favoring manned efforts and increased unmanned missions as well.[54]

Political conditions in the USSR inhibit this kind of debate from emerging publicly. Until recently there was little or no public evidence in the Soviet Union of differences among scientists or others on the amount of effort that ought to be devoted to manned and unmanned flights. There seemed to be a consensus among Soviet scientists favoring manned operations both in terms of increased reliability and in keeping with the long-range goal of manned exploitation and colonization of the solar system.[55]

At least until 1967 most Soviet public statements about space flight by scientists and cosmonauts emphasized the importance of manned flights and the Soviet intention to mount some kind of manned mission to the moon.[56] Beginning in 1967, Soviet scientists and cosmonauts emphasized both manned and unmanned activities. Later, especially after the success of Apollo 11 in 1969, Soviet commentators focused on unmanned explorations of the moon and planets and the role of man in doing scientific work in stations in earth orbit.[57] Thus the Soviet reaction to Apollo was not so much to downgrade manned operations as to emphasize their expense insofar as the moon was concerned and to defend Soviet efforts for unmanned exploration of the moon with manned exploration to come later, and in the meantime, a program of manned activities in earth orbit of which the tragedy-struck *Soiuz* 11 was evidently part of a long series, as evidenced by the launchings of *Soiuz* 12 and 13 in 1973.

PREPARATION, SAFETY, RISK AND HASTE

Part of the policy style of any power is its tendency to exercise risk or caution.* In space policy haste and risk are indicated by attempts to cut technological corners to achieve early launch dates and by tolerance of failure and danger to human life through inadequate safety precautions and preparations. The adequacy of early Soviet orbital missions in terms of preparation is indicated by the success of Sputniks 1-3, just as the haste and lack of preparation of the Vanguard Project after Sputnik is evidenced by the long string of American failures before Explorer 1. Repeated failures of American attempts to impact the moon and of both countries' attempts to impact the planets also suggest some measure of risk and haste. Of course, the Soviet policies of minimizing prelaunch publicity and almost never admitting space failures do serve to decrease the risks of adverse publicity to which the American program is wide open. It is generally possible without recourse to secret methods of data gathering to distinguish failures from successes among those spacecraft that have been partially successful in the boost phase. Whatever boost-stage failures the USSR has suffered have not been made public by the Soviets or anyone else in a position to know about them. Except for some military launches, American boost failures have been public events.

Much has been made in the United States of Soviet technological short-

*See Jan F. Triska and David D. Finley's Chapter IX in *Soviet Foreign Policy* (New York: Macmillan, 1968), pp. 310-349.

cuts on the theory that the USSR used its superior booster capacity in the late 1950s and early 1960s to launch large but crude vehicles and experiments, while the USA, working with smaller boosters, made much greater advances in the "sophistication" and "miniturization" of its payloads.[58] The Soviet defector Penkovsky described this situation as "a big lag in electronics."[59] Although it is certainly true that the USSR was behind and still is far behind in the miniturization of electronic circuitry, this in itself has probably not produced any great hardships for the Soviet space program, nor is it in itself evidence of any undue haste or risk in Soviet launchings. Indeed, the early Soviet successes as well as the more recent American ones indicate that it is booster capacity more than any other factor that sets the limits to what a country can accomplish in space in terms of missions, while electronics technology sets limits to the amount of scientific data returned. Perhaps it is only in the area of planetary probes that the relative unreliability of Soviet electronic components caused great difficulties, occasioned risk and haste, and effectively eliminated much of the Soviet advantage in booster capability in the early 1960s. Otherwise the Soviets have not been inhibited specifically by the "electronic lag" in achieving most of their goals in space.

Perhaps it is not only haste and lack of preparation that has led the Soviets to make so much use of an obsolete booster employing a configuration of clustered small rocket motors and of off-the-shelf electronic components, but a combination of haste and economy.[60] With the use of satellites for economic applications such as communications and resource location, this original savings may turn into false economy, since more "sophisticated" components increase capacity and prolong useful life. By exploiting to the full the booster technology of Sputnik, the Soviets were able to make many impressive accomplishments in the late 1950s and early 1960s. Meanwhile, the United States developed the booster and spacecraft technology of Gemini, Apollo and Skylab, leaving the USSR far behind in the sixties and seventies.

As regards manned flights, both the USSR and the USA have gone to considerable lengths in training and assuring the safety of their cosmonauts and astronauts.[61] Two Soviet flights involving four men, and one American ground test involving three men, have resulted in fatal accidents for the crews. The contention of Penkovsky that several Soviet cosmonauts perished in launch and in orbit before Gagarin's successful flight appears to be unconfirmable and is probably incorrect.[62] If the contention of one Soviet defector is correct, the Chief Designer, Korolev, was pressured by Khrushchev to cut corners to expand the one-man *Vostok* into the three-man *Voskhod*.[63] Whether this is or is not so, there were no recorded fatalities in either the *Vostok* or *Voskhod* series of manned flights.

Both the Soviet Union and the United States have tested their manned spacecraft in unmanned flight previous to their use with crew, the only exception being the Apollo moon landing with no prior unmanned tests of orbiting or landing on the moon by the same vehicle. Thus the USSR flew five unmanned precursors to *Vostok* before Gagarin's flight, two precursors to *Voskhod* and two precursors to *Soiuz*. Much as was the American's case after the 1967 Apollo fire on the launch pad, the USSR flew unmanned tests after the accident on *Soiuz* 1. Three flights in 1970 and 1971 were also possibly unmanned precursors to the *Soiuz-Saliut* space-station missions carried out by *Soiuz* 10 and the ill-fated *Soiuz* 11.[64]

If good luck averted the dangers occasioned by haste in the case of *Voskhod*, Apollo was not so fortunate. Erlend Kennan and Edmund Harvey, Jr., argue persuasively that the January, 1967, fire on board the Apollo spacecraft was in part the result of the choice of a pure oxygen atmosphere made partially for reasons of haste, and that the mechanisms for safety, escape and rescue in the spacecraft were inadequate because of haste.[65] These authors note that Designer Korolev rejected a pure oxygen atmosphere after an electrical fire took place in a Soviet simulator.[66] According to NASA Deputy Administrator George Low, the choice of a pure oxygen atmosphere for Mercury, Gemini, Apollo (and Skylab) was an "expediency." American plans for the space shuttle involve the use of normal air at terrestrial surface pressure.[67] The Soviets have also criticized the Apollo Project for what they see as undue risks to the lives of the astronauts.[68] Oddly enough, however, the Soviets have not made a great deal of their own safety precautions. They seem to wish to give the impression that their techniques are so advanced and reliable that extensive testing and the use of precursor craft are unnecessary. At least this was the case until the death of Vladimir Komarov aboard *Soiuz* 1 in April 1967. Previously all precursor flights had carried the labels *Tiazhelii Sputnik or Kosmos* and were not therefore publicly identified as precursors to manned flights. *Soiuz* 2 was the first and, to this date, only Soviet craft explicitly identified as an unmanned test vehicle to test safety routines before manned flight. Even the precursor flights subsequent to the October, 1968, flight of *Soiuz* 2 bore the vague *Kosmos* label.[69]

STRATEGIES USED TO OBTAIN PROPAGANDA ADVANTAGES

Both the Soviet Union and the United States have sought to achieve the greatest propaganda advantages for their own space programs through various public-relations strategies aimed at domestic and foreign audiences. For the USSR the principal tactic used is to identify the Soviet program as wholly peaceful in aim and nonmilitary in contrast to the American space program, which is labeled as militarist and aggressive in intent.[70] Some examples follow.

Academician Blagonravov has expressed his sympathy for American space scientists who must work under military direction. By contrast, Soviet programs are wholly "scientific." This fiction is maintained by the Soviet policy of refusing to acknowledge any of their own military space activities.[71] In large measure, American military programs are a matter of public record.[72] Although NASA was originally planned as a civilian space agency with (for Eisenhower) a rigorous separation of military and civilian functions, this fast distinction was not made by subsequent administrations. Because the USSR evidently does not have a civilian space agency, the distinction between its military and nonmilitary space activities is even harder to make. NASA has at least succeeded in keeping its international programs free of any military taint. All proposals from foreign countries must be submitted by civilian agencies with some promise of scientific payoff and some investment by both the foreign country and the USA. This has made it easier for NASA to acquire tracking stations overseas since there is no military stigma attached. Moreover, NASA can use foreign scientists to refute Soviet propaganda charges that those programs in which NASA cooperates with foreign countries serve military purposes.[73]

The Soviets have also charged that since the American space program must serve the interests of the military and/or the profit motives of the corporations involved, it cannot make the same progress in fundamental scientific research as can the Soviet program.[74] The charge is also made that American international space programs are designed to gain control of the space programs of other countries.[75]

Soviet journalists, military figures and legal specialists have attempted to portray American reconnaissance, geodetic and navigation satellites as aggressive weapons designed to make possible a surprise attack on the USSR. They have also credited the United States with plans to shoot down Soviet satellites, orbit nuclear weapons and set up military bases on the moon.[76]

Although not so pronounced as Soviet propaganda nor so oblivious to the facts, propaganda on the question of the military use of space is also part of American policy. Because the United States *does* acknowledge a military space program, it cannot claim that the USSR is militarizing space and the USA is not. Instead, American programs are identified as being "defensive" and "nonaggressive" in contrast to "aggressive" Soviet programs.[77] One way of doing this is to contrast (and exaggerate) the openness of the American space program with the secrecy of the Soviet program on the assumption that what is done in secret must be dangerous or malicious.[78] Moreover, the claim is made without much substantiation that the military plays a much greater role in the Soviet program than in the American.

A number of other tactics have been employed by Soviet spokesmen to make invidious comparisons of the two nations' space programs. Khrushchev ridiculed the smallness of the early American satellites, comparing them to oranges.[79] Other Soviet spokesmen have claimed that in the haste to catch up with Soviet space feats, the Americans have subjected their astronauts to undue risks.[80] The Soviets have also sought to portray their own cosmonauts as selfless heroes and dedicated Communists, while imputing the American astronauts with the motives of getting "space fees" and becoming "bourgeois exploiters."[81] Their association with the space program has in fact been a bonanza for the American astronauts. The original seven astronauts signed a half-million-dollar contract for their personal stories with *Life* magazine. *Life* and *World Book Encyclopedia* pledged another million to the next team of sixteen astronauts. Astronauts have frequently emerged as owners and officers in banks, motels and corporations in the Houston and Cape Canaveral areas and have, after "retirement," lent their endorsements to various products and enterprises.[82] Although Soviet astronauts cannot find these sorts of rewards and opportunities in the socialist USSR, it should be noted that they have been afforded the highest public honors and often been given rapid promotions in the ranks of armed forces and the Soviet space program. Of the 25 Soviet cosmonauts through 1972, three held the rank of major general, four were colonels and nine lieutenant colonels. The other two Red Air Force members remaining at the rank of lieutenant were a woman and a physician. At least two cosmonauts, Gagarin and Feoktistov, have held high administrative posts in the space program.[83] Cosmonauts have also been honored by election to the Supreme Soviet and by burial in the Kremlin Wall.

Soviet criticism of the American space program has been moderated in recent years. Space scientists and cosmonauts have generally praised American achievements, and political leaders have had relatively little to say about space except at ceremonial occasions and in making economic justifications of space spending. Some criticism of American programs still goes on, generally attributing military purposes to such projects as MOL (canceled by Nixon). One usually finds such articles in the military press or in popular periodicals like *New Times* and *International Affairs*.[84] In line with the recent Soviet policy of emphasizing the economic return on space spending, and noting the comparative cheapness of unmanned flight, some mild criticisms of the Apollo program have been made which may take note of the reservations of American scientists and claim that Apollo was economically inefficient and retarded the development of other space projects.[85]

It has been suggested that Soviet space shots are timed to achieve maximum propaganda advantage. This is difficult to prove, but a number of "coincidences" can be cited in evidence. Many Soviet flights seem to have been timed to accomplish certain missions before, sometimes only days or weeks before, the American attempts, which are announced months in advance. Other Soviet flights seem to have been timed for special occasions. *Luna* 2 impacted the moon just before Khrushchev set off on his 1959 visit to the United States, where the Soviet leader presented Eisenhower with replicas of Soviet insignia deposited on the lunar surface. Sputnik 4 coincided with the 1960 summit conference and *Vostok* 2 with the Berlin crisis and Warsaw Pact meeting of 1961.[86] *Luna* 10 coincided with the 23rd Party Congress of the CPSU, and NASA expected a Soviet attempt at a manned lunar landing to coincide with the fiftieth anniversary of Soviet rule in November, 1967.[87] It did not.

The selection of the very first Soviet cosmonaut was evidently made with propaganda considerations in mind. Among Gagarin's "qualifications" were that he was a pure Great Russian and of humble peasant origins. Similarly, the Soviet leaders may have hoped to gain some propaganda advantage by orbiting a woman in *Vostok* 6 and a physician and a scientist in *Voskhod* 1. In the case of the woman, Valentina Tereshkova, this supposition is reinforced by the fact that Penkovsky predicted the flight in advance and cited propaganda motives as being the basis for it. Moreover, Miss Tereshkova returned from space just in time to speak to the International Congress of Women in Moscow.[88]

As will be discussed in a subsequent chapter, both space powers have steadfastly maintained their public support of international cooperation in space. A good part of this public commitment is pretense maintained for propaganda purposes. In particular, the USSR has been very reticent in actually implementing cooperation, even with other socialist countries.[89] The United States, which does maintain a very considerable program of cooperation with various countries, also pays lip service to cooperation as well. Perhaps the best example of this was Kennedy's proposal for a joint US-USSR lunar mission, an apparent reversal of the spirit in which Apollo was planned and one that would have been impossible in light of its reception by the Congress.[90]

Another Soviet tactic, evidently designed to express national pride and secure a lasting propaganda advantage, was tried out in October, 1959, after the moon flyby mission of *Luna* 3, which returned photographs of the far side of the moon. The Soviets proceeded to give Russian names to all the identifiable features of the lunar geography in violation of the international custom of having the International Astronomical Union name newly discovered celestial objects. It was not until 1967 that a compromise settlement was reached which apportioned lunar place names to reflect both Soviet and American discoveries on the moon.[91]

Both space powers have extensively employed tours of astronauts and cosmonauts and space exhibits at various international expositions to bolster the propaganda impacts of their space programs. In the United States the information agency, USIA, works very closely with NASA to maximize propaganda benefits. The Apollo 11 mission in particular was accompanied by an orgy of propaganda conducted by the USIA. Through 1970 18 American astronauts visited 57 countries in ten separate tours with deliberate attempts to gain access to and influence special foreign publics.[92] Soviet cosmonauts have also gone on world tours and been employed to proclaim the official positions of Soviet foreign policy in Vietnam and the Middle East.[93]

NASA has used the astronauts for propaganda supporting the United States and its space program by permitting and even encouraging the superhero images thrust on the astronauts by the public press. In January through November of 1966 the astronauts selected for Apollo missions made 810 public appearances and 314 formal presentations. They conducted 137 press interviews and processed over 73,000 items of correspondence.[94] The astronauts have also played a large part in the extensive publicity campaigns conducted by the USIA abroad.[95] The Soviet cosmonauts have been similarly idolized, publicized and sent on good-will junkets throughout the USSR and the world. In particular, Iurii Gagarin enjoyed unrivaled status as a Soviet superhero until his accidental death in March, 1968. Although the Soviets have not matched the publicity campaigns of the Apollo program, their biggest splash occurred at the Paris Air Show of July and August, 1967. There they displayed a complete *Vostok* spaceship together with its entire launcher, the first public display of the standard booster. Previously the Soviets had never published photos or accurate artists' conceptions of this (obsolete) launcher.[96]

APPLICATIONS OF SPACE PROPAGANDA TO INTERNATIONAL POLITICS

Previous chapters have shown how Khrushchev and other Soviet leaders attempted to extract the greatest possible international political advantages from Soviet space successes.* Khrushchev sought to convey the impression that the USSR was technologically and strategically ahead of the USA, as proved by the Sputniks.[97] Claims made for Soviet weapons systems by Khrushchev and others before 1962 were often exaggerated in line with this policy.[98] The Soviet policy of maintaining a strict cloak of secrecy over their space and missile programs made it possible for Khrushchev to hope that his boasts would be believed outside the USSR. Perhaps it is because the photos taken by U-2 aircraft showed that Khrushchev was bluffing, that measures were taken to halt the flights of U-2s over the USSR. Perhaps Khrushchev's bluff worked a little better in the short space between the U-2 incident in May, 1960, and the development of reliable satellite reconnaissance by the USA. In the case of space vehicles, some attempts were made to minimize the provocativeness of the policy of bluff and bluster by claiming that the Soviet program was wholly peaceful and scientific and that, not the space vehicles themselves, but the technology behind them, reflected the strategic might of the Soviet Union.[99]

Khrushchev's policy, a kind of reversal of Teddy Roosevelt's admonition to "speak softly and carry a big stick," proved to have undesirable consequences, as already described. Not only has there been a precipitous decline in Soviet statements linking space achievements to strategic might,[100] but the USSR has permitted its veil of secrecy to be penetrated by not attempting to shoot down American spy satellites, despite the fact that it has developed the capability to do so in recent years.[101] Moreover, both the US and the USSR are bound by the Strategic Arms Limitation Agreements of 1972 not to interfere with or attempt to use concealment techniques to thwart the satellite systems used to police the agreements. Since the Khrushchev period, the USSR has muted criticism of American military activities in space and ceased to impute military purposes to all American space flights.[102] It has also developed numerous military space systems of its own without acknowledging them.

If the USSR has become less critical of the American space program, it does not take such a view of the first space efforts of the People's Republic of China. The Soviets have claimed that it is foolish for the Chinese to build rockets and missiles since they cannot afford the expense and are evidently afraid that the "main target" of Chinese nuclear and missile development is the USSR.[103]

---

*For example, by choosing radio frequencies for Sputnik that could be heard on home receivers around the world.

SECRECY, PUBLICATION AND ANNOUNCEMENT POLICIES

There is no aspect of the Soviet space program in which the influence of political considerations is more evident than in Soviet information policies about that program. Political considerations are particularly obvious in cases of deliberate misinformation and withholding of the facts by those in positions of authority. The most blatant instance of the Soviet credibility gap in space, the pretense that the USSR does not carry out any military activities in space, has already been described. The use of space achievements in inculcating exaggerated estimates of Soviet technological and strategic development has also been dealt with above. Other aspects of space information policy will be detailed here.

Secrecy

Neither the United States nor the Soviet Union made any great secret of its intentions to mount some kind of space program. As noted in Chapter 1, the president of the USSR Academy of Sciences indicated the Soviet intention to launch an earth satellite as early as November, 1953.[104] The actual announcement of plans to launch satellites came in 1954 in the US and in 1955 in the USSR.[105] Leonid Sedov's 1955 statement that the USSR would launch a satellite in about two years was only two months off.[106] A public description of the *Lunakhod* vehicle which explored the moon in 1971 was made as far back as 1955.[107] At least in nonmilitary programs, the Soviets have given good indications of what they were planning to do, although on occasion they have either achieved certain goals ahead of time or else delayed or canceled the carrying out of plans that have encountered difficulties. In general US programs from planning to execution stages are open to the glare of publicity, including even military developments.

Despite the many indications of its coming, the launch of the first Sputnik was surrounded with secrets. The launch vehicle itself, although later acknowledged to have been the first-generation Soviet ICBM, was never publicly depicted in accurate form until the modified version used to launch *Vostok* was put on display at the Paris Air Show in 1967.[108] The Soviets indicated that the vehicle was kept secret because it was also used as a Soviet ICBM.[109] This explanation is found wanting. To begin with, the USSR has regularly displayed its military weapons including ICBMs on May Day and during other celebrations. Moreover, the USSR has never publicly displayed or accurately depicted its heavy (*Proton*) and very heavy launch vehicles which are not used for military missions, nor have they acknowledged the use as space boosters of the SS-9 ICBM or the SS-5 IRBM.[110]

Finally, the SS-6, which launched the early Sputniks, was obsolete as a military launcher years before its display as *Vostok* in 1967. Perhaps it was the very obsolescence of this craft which the Soviet leaders chose to conceal. Instead of a simpler configuration based on one or a few engines, which would be more reliable, efficient and technically advanced, the standard Soviet booster consists of *five* rocket bodies strapped together with four main engines in each, combining a *total* thrust less than that of a single F-1 engine, five of which are contained in the single body of the US Saturn V booster. Another result of keeping the design secret was to encourage years of speculations and fantastic artists' conceptions which would have been put to rest by a single photograph. As it turned out, the thrust of each of the *20* engines of the SS-6 was about that of the "primitive" engine employed in the US Army's Redstone MRBM first launched in 1953.[111] Being unaware of this, those outside the USSR were inclined to overestimate the development of Soviet space and missile technology vis-a-vis the US. It is speculated even now that the heavy and very heavy Soviet launch vehicles, still secret, employ a similar strap-on configuration with multiple engines of relatively low thrust.[112] The contention that the USSR has used secrecy to conceal the backwardness of its space technology is supported by recent Soviet publication of details of the cabin of the *Soiuz* spacecraft for the forthcoming Apollo-*Soiuz* mission. Western observers were surprised at the lack of instrumentation and pilot control of the vehicle, which is the Soviet's latest manned capsule.[113]

The Soviet Union maintains three separate space launch sites. Of these only one was publicly acknowledged in the early years of the Soviet space program. The location of that launch site is in fact some 230 miles from the place publicly indicated.[114] A second launch site, which has only been used in recent years for a few cooperative Communist-bloc launches, was acknowledged in 1971, and its general location was identified.[115] The organization of the Soviet space program, along with names of the leading officials and the budget figures, are also kept secret. Khrushchev claimed the names of space officials were kept secret to prevent their assassination by foreign agents, but this seems openly disingenuous. Instead one suspects that the top Soviet leaders prefer that the credit for the space program goes to themselves.[116] Moreover, the leading figures in space research may not enjoy the trust of the political leaders, who fear what they might say to foreign journalists. Soviet secrecy seems to contradict the pretense that the Soviet program is conducted for purely scientific ends. In fact, the indications are that many of the scientists who work in the Soviet space program are not permitted to participate in the launching of space vehicles and that they are often kept in the dark about the hardware and operational characteristics of the spacecraft themselves.[117] Undoubtedly, this Soviet penchant for secrecy and red tape has retarded the space program, just as it has retarded technical progress in other areas.

Although the United States does not publicize every aspect of its space program, the difference in policy compared to the USSR is extreme. US secrecy is largely confined to military space activities. Even here, the United States acknowledges its military space activities, while the USSR does not. Until late 1961 the United States retained a military program almost as open as its civilian space program. Afterward, although security was tightened, the nature of various US military space activities were "open secrets," with the purposes of various programs discussed in advance along with the general nature of the equipment involved. The nature of ongoing military systems can be divined from censored congressional testimony and from nongovernmental publications. The organization of the American space program was worked out in public congressional debates culminating in the National Aeronautics and Space Act of 1958. Leading space scientists and administrators are public figures whose identity is well known. The location of American spacedromes is public information. Millions have toured Cape Canaveral and witnessed launchings from there. Vandenberg Air Force Base is not generally open to the public, but several launchings from there have been filmed by the public media. The US has even invited Soviet cosmonauts and others to tour Cape Canaveral and view the launchings, probably to propagandize the relative openness of the American program. The Soviets have turned down these invitations, not wishing to help the US score a propaganda victory and because they are evidently unwilling to reciprocate. Representatives of other socialist countries who have no space programs of their own have visited Cape Canaveral.[118] As part of the scheduled Apollo-*Soiuz* mission of 1975, American astronauts have visited "Star City" near Moscow and Soviet cosmonauts the Johnson Manned Spacecraft Center in Houston. Recently the Soviets finally agreed to permit technicians of each to be present as advisors and observers when Apollo and *Soiuz* are launched from Canaveral and Tiuratam in 1975.[119]

All of the American space launch vehicles have been filmed, photographed and diagramed by the public information media and described according to such general technical details as thrust, size, staging, fuels, etc. This has even been done well in advance of the operational use of these vehicles. Aerospace firms, anxious to secure support for the purchase of their latest hardware, furnish glossy pictures and semitechnical descriptions in the general media and aerospace trade press. For every single space launch, including military launches, the US makes public the launching vehicle employed.

The Vanguard Project was classified for a time preceding its transfer to NASA in 1958.[120] So, for a time, was the "Anna" geodetic satellite, which resulted in NASA's withdrawing from the project.[121] NASA does exercise its authority to classify anything of military significance, applying this largely to the technical details of military launch vehicles. NASA has also been somewhat reluctant to publicize certain kinds of technical information so long as the USSR is unwilling to provide anything similar. By contrast, the US National Academy of Sciences generally favors a policy of full disclosure.[122]

Almost all American military flights have been missions of reconnaissance, communications, navigation, monitoring for nuclear tests and electronic ferreting. The US Government has not openly publicized its reconnaissance satellites since the fall of 1961.[123] In choosing not to do so, it was probably not attempting to hide these programs from domestic and international publics or to perpetuate the myth of a completely nonmilitary space program, but rather to avoid belaboring the fact that the United States is spying on the Soviet Union, which would be embarrassing to Soviet and international publics. The withholding of technical details is probably more of a conventional military secret intended to prevent those spied on from perfecting active and passive countermeasures.[124] The earlier secrecy surrounding communications satellites, nuclear test monitors and navigation satellite systems has been dropped, although the more politically sensitive ferret missions remain secret.[125] Once again, it was probably to avoid offending Soviet sensibilities that the United States in 1968 for the first time withheld from public disclosure the orbital elements of a new reconnaissance satellite that traced a fixed orbit over Soviet territory, although the orbit was filed with the UN Secretariat and thus known to the USSR and confirmable by its radars.[126]

Launch Announcement Policies

Soviet launch announcement policies are more notable for what they conceal than what they reveal. The largest number of Soviet space flights whose nature is concealed by launch announcement policies are the military flights, including satellites for reconnaissance, military communications, navigation, ferreting, and test of FOBS and the satellite interceptor. All military flights which reach orbit are classified as *Kosmos* satellites with a serial number designator. Reference to the purpose of the flights is made to the original announcement of the *Kosmos* series in March, 1962. The text of this announcement is found in Appendix A at the end of this study. The announcement briefly describes the purpose of the series as one of investigating electromagnetic phenomena in near-earth space, testing vehicles and materials, and studying earth's atmosphere.[127] The *Kosmos* cover for military flights makes possible the Soviet pretense to a purely scientific and nonmilitary space program. It also makes possible the deception of the Soviet public and sections of the international public who cannot be apprised of Soviet military programs by citation of official Soviet sources. It has already been shown how Soviet propagandization of American military space activities relies heavily on quotation of American officials. This policy does *not* conceal Soviet military space activities from American decisionmakers and others who can generally deduce the nature of the missions from radar, radio and optical tracking and by monitoring of orbital characteristics, maneuver, reentry and telemetry. At best any strategic advantage in terms of foreign uncertainty is soon overcome when patterns of repetition make the missions of Soviet flights quite clear. Since the *Kosmos* label is also used for other purposes, including those mentioned in the official announcement, the military nature of the flights is not immediately clear to those interested parties who lack all the necessary tracking equipment, and they must often wait until such time as the failure to announce any scientific findings indicates a high probability of some kind of military mission.

By using the *Kosmos* cover, the Soviet Union is able to specify, albeit in the vaguest way, the name of the payload of almost all its flights, even though it is often practicing deception in doing so. This constitutes a certain advantage for propaganda purposes over the USA, which usually has not named the payloads of its military launches since the fall of 1961. The US does list with the UN the names of all launch vehicles including military boosters and boosters with military payloads. The USSR does not name the launching vehicle for most of its launches including *all* military flights.[128] Hundreds of Soviet military flights have been obscured under the *Kosmos* label since the first ones in 1962.

In 1966 the USSR broke this pattern briefly with the launch of two satellites which were probably tests of FOBS or an orbital interceptor. No announcement whatsoever was made about these flights by the USSR at the time or later. Since this was the first time that any country had tested such a weapons system, the Soviet leaders evidently felt that the less said about it the better. They may have hoped that the tests would not be understood for what they were in the West at least for a time, and feared that the sudden discovery of them by a wide public, if the orbital characteristics were openly published, would have set off a hostile reaction in the United States and elsewhere. One last possibility is that the flights failed so miserably that the Soviet leaders did not want to help Western experts in any way to detect the relative unreliability of their new weapons system. Some attempt at concealment may also be indicated by what appears to have been the explosion of both of these payloads. Both may have failed in reentry attempts after partial completions of their orbits.[129] Other test flights of the Soviet FOBS and orbital interceptor, beginning in early 1967, have all come under the *Kosmos* designation as "scientific" flights. In keeping with established Soviet policy, their obital inclinations, apogees and perigees have been announced. The Soviets do not announce orbital periods for FOBS flights, since they fly only partial orbits. This makes their identification as successful FOBS tests even easier.[130] "Scientific findings" from FOBS and interceptor flights have not been published.

With respect to navigation satellites which serve largely military purposes, the Soviets have acknowledged possessing such a system, but they have never identified the specific launches of navigation satellites.[131] The situation here is somewhat similar to that of FOBS, which the Soviets have long claimed to possess without ever acknowledging the test flights.

Policies of using the *Kosmos* label or failing to announce flights altogether in violation of established international conventions have also been used by the USSR to cover up space flights which have been failures. Presumably, the Soviet space launchings are subject to the same failings of men and equipment as American flights. However, while the United States has acknowledged and publicized several boost-phase failures of space launchings, the USSR has not acknowledged a single one.[132] Since the United States announces planned launch dates in advance and permits public observation of launches, it would be impossible to hide boost-phase failures except in the case of the semisecret military flights in which such failures may have occurred in the unannounced Soviet style.

The Soviet policy of delaying the announcement of space flights until they have reached earth orbit is probably intended largely to make it possible to cover up failures and thus maintain a propaganda image of technical infallibility.[133] In the early days of the Soviet space program, high-ranking Soviet officials even claimed that their program had no failures.[134] By delaying the announcement of various activities to be carried by and aboard space vehicles until these have actually taken place, the Soviets also avoid the risk of having failed to accomplish everything they said they would do.[135] This also has made it possible for the Soviets to convey the impression that the United States' space failures reflect a policy of boasting without solid accomplishment, in contrast to Soviet modesty and success. The Soviet policy of withholding announcements and leaving out many of the details also has had the effect of maximizing international curiosity and, hence, press coverage of the major Soviet missions, especially manned missions.[136]

Aside from any early boost-phase launch failures, which may or may not have occurred, the first Soviet space failure of note was *Luna* 1 in May,

1959. *Luna* 1 was evidently intended to strike the moon, as did *Luna* 2. Until very recently Soviet spokesmen did not indicate that *Luna* 1 was an attempt to impact the moon.[137] The next identifiable attempts to hide Soviet failures occurred with the failure of two attempts to send Soviet spacecraft into the vicinity of Mars in October, 1960, and were revealed by US officials in September, 1962.[138] Neither succeeded even in reaching earth orbit. Three Venus attempts in 1962 reached earth orbit but failed to depart for Venus, and the same occurred with two of three Mars attempts that same year. In January, 1963, the Soviets attempted to launch an improved vehicle to the moon that also remained stuck in earth orbit. All eight of these deep-space failures have never been named or announced by Soviet spokesmen. The United States did not immediately attribute these flights to the USSR or disown them itself, resulting in rumors and charges that the six which reached orbit were secret American military missions. These charges were even concurred in by the Soviet representative to the United Nations. It was only after a heated debate in 1963 marked by Soviet charges concerning secret American flights that the American representative to the UN revealed that the six flights of 1962 and 1963 were Soviet lunar and planetary failures.[139] Except for the two unannounced flights of 1966, no other unacknowledged Soviet space flights have been detected by unclassified sources.

Another technique employed by the Soviet Union to obscure the failure of its space flights is the naming of Soviet spacecraft in such a way that their missions remain so vague that failures can be passed off as successes. Probably the first instance of the use of this technique was the launch of *Tiazhelii Sputnik* (heavy Sputnik) 4 in February, 1961. Because the USSR uses the technique of launching deep-space probes from earth-parking orbits, the delayed announcement policy makes it possible to pass off lunar and planetary failures as earth-orbital successes. This was probably the case with *Tiazhelii Sputnik* 4, which was launched during a launch window for the planet Venus. Generally, when early indications reveal the real intent of a Soviet deep-space mission and promise at least partial success, a name is given such as Mars, *Venera* or *Luna*, which makes clear the intended mission. Thus, the probe successfully launched toward Venus by *Tiazhelii Sputnik* 5 was named *Venera* 1.[140]

In March, 1964, a diagnostic flight designed to overcome previous difficulties evidently failed and was given the name *Kosmos* 21, the same technique already employed to disguise military flights. Likewise, a Venus failure in November was dubbed *Kosmos* 27 to cover a similar failure. These two flights mark a shift from the previous policy of refusing to name planetary failures at all. Both were passed off as scientific earth-orbital missions and their parameters filed with the UN.[141] Similar use of the

Kosmos label has been made ever since, including *Kosmos* 60, 96, 111, 167, 300, 305, 359, 419 and 482, all of which attained earth orbit but failed to depart for the moon, Venus or Mars.[142]

The repeated failure of Soviet deep-space probes to the moon and beyond also engendered a different tactic for the naming of vehicles. Although those missions which failed to get out of their earth-parking orbits could be passed off as scientific flights, several craft which had been successfully propelled toward Mars and Venus and given the appropriate names suffered communications failures en route. Since the telemetry of these vehicles was monitored by non-Soviet radio telescopes at England's Jodrell Bank and elsewhere, their failure was soon evident and broadly publicized in the West, although never officially by the US Government except in the cases of the two delayed announcements mentioned above. The USSR, fearing such a possible failure of a space probe launched during the Venus window of 1964, gave it the name *Zond* 1 rather than *Venera* 2.[143] As with prior planetary probes, communication from *Zond* 1 was lost. Although the *Zond* name (meaning probe) made possible a public claim of success, the failure of the mission to return data from Venus was obvious and made public in the West. Communications also failed with *Zond* 2, launched toward Mars by a *Tiazhelii Sputnik*. *Zond* 3, launched toward the orbit of Mars when there was no Mars window, was a diagnostic flight for communications tests which returned photographs of the far side of the moon. In 1967 the USSR returned to the use of the *Venera* label for two craft, probably encouraged by the success of *Zond* 3. Although *Venera* 3 did successfully impact Venus, the USSR revealed that communications failures occurred with it and with *Venera* 2 shortly before they reached Venus.[144] Since *Zonds* 2 and 3, the *Zond* name has not been used for flights relating to the planets. The Mars and *Venera* names have been used instead, with the *Kosmos* label applied to those flights that get stuck in earth orbit.

The USSR attempted at least two planetary flights to Mars or Venus at every window of these planets between 1960 and 1967. During Mars windows in 1967 and 1969 no Mars flights, *Zond* flights aimed at Mars, or *Kosmos* flights identifiable as Mars faliures took place. That Soviet flights to Mars had not been permanently canceled is indicated by the flights of Mars 2 and 3 and *Kosmos* 419 in 1971 and by Mars 4-7 in 1973.[145] If any attempts to fly to Mars were made by the USSR in the earlier period, they were evidently boost-phase failures and never announced.

In naming its spacecraft the USSR has on several occasions concealed the nature of precursor spacecraft designed to test equipment and techniques before operational flights are made. Thus, for example, *Korabl Sputniki* (satellite ships) 1-5 tested orbit and reentry of the *Vostok* vehicle. *Vostok*

1 carried Iurii Gagarin. In this case the name and Soviet publicity *did not* in fact conceal the nature of these flights as manned precursors. However, *Kosmos* 47, 57, 119, 133, 140, 146, 154, 186, 188, 212, 213, 238, 379, 382, 398, 434, 496, 557 and 573 have all been probable manned precursors or failures whose nature has been obscured by the *Kosmos* name and whose relation to manned flights has not been identified by Soviet officials.[146] The second series of *Zond* flights has been identified by Soviet spokesmen as man-related (*Zonds* 4-8).[147] Although Penkovsky and others have contended that the USSR experienced failures of manned flights which have resulted in secret cosmonaut fatalities, these contentions are unconfirmed and seem improbable from previous monitoring of Soviet telemetry and the fact that cosmonaut fatalities in 1967 and 1971 were not concealed.[148] The Soviets could have concealed cosmonaut deaths in a boost-phase or ground-test accident such as the Apollo fire in the United States. It is possible that they have done so.

The purpose of various unmanned spacecraft has also been obscured by Soviet policies of naming their spacecraft. Meteorological and communications satellites known as *Meteora* and *Molniia* (Lightning) both had their precursors which flew under the *Kosmos* label. The first *Meteora* did not fly until 1969. Several satellites exhibiting the characteristics of weather satellites had flown since 1963, and announcement of the nature of these flights was either delayed or avoided altogether, indicating substantial difficulties and perhaps some military purposes. It also appears that *Kosmos* has also been used to cover up failures of communication and weather satellites that would be evident if their purpose had been identified.[149]

Such use of the *Kosmos* label represents an attempt, for political purposes, to cover up any difficulties and failures which might tarnish the propaganda impact of Soviet space successes and raise questions about the reliability of Soviet strategic weapons and spaceborne systems for military reconnaissance, communications, targeting and meteorology. Obscuring precursor flights with *Kosmos* and other labels is evidently intended to convey the impression that Soviet space technology is so reliable that precursor tests are unnecessary, and to maximize the drama and public impact of various flights by withholding information about them until they actually occur.

The announcement policies of the United States for prelaunch and post-launch phases are considerably more open than those of the Soviet Union. The United States does not obscure the nature and purposes of its non-military spacecraft and divulges considerable information about all these craft *in advance*. The USA identifies the launch vehicles employed for its military flights but does not divulge the name of the spacecraft themselves.

The USSR does not generally name its launch vehicle for military or non-military flights but names the spacecraft launched, applying the vague or, in the case of military craft, deceptive *Kosmos* label to them.

The United States classified its military space flights in the fall of 1961.

Two military flights which occurred in November and December of 1961 were not announced by the US to the UN or in the President's annual report to Congress. Since February of 1962, the US has filed semimonthly reports with the UN Secretary-General which cover all spacecraft in orbit. Those which have been orbited and deorbited during the intervals are described in supplementary reports. The US delays the public announcement of the orbital elements of its military flights, and in the case of those in fixed positions and cycling orbits over the USSR, does not openly publicize them at all.[150] The Soviet Union, however, makes an immediate public announcement of the orbital elements of all its flights, which have not to date included any reconnaissance craft of fixed cycling or synchronous orbits over the USA. The USSR also files this information with the UN, although, unlike the US, it does not file the orbital elements of or in any other way acknowledge the existence of debris in orbit associated with the explosion, malfunction or staging of its space vehicles.[151]

Publication of the Findings of Space Research and Exploration

The Soviet Academy of Sciences has been publishing the results of its theoretical and experimental space research since 1958.[152] The leading source of such information is the irregularly published *Izkustvennie Sputniki Zemli* (Artificial Earth Satellites). The publication of data and findings from research is generally delayed for a period of up to a few years. In volume the USA publishes several times as much material.[153] American technical information, besides appearing in scientific and technical journals, is also found in the popular press.[154] Soviet space scientists have been eager to discuss their work with their American counterparts, even before they have published it themselves. Nonetheless, delays in exchange of Soviet satellite data and imagery have complicated cooperative Soviet-American efforts in meteorology and mapping the earth's magnetic field.[155] Unlike the US, the Soviet Union has not published any map quality photographs from its satellites, nor has it provided the details of its procedures for sterilization of lunar and planetary probes.[156] The US has published thousands of pictures from its nonmilitary satellites, although some have been classified by the State Department, the Pentagon and the AEC.[157] As mentioned above, the failure of certain Soviet missions has in many cases been obscured or kept secret for long periods. In one case the physical condition of a cosmonaut was described as being normal, it being admitted later that he had felt quite ill.[158]

POLICIES TO PRESERVE THE COSMOS FROM CONTAMINATION

One problem that the space-faring nations must face in their exploration of the cosmos is the possibility that they will permanently alter space and the celestial bodies *before* they discover what they are like. To minimize this risk, the Committee on Space Research of the International Council of Scientific Unions (COSPAR) recommended that those nations launching space probes avoid compromising the search for extraterrestrial life by adopting techniques of sterilizing their spacecraft, which would reduce risks of contamination to acceptable levels.[159]

Although the USSR has participated in international symposia devoted to discussion of sterilization of space vehicles and published technical articles dealing with the problem, the Soviets have not specifically described the techniques they employ, as has the United States, nor does it appear that they go to the lengths of thoroughness or expense that the Americans do.[160]

The Soviet Union's deep-space probes have impacted Mars and Venus, while American probes have not. Moreover, the Soviet Union employs a technique known as bus deflection whereby not only the capsule lander but the entire planetary vehicle are aimed at the planet, and the capsule carrier or bus is maneuvered from impact trajectory in the final stage of approach. Should the command from earth to execute bus deflection fail to produce the desired maneuver, the entire vehicle, including parts probably not so thoroughly sterilized as the capsule itself, may impact the planet, greatly increasing the chance of contamination. This occurred with the Soviet *Venera* 3 and possibly the *Zond* 2 probe of Mars. The United States does not use or plan to use the bus-deflection technique but rather the safer capsule-deflection technique whereby the whole vehicle is aimed away from the planet and the capsule deflected toward it.[161]

The Soviet Union has on three separate occasions charged the United States with conducting military activities in space which have disrupted and contaminated the environment of space and interfered with the scientific study of the cosmos. The first instance was the West Ford Project, the October, 1961, launch by the US Air Force of 350 million tiny needles intended to form a belt in space capable of reflecting radio waves as part of an experimental defense communications system. Soviet spokesmen made the most of the protests voiced by scientists in the West and themselves lodged an official protest at the UN.[162] Mstislav Keldysh, president of the Soviet Academy of Sciences, also sent a formal protest to the president of the US National Academy of Sciences.[163] The Soviet Union reacted similarly to West Ford II, a repeat of the original experiment, which had not been entirely successful. A TASS (Soviet News Agency) statement was presented to the UN Secretary-General asserting that the charged needles or dipoles of West Ford II would interfere with communication between the earth and planetary probes (presumably the USSR's). The United States contended that there was no such danger and that the dipoles would eventually decay and be burned up in reentry.[164]

The USSR in July, 1962, reacted to a US high-altitude (400 km), one-megaton nuclear test by noting the complaints of the world's scientists and lodging its own protests with the UN's Committee on the Peaceful Uses of Outer Space (COPUOS). It was charged that the US nuclear test would hamper earth communications, create risks for astronauts and spacecraft, and lastingly alter the earth's radiation belts. The Soviets contended that the test had in fact damaged US and French satellites and would interfere with research during the International Year of the Quiet Sun.[165]

Since West Ford II the USSR has not made any such voluble complaints about American disruption or contamination of space. Probably both the US and the USSR now take considerable precautions to avoid the contamination of space, but nonetheless, as the two great space polluters, each now fears being the one to cast the first stone.

NOTES

[1]William Shelton, *Soviet Space Exploration* (New York: Washington Square Press, Inc., 1968), pp. 13-14.

[2]*Satellites, Science and the Public: A Report of a National Survey of the Public Impact of Early Satellite Launchings—for the National Association of Science Writers and New York University Survey Research Center,* University of Michigan, 1959, as cited in Joseph M. Goldsen, *Public Opinion and Social Effects of Space Activity,* Research Memorandum #RM-2417, Rand Corporation, July 20, 1959, p. 2.

[3]Sam Lubell, "Sputnik and American Public Opinion," *Columbia University Forum, 1* (1), Winter 1957, pp. 15-21.

[4]B. W. Augenstein, *Policy Analysis in the National Space Program,* Paper #P-4137, Rand Corporation, July 1969, p. 22.

[5]Reported by Raymond A. Bauer with Richard S. Rosenbloom and Laure Sharp *et al., Second-Order Consequences, A Methodological Essay on the Impact of Technology* (Cambridge, Mass.: MIT Press, 1969), pp. 85, 86.

[6]*Izvestiia,* January 24, 1960, as cited in Wolfgang Leonard, *The Kremlin Since Stalin* (New York: Praeger, 1962), pp. 370-371. See also *Komsomolskaia Pravda,* June 11, 1960, as cited in Herbert Ritvo (annotator), *The New Soviet Society* (New York: *The New Leader,* 1962), p. 41.

[7]*Washington Post,* March 5, 1971, p. A-11 as cited in Whelan, "Political Goals and Purposes of the USSR in Space," *Soviet Space Programs, 1966-1970, op. cit.,* p. 35.

[8]See *New York Times,* February 28, 1971, p. 20.

[9]See Amitai Etzioni, *The Moon-Doggle, Domestic and International Implications of the Space Race* (Garden City, N.Y.: Doubleday and Co., Inc., 1964), xii.

[10]See John M. Logsdon, *The Decision To Go to the Moon: Project Apollo and the National Interest* (Cambridge, Mass.: MIT Press, 1970), pp. 103-104, 129; and Charles S. Sheldon II, "An American 'Sputnik' for the Russians?" in Eugene Rabinowitch and Richard S. Lewis, editors, *Man on the Moon—The Impact on Science, Technology and International Cooperation* (New York: Basic Books, Inc., 1969), p. 55.

[11]John R. Walsh, "NASA: Talk of Togetherness with Soviets Further Complicates Space Politics for the Agency," *Science, 142* (3588), October 4, 1963, pp. 35-38.

[12]Edwin Diamond, *The Rise and Fall of the Space Age* (Garden City, N.Y.: Doubleday and Co., Inc., 1964), p. 102.

[13]Logsdon, *op. cit.*, pp. 90-91.

[14]Erlend A. Kennan and Edmund H. Harvey, *Mission to the Moon, A Critical Examination of NASA and the Space Program* (New York: William Morrow and Co., 1969), p. 483, who argue that this is why the committees did not look too deeply into the Apollo fire.

[15]Hugh Dryden (Deputy Administrator, NASA) in Jerry and Vivian Grey, eds., *Space Flight Report to the Nation* (New York: Basic Books, 1962), p. 182.

[16]Nicholas Daniloff, *The Kremlin and the Cosmos* (New York: Alfred A. Knopf, 1972), p. 208.

[17]Unpublished address delivered by Dr. Sergei P. Fedorenko at Illinois State University, December 6, 1972.

[18]Cited in Charles S. Sheldon II, *Review of the Soviet Space Program, With Comparative US Data* (New York: McGraw-Hill Book Company, Inc., 1968), p. 84.

[19]See Jan F. Triska and David D. Finley, *Soviet Foreign Policy* (New York: Macmillan, 1968), pp. 284-287, 298-309.

[20]Sheldon, "Overview, Supporting Facilities and Launch Vehicles of the Soviet Space Program," *Soviet Space Programs, 1966-1970, op. cit.*, p. 147; and Sheldon, "Program Details of Manned Flights," in *ibid.*, p. 251.

[21]Etzioni, *op. cit.*, p. 36.

[22]See *ibid.*, pp. 135, 137.

[23]Edward Crankshaw, *Krushchev: A Career* (New York: Viking Press, 1966), p. 266.

[24]Diamond, *op. cit.*, p. 27.

[25]See Logsdon, *op. cit.*, p. 144.

[26]See Frederick C. Barghoorn, *Politics in the USSR* (Boston: Little, Brown and Co., Inc., 1972), pp. 340-341.

[27]Seth T. Payne and Leonard S. Silk, "The Impact on the American Economy," in Lincoln P. Bloomfield, editor, *Outer Space, Prospects for Man and Society* (New York: Frederick A. Praeger, Inc., 1968), p. 97.

[28]Logsdon, *op. cit.*, pp. 87-88.

[29]See Etzioni, *op. cit.*, pp. 7-8; and Shelton, *Soviet Space Exploration, op. cit.*, pp. 88, 300-301.

[30]Robert E. Marshak, "Reexamining the Soviet Scientific Challenge," *Bulletin of the Atomic Scientists, 19* (4), April 1963, pp. 13-14.

[31]See Joseph G. Whelan, "Political Goals and Purposes of the USSR in Space," in *Soviet Space Programs, 1966-1970, op. cit.*, pp. 17-18, 28-29, 35-37.

[32]Bauer *et al.*, *op. cit.*, pp. 156, 162.

[33]Vernon Van Dyke, *Pride and Power, The Rationale of the Space Program* (Urbana: University of Illinois Press, 1964), p. 190.

[34]Logsdon, *op. cit.*, ix-x, pp. 13, 35-36.

[35]See *ibid.*, p. 20; and T. Keith Glennan, "The Task for Government," in American Assembly, Columbia University, *Outer Space Prospects for Man and Society* (Englewood Cliffs, N.J.: Prentice-Hall Inc., 1962), p. 94; and James R. Killian, Jr., "Shaping a Public Policy for Outer Space," in the same work, pp. 184-188.

[36]Logsdon, *supra*, p. 155.

[37]Etzioni, *op. cit.*, xii, xiv; and Logsdon, *supra*, x.

[38]Logsdon, *supra*, x, pp. 94-99.

[39]Etzioni, *op. cit.*, xii.

[40]*Ibid.*, xiv-xv.

[41]Lyndon B. Johnson, "The Politics of the Space Age," in Lillian Levy, editor, *Space: Its Impact on Man and Society* (New York: W. W. Norton & Co., Inc., 1965), pp. 4-5.

[42]Shelton, *Soviet Space Exploration, op. cit.*, p. 41.

[43]See Sheldon, *Review of the Soviet Space Program . . . , op. cit.*, p. 84; and Whelan, "Political Goals and Purposes of the USSR in Space," in *Soviet Space Programs, 1966-1970, op. cit.*, pp. 35-37.

[44]Charles S. Sheldon II, *United States and Soviet Progress in Space: Summary Data Through 1973 and a Forward Look*, Library of Congress, Congressional Research Service, January 8, 1974, pp. 9-10.

[45]Odishaw introduction in Hugh Odishaw, editor, *The Challenges of Space* (Chicago, The University of Chicago Press, 1963), pp. 157-158.

[46]Alton Frye, "Soviet Space Activities: A Decade of Pyrrhic Politics," in Bloomfield, *op. cit.*, p. 193.

[47]Augenstein, *op. cit.*, p. 42.

[48]See Shelton, *Soviet Space Exploration, op. cit.*, p. 312; and Sheldon, *United States and Soviet Progress* (1973), *op. cit.*, pp. 60-62.

[49]Kennan and Harvey, *op. cit.*, p. 75.

[50]James R. Killian, Jr., speech to MIT Club of New York, December 13, 1960 in *Science, 133*, January 6, 1961, pp. 24-25.

[51]See the remarks of Urey and Kistiakowskii in Shelton, *supra*, pp. 300-301. See also pp. 306-307 in *ibid.*

[52]Van Dyke, *op. cit.*, pp. 90-91.

[53]See Logsdon, *op. cit.*, p. 75.

[54]See Loyd S. Swenson, Jr., James M. Grimwold and Charles C. Alexander, *This New Ocean: A History of Project Mercury* (Washington, D.C.: NASA, 1966), p. 307; and Jay Holmes, *America on the Moon: The Enterprise of the Sixties* (Philadelphia: J. B. Lippincott Co., 1962), p. 194.

[55]Shelton, *Soviet Space Exploration, op. cit.*, pp. 90-91, 190.

[56]See Sheldon, "Projections of Soviet Space Plans," in *Soviet Space Programs, 1966-1970, op. cit.*, pp. 359-367.

[57]See *ibid.*, pp. 364-384. See also Khlebsevich in Sergei Gouschev and Mikhail Vassiliev, editors, *Russian Science in the 21st Century* (New York: McGraw-Hill Book Co., Inc., 1960), p. 207.

[58]Frye, "Soviet Space Activities: A Decade of Pyrrhic Politics," *op. cit.*, p. 192.

[59]Oleg Penkovsky, *The Penkovsky Papers*, translated by P. Denatin (London: Collins Clear Type Press, 1966), p. 243.

[60]Sheldon, "An American 'Sputnik' for the Russians?" *op. cit.*, pp. 59-60.

[61]See Shelton, "Summary" in *Soviet Space Programs, 1966-1970, op. cit.*, xxx; and Shelton, *supra*, pp. 7-8.

[62]Penkovsky, *op. cit.*, p. 243.

[63]Leonid Vladimirov in *Posev* (Munich), September 1969, pp. 47-51. See also Leonid Vladimirov, *The Russian Space Bluff* (New York: Dial Press, 1973), pp. 123-133.

[64]Appendix A in *Soviet Space Programs, 1966-1970, op. cit.*, pp. 532, 536, 537, 541, 544, 545, 547, 556, 557.

[65]Kennan and Harvey, *op. cit.*, p. 136; and introduction by Ralph Lapp in the same work, x-xii.

[66]See also William Shelton, "Neck and Neck in the Space Race," *Fortune, 76* (5), October 1967, p. 168.

[67]Testimony of George Low before the Senate Committee on Aeronautics and Space Sciences on *Space Agreements with the Soviet Union*, June 23, 1972, pp. 20-21.

[68]See the statements of General Kamanin (p. 364) and Academician Petrov (p. 370) in Sheldon, "Projections of Soviet Space Plans," *op. cit.* Also Joseph G. Whelan, "Political Goals and Purposes of the USSR in Space," in *Soviet Space Programs, 1962-1965, op. cit.*, p. 48; and G. I. Pokrovskii, "Competition in Space," *New Times*, No. 20, May 17, 1961, pp. 3-4.

[69]Sheldon, "Program Details of Manned Flights," in *Soviet Space Programs, 1966-1970, op. cit.*, p. 233.

[70]Arnold L. Horelick, "The Soviet Union and the Political Uses of Outer Space," in Joseph M. Goldsen, editor, *Outer Space in World Politics* (New York: Frederick A. Praeger, 1963), pp. 62, 64.

[71]*Ibid.*, pp. 64-66.

[72]Paul Kecskemeti, "Outer Space and World Peace," in Goldsen, *op. cit.*, p. 33.

[73]Don E. Kash, *The Politics of Space Cooperation* (Purdue University Studies, Purdue Research Foundation, 1967), p. 40.

[74]Y. Listvinov, "Space Research American Style," *International Affairs*, No. 11, November 1968, pp. 46-47; J. Y. Sejnin, "Space and Earth," *New Times*, No. 14, April 1, 1960, p. 28; "Probing the Secrets of the Universe" (editorial), *New Times*, No. 33, August 15, 1962, pp. 1-2; E. A. Korovin and G. P. Zhukov, editors, *Sovremennye Problemy Kosmicheskogo Prava* (Moscow, 1963), pp. 6-7.

[75]G. Khozin, "Pentagon Seeking Control of Space Research in Asia and Africa," *International Affairs*, No. 2, February 1968, pp. 30-34.

[76]G. I. Pokrovskii, "On the Problems of the Use of Cosmic Space," *International Affairs*, No. 7, July 1959, pp. 105-107; B. L. Teplinski, "Space Maniacs," *New Times*, No. 35, September 1, 1965, pp. 7-8; G. Zhukov, "Space Espionage and International Law," *International Affairs*, No. 10, October 1960 pp. 53-57; B. Teplinskii "Pentagon's Space Programme" *International Affairs*, No. 1, January 1963, pp. 56-59; V. Glasov, "Cannibals in Space," *New Times*, No. 24, June 19, 1963, pp. 12-13; Lt. Colonel V. Larionov, "The Doctrine of Military Domination of Outer Space," *International Affairs*, No. 10, October 1964, pp. 25-27.

[77]See Richard N. Gardner (Deputy Assistant Secretary of State), "Cooperation in Outer Space," *Foreign Affairs, 41* (2), January 1963, p. 359.

[78]Kash, *op. cit.*, pp. 60-61; and Van Dyke, *op. cit.*, p. 132.

[79]Whelan, "Political Goals and Purposes of the USSR in Space," in *Soviet Space Programs, 1962-1965, op. cit.*, p. 47.

[80]*Ibid.*, pp. 48, 52; and Georgii *Pokrovskii*, "Competition in Space," *op. cit.*, pp. 3-4.

[81]Whelan, "Political Goals and Purposes of the USSR in Space," in *Soviet Space Programs, 1962-1965, op. cit.*, p. 4; and Observer, "Space, Science and Peace," *New Times*, No. 35, August 24, 1962, pp. 1-2.

[82]Frank Gibney and George J. Feldman, *The Reluctant Spacefarers. A Study in the Politics of Discovery* (New York: New American Library, 1965), pp. 75-77.

[83]See Barbara M. DeVoe, *Astronaut Information: American and Soviet (Revised),* Library of Congress, Congressional Research Service, February 12, 1973, pp. 29-34.

[84]See Whelan, "Political Goals and Purposes of the USSR in Space," in *Soviet Space Programs, 1966-1970, op. cit.,* pp. 39-40, 43-45; and Sheldon, "Soviet Military Space Activities," in the same work, pp. 346-348. Also see Donald G. Brennan, "Arms and Arms Control in Outer Space," in Bloomfield, *op. cit.,* p. 160.

[85]See S. Nevskii, "Apollo 12," *New Times,* No. 48, December 3, 1969, pp. 9-10; and Sheldon, "Projections of Soviet Space Plans," in *Soviet Space Programs, 1966-1970, op. cit.,* pp. 375, 376, 379, 380, 383.

[86]Daniloff, *op. cit.,* p. 105; and Horelick, "The Soviet Union and the Political Uses of Outer Space," *op. cit.,* p. 59.

[87]Whelan, "Political Goals and Purposes of the USSR in Space," in *Soviet Space Programs, 1966-1970, op. cit.,* p. 17.

[88]Leonid Vladimirov, *The Russian Space Bluff, op. cit.,* pp. 92-93; Shelton, *Soviet Space Exploration, op. cit.,* pp. 147-148, 162; Penkovsky, *op. cit.,* p. 245; and Daniloff, *op. cit.,* pp. 109-110.

[89]See Sheldon, "Summary" of *Soviet Space Programs, 1966-1970, op. cit.,* xxxvi-xxxviii; and Whelan, "Soviet Attitude Toward International Cooperation in Space," *Soviet Space Programs, 1966-1970, op. cit.,* pp. 399-406.

[90]Van Dyke, *op. cit.,* p. 160.

[91]Bloomfield, "The Quest for Law and Order," in Bloomfield, *op. cit.,* p. 121.

[92]See Van Dyke, *op. cit.,* p. 259; and Simon Bourgin (Science and Space Advisor, USIA), "Impact of US Space Cooperation Abroad," in *International Cooperation in Outer Space, Symposium, op. cit.,* pp. 167-172.

[93]See Whelan, "Political Goals and Purposes of the USSR in Space," in *Soviet Space Programs, 1966-1970, op. cit.,* pp. 50-53.

[94]Kennan and Harvey, *op. cit.,* p. 54.

[95]Bourgin, *op. cit.,* pp. 167-171.

[96]Fermin J. Krieger, *The Space Programs of the Soviet Union,* Rand Corporation, Paper #P-3632, July 1967, p. 5.

[97]See J. Goldsen and L. Lipson, *Some Political Implications of the Space Age,* Rand Corporation, Paper #P-1435, February 24, 1958, pp. 6-9; Sheldon, *Review of the Soviet Space Program ..., op. cit.,* p. 85; Penkovsky, *op. cit.,* pp. 238-239; and the interview with Khrushchev by the editor of *Dansk Kolkestyre* on January 4, 1958, in N. S. Khrushchev's *For Victory in Peaceful Competition with Capitalism* (New York: E. P. Dutton and Co., 1960), pp. 23 ff.

[98]For example, see Horelick, "The Soviet Union and the Political Uses of Outer Space," *op. cit.,* pp. 46-47, 53; and Klass, *op. cit.,* p. 70.

[99]See Kecskemeti, *op. cit.,* pp. 31-32; "Earth-Moon," *New Times,* No. 38, September 1959, pp. 3-4; and "Leading on Earth and in Space," *International Affairs,* No. 9, September 1962, p. 5.

[100]F. S. Nyland, *Space: An International Adventure?,* Rand Corporation, Paper #P-3043, December 1964, p. 13.

[101]Sheldon, *Review of the Soviet Space Program ..., op. cit.,* p. 85.

[102]Sheldon, "Soviet Military Space Activities," *op. cit.,* p. 326.

[103]See Whelan, "Soviet Attitude Toward International Cooperation in Space," in *Soviet Space Programs, 1966-1970, op. cit.,* p. 441n; and M. Dubovsky, "What For and Against Whom?" *New Times,* No. 30, July 29, 1970, pp. 9-11.

[104]Shelton, *Soviet Space Exploration, op. cit.,* p. 40.

[105]*Soviet Space Programs, 1962, op. cit.,* p. 90.

[106]*New York Times,* August 3, 1955, p. 8.

[107]*Soviet Space Programs, 1962, op. cit.,* pp. 73, 74.

[108]Frye, "Soviet Space Activities: A Decade of Pyrrhic Politics," *op. cit.,* p. 182.

[109]Sheldon, "Overview, Supporting Facilities and Launch Vehicles of the Soviet Space Program," in *Soviet Space Programs, 1966-1970, op. cit.,* p. 130; and Sheldon, "Soviet Military Space Activities," in the same work, p. 323.

[110]Sheldon, "Overview . . . ," *supra,* pp. 140, 141, 146; and Sheldon, "Soviet Military Space Activities," *supra,* p. 342.

[111]Frye, "Soviet Space Activities: A Decade of Pyrrhic Politics," *op. cit.,* pp. 182-183.

[112]See Sheldon, Appendix B of *Soviet Space Programs, 1966-1970, op. cit.,* pp. 561, 562.

[113]See *Aviation Week and Space Technology, 100* (3), January 21, 1974, pp. 38-43; and *Aviation Week and Space Technology, 100* (4), January 28, 1974, pp. 36-41.

[114]Sheldon, "Overview . . . ," *op. cit.,* pp. 126-128.

[115]This is the "Volgograd" cosmodrome in nearby Kapustin Iar. See *Soviet Space Programs, 1971, A Supplement to the Corresponding Report Covering the Period 1966-1970,* Staff Report Prepared for the Use of the Committee on Aeronautics and Space Sciences, US Senate, Congressional Research Service, Library of Congress, April 1972, 92nd Congress, 2d Session, p. 62.

[116]Leon M. Herman, "Soviet Economic Capabilities for Scientific Research," in *Soviet Space Programs, 1962-1965, op. cit.,* p. 418; and House Committee on Science and Astronautics, *Review of the Soviet Space Program,* 90th Congress, 1st session, November 10, 1967 (Washington, D.C.: Government Printing Office, 1967), p. 79; and Fermin J. Krieger in Grey, *op. cit.,* p. 182.

[117]Sheldon, *Review of the Soviet Space Program . . . , op. cit.,* p. 79; and Arnold W. Frutkin, *International Cooperation in Space* (Englewood Cliffs, N.J.: Prentice-Hall, 1965), pp. 102-103.

[118]Kash, *op. cit.,* pp. 60-61.

[119]Sheldon, *United States and Soviet Progress in Space* (1973), *op. cit.,* p. 17; and *Aviation Week and Space Technology, 99* (19), November 5, 1973, pp. 20-21.

[120]Shelton, *Soviet Space Exploration, op. cit.,* p. 50.

[121]Kash, *op. cit.,* p. 35.

[122]*Ibid.,* pp. 52, 55.

[123]Klass, *op. cit.,* p. 109.

[124]*Ibid.,* pp. 109-110.

[125]Sheldon, *Review of the Soviet Space Program . . . , op. cit.,* p. 123.

[126]Klass, *op. cit.,* pp. 180, 187.

[127]See Sheldon, "Program Details of Unmanned Flights," in *Soviet Space Programs, 1966-1970, op. cit.,* p. 173.

[128]US Congress, *Review of the Soviet Space Program,* Committee on Science and Astronautics, House, 90th Congress, 1st session, November 10, 1967, p. 123 (hereinafter *Review of the Soviet Space Program, 1967*).

[129]See Sheldon, "Soviet Military Space Activities," *op. cit.,* pp. 334-336. See also Sheldon and DeVoe, Appendix A of *Soviet Space Programs, 1966-1970, op. cit.,* p. 541. In 1972 Sheldon, who originally interpreted these flights as FOBS tests, concluded they were more probably tests of an orbital interceptor. See *Soviet Space Programs, 1971, op. cit.,* p. 51.

[130]Sheldon, "Soviet Military Space Activities," *op. cit.*, pp. 336-337.

[131]Sheldon, "Summary," *op. cit.*, xxvii-xxviii.

[132]Sheldon, "Overview, Supporting Facilities and Launch Vehicles of the Soviet Space Program," *op. cit.*, p. 118; and Sheldon, *Review of the Soviet Space Program . . . , op. cit.*, p. 119.

[133]House, *Review of the Soviet Space Program, 1967, op. cit.*, pp. 117-118.

[134]Sheldon, *Review of the Soviet Space Program . . . , op. cit.*, p. 121.

[135]*Ibid.*, pp. 117-118; and Horelick, "The Soviet Union and the Political Uses of Outer Space," *op. cit.*, pp. 60-61.

[136]Shelton, *supra*, p. 142; Goldsen and Lipson, *Some Political Implications of the Space Age, op. cit.*, p. 10.

[137]Sheldon, "Program Details of Unmanned Flights," in *Soviet Space Programs, 1966-1970, op. cit.*, p. 162.

[138]*Ibid.*, p. 163.

[139]Letter of June 6, 1963, from Ambassador Adlai E. Stevenson, Jr., to the Secretary-General of the UN. See also *ibid.*, pp. 163-167; and Sheldon, *Review of the Soviet Space Program . . . , op. cit.*, p. 124.

[140]Sheldon, "Program Details of Unmanned Flights," *op. cit.*, p. 164.

[141]*Ibid.*, p. 165.

[142]*Ibid.*, pp. 166, 167, 174, 192, 197-198, 214, 217; and Sheldon, "Postscript," in *Soviet Space Programs, 1966-1970, op. cit.*, p. 527; and Sheldon, *United States and Soviet Progress in Space* (1973) *op. cit.*, p. 58.

[143]Sheldon, "Program Details of Unmanned Flights," *op. cit.*, p. 165.

[144]*Ibid.*, p. 166.

[145]Sheldon and DeVoe, Appendix A of *Soviet Space Programs, 1966-1970, op. cit.*, p. 558; and Sheldon, *United States and Soviet Progress in Space* (1973), *op. cit.*, p. 58.

[146]Sheldon, "Program Details of Manned Flights," *op. cit.*, pp. 224-225, 227, 229-230, 232-233, 240; Sheldon, "Projections of Soviet Space Plans," *op. cit.*, p. 390; Sheldon, "Postscript," *op. cit.*, pp. 513-516; *Soviet Space Programs, 1971, op. cit.*, p. 67; Sheldon, *United States and Soviet Progress in Space* (1972), *op. cit.*, p. 9; and Sheldon, *United States and Soviet Progress in Space* (1973), p. 9.

[147]Sheldon, "Program Details of Manned Flights," *supra*, pp. 240-245; and "Projections of Soviet Space Plans," *supra*, pp. 390-391.

[148]Penkovsky, *op. cit.*, p. 243; and Shelton, *Soviet Space Exploration, op. cit.*, pp. 6-7.

[149]Barbara M. DeVoe, "Soviet Application of Space to the Economy," *Soviet Space Programs, 1966-1970, op. cit.*, pp. 298-310.

[150]Klass, *op. cit.*, pp. 180, 187.

[151]*Review of the Soviet Space Program, 1967, op. cit.*, p. 122; and Sheldon, *Review of the Soviet Space Program . . . , op. cit.*, pp. 122-123.

[152]Robert W. Buchheim and the Staff of Rand Corporation, *New Space Handbook: Astronautics and Its Applications* (New York: Vintage Books, 1963), p. 310.

[153]F. J. Krieger (Rand Corporation) in Grey, *op. cit.*, p. 178.

[154]*Loc. cit.*; and Whelan, "Political Goals and Purposes of the USSR in Space," *Soviet Space Programs, 1966-1970, op. cit.*, p. 31. James Van Allen as cited by Whelan, "Soviet Attitude Toward International Cooperation in Space," *op. cit.*, p. 429.

[155]Whelan, 'Soviet Attitude Toward International Cooperation in Space," *op. cit.*, pp. 406-410.

[156]*Ibid.*, pp. 410-412; and Sheldon, "Soviet Military Space Activities," *op. cit.*, p. 330.

[157]Kennan and Harvey, *op. cit.*, p. 228.

[158]Oskar Morgenstern (Professor, Econometric Research Program, Princeton), "Political Effects," in Grey, *op. cit.*, p. 134.

[159]US Department of State, Appendix E of *Soviet Space Programs, 1966-1970, op. cit.*, p. 610.

[160]James M. McCullough, "Soviet Bioastronautics: Biological Behavioral and Medical Problems," in *Soviet Space Programs, 1966-1970, op. cit.*, p. 295, n. 118; and Whelan, "Soviet Attitude Toward International Cooperation in Space," *op. cit.*, pp. 410-413.

[161]B. C. Murray, M. E. Davies and P. K. Eckman, *Planetary Contamination II: Soviet and US Practices and Policies*, Rand Corporation, Paper #P-3517, March 1967, pp. 6-15.

[162]Gyula Gal, *Space Law* (Leyden: A. W. Sijthoff and Dobbs Ferry, N.Y.: Oceana Publications, Inc., 1969), p. 146.

[163]Commission on Legal Problems of Outer Space of the Soviet Academy of Sciences, "American Diversion in Space," *International Affairs*, No. 12, December 1961, pp. 117-118.

[164]Gal, *op. cit.*, pp. 146-147; Domas Krivickas and Armins Rusis, "Soviet Attitude Toward Space Law," in *Soviet Space Programs, 1962-1965, op. cit.*, pp. 502-503; and "Space Crime" (editorial), *New Times*, No. 21, May 24, 1963, p. 2.

[165]Gal, *supra*, pp. 149-151; Georgii Pokrovskii, "Crime in Space," *New Times*, No. 25, June 20, 1962, pp. 9-11; and V. Glasov, "Cannibals in Space," *op. cit.*, pp. 12-13.

6 Politics and the Moon

It was to be expected that after their first successes in orbiting spacecraft around the earth, the space powers would next set their sights on earth's nearest celestial neighbor, the moon. Missions intended to impact or orbit the moon were within the capability of existing ICBM and IRBM boosters such as the "Sapwood," Atlas and Thor with the addition of upper stages used to accelerate lunar vehicles from orbital to earth-escape velocity. Because the "Sapwood" ICBM had this capability, it may be presumed that the USSR had begun planning flights to the moon, including development of the required technology about the same time as the announcement of what became the Sputnik program in 1953. The first American satellite program, Vanguard, employed a launch vehicle without the capacity for lunar missions through the use of additional stages. Thus, Soviet planning and development had gone further toward implementing lunar missions by the time of Sputnik than had similar plans and developments in the United States.

WHY GO TO THE MOON? MOTIVES AND JUSTIFICATIONS

The motives and goals for spaceflight examined generally in Chapter 4 apply to moon flight as well. Sending spacecraft to the moon requires the development of certain kinds of hardware and techniques and promises its own scientific and possible military benefits, although the last are rather dubious and in many cases now illegal under the terms of the 1967 Space Treaty. Other than technological spinoffs and economic pump-priming, the practical applications of lunar space flight are few. Therefore, the benefits most often cited are in terms of prestige, scientific discovery and exploration of the moon as a stepping stone to later economic exploitation and exploration deeper into the universe.

As already described, the US attached immense prestige value to its manned lunar program. The Apollo mission was chosen because it promised a significant space achievement which the United States, with a concerted effort, could probably accomplish *before* the Soviet Union.[1] The scientific and economic benefits of Apollo were simply not worth the cost.[2] Although it is somewhat risky to make comparisons and impossible to cite the actual costs borne by both sides in conducting various lunar missions,[3] one may conclude that the $35 billion total cost of the Apollo program is way beyond whatever the USSR has spent in its various unmanned and man-related flights to the moon.[4]

These immense costs have, as is well known, produced much criticism of the Apollo program. NASA has therefore been quite defensive about Apollo and sought to justify the program any way it could. For example, NASA Administrator Webb emphasized the military benefits of Apollo, claiming that 80 per cent of the mission could have military uses.[5] He also claimed (perhaps disingenuously) that the technology developed in the course of Project Apollo was "more important than the lunar landing itself."[6] After the landing of the unmanned lunar rover *Lunakhod* by the USSR, the USIA extensively propagandized abroad the relative benefits of the manned approach to the moon.[7]

Right up to the successful landing of Apollo 11 in the summer of 1969, NASA insisted that the USSR was going ahead with its own plans for a manned lunar landing, suggesting that any slackening of pace or budget cutbacks for Apollo would award the lunar prize to the Soviets. It may even be the case that the USSR encouraged statements by its own cosmonauts and scientists to support such claims, possibly with the intention of forcing the United States into further expenditures and, perhaps, accidents.[8] Whatever Soviet motives, both NASA and such highly placed Soviet spokesmen as Khrushchev and Academician Keldysh sought to dispel public reports that the USSR did not have its own program to land men on the moon.[9]

DEVELOPMENT OF SOVIET LUNAR PROGRAMS

A systematic plan for the development of Soviet space research was put forward in 1959 by G. V. Petrovich (pseudonym for V. P. Glushko, "Chief Designer of Rocket Engines"). Petrovich's plan followed that suggested by K. E. Tsiolkovskii at the turn of the century. It divided the conquest of space into three phases. The first involved operations in earth orbit; the second, impact, orbiting, unmanned and manned landing on the moon; the third phase was to be the exploration of the planets.[10]

Both space powers made several attempts to send spacecraft to the moon at an early stage in the history of space exploration. This eagerness is demonstrated by the large number of failures in the lunar flights of both the

USA and the USSR. It would appear that the prestige associated with being the first to accomplish this or that space feat underlay this eagerness. So, for example, the United States, still lagging significantly in space technology, made the first attempts to orbit the moon and one attempt at lunar impact in 1958, all of them failures. American haste to strike, orbit or return pictures from the moon was further evidenced by eleven more failures before the successful flight of Ranger 7 in July, 1964. Soviet preparations were evidently better in that the second acknowledged attempt to hit the moon by the USSR was a success. This flight, *Luna* 2, was launched on the eve of Khrushchev's visit to the USA, and *Luna* 3, which photographed the moon's far side, was launched on the second anniversary of the first Sputnik. Soviet haste to *soft-land* on the moon is demonstrated by the fact that at least seven failed attempts at a soft landing were made by the USSR before *Luna* 9 accomplished this feat in 1966, two months in advance of the American Surveyor 1, the first American attempt at a soft landing.[11]

The haste of both space powers to achieve lunar impact, orbit and soft landing, in addition to being aimed at achieving direct space firsts, was also intended to achieve and test the procedures required for subsequent missions, especially manned lunar landing and return. So, for example, both Soviet and American flights monitored hazardous radiation, mapped possible landing sites and tested landing techniques. In the Soviet timetable a manned lunar landing in 1968–1970 was planned to follow unmanned soft landing in 1964–1965 and manned orbit in 1966–1967. The delay or cancellation of the timetable is evident in that soft landing occurred one to two years late, and in that the USSR has yet to send a man to a lunar orbit or landing.[12] Successful unmanned orbiting of the moon was achieved by the USSR (*Luna* 10) and the United States (Lunar Orbiter 1) in 1966.

After the orbiting of the moon by *Luna* 13 in December, 1966, a period of sixteen months passed without an announced or verifiable Soviet moon launch. This, along with the skipping of Mars windows in 1967 and 1969, is a good indication of some cutback in Soviet budgeting of deep-space missions, perhaps in part the deferred result of the change in political leadership that took place in 1964–1965. Between *Lunas* 13 and 14, nine American flights to the moon took place.[13]

Further Soviet moon missions depended on the use of larger launch vehicles. The heavy *Proton* launch vehicle which had first flown in 1965 was employed for lunar work only in 1968 with the successful circumflight of the moon and return to earth of *Zonds* 5 and 6. The *Zonds* were unmanned tests of vehicles capable of carrying men and were shortly followed by the orbiting of the moon and return to earth by the American astronauts in Apollo 8.[14]

In spite of the evident success of *Zonds* 4-6, which were belatedly acknowledged as precursors to manned circumlunar flight, Academician Blagonravov maintained that further tests would be necessary. *Zonds* 7 and 8 were in this category and were evidently quite successful. American expectations for a *Zond* 9 or subsequent *Zonds* to fly men around the moon had not been fulfilled as of July, 1974.[15] The United States had accomplished manned circumlunar flight with Apollo 8 and landed men on the moon with Apollo 11 even before the flights of *Zonds* 7 and 8. It is quite possible, therefore, that political considerations may have forced a cancellation or postponment of further *Zond* missions beyond *Zond* 8 even before these flights took place. There was no longer a chance of "beating" the Americans after Apollo 8, and the *Zond* missions would be open to invidious comparisons with the spectacular achievements of Apollos 11, 12 and 14-17. It may be that the commitment to the unmanned series of *Zond* circumlunar flights was undertaken during the Khrushchev period and that manned *Zond* flights and possibly manned landings which were meant to follow *Zond* were canceled by the new Soviet leaders in 1964–1965.

As things turned out, it was in the field of *unmanned* lunar exploration that the USSR scored firsts of a kind surpassing the USA. These accomplishments, paled considerably in the light of Apollo, were substantial achievements nonetheless. Continuation of the series of unmanned *Luna* flights offered the USSR a chance of getting to the moon and accomplishing a return before Apollo 11; and barring that, they could record unmanned firsts and create a solid basis, along with American findings, for later manned flights to the moon and planets.

Luna 15 was launched a bare three days before Apollo 11. Had it not failed in its attempt in a soft landing on the moon, it almost certainly would have carried out the mission of *Luna* 16, which, over a year later, returned a sample of lunar soil to the USSR, or possibly that of *Lunas* 17 and 21, which landed automatic roving vehicles, or *Lunakhods*, on the moon. *Kosmos* 300 and 305, launched two and three months after Apollo 11, were also attempts to perform similar missions.[16] *Lunas* 15-21 were the first in the *Luna* series to employ the heavy *Proton* launch vehicle also used in the later *Zond* flights.

Luna 16 accomplished its mission at a much lower cost than an Apollo flight, although its soil sample and flexibility were much less, making the Apollo a better bargain by the pound of soil or amount of scientific return.[17] However, as demonstrated by *Luna* 17, the *Lunakhod* vehicle can be operated for a much longer time than an Apollo mission.[18] After Apollo 11, Soviet spokesmen made the most of the relative cheapness and durability of unmanned craft and downplayed their disadvantages. One Soviet engineer even claimed that the scientific return from unmanned craft is *greater*, which is patently false.[19] Highly placed spokesmen like Academicians Blagonravov, Sedov and B. Petrov as well as some cosmonauts, who had a personal stake in manned flight, were inclined to admit the advantages of manned operations.[20] Since the *Lunakhod* vehicle had been roughly described as much as sixteen years previous to its launch, there is little reason to consider it simply a last-minute effort to steal some prestige from the US.[21]

SOVIET COMMITMENT TO THE MOON RACE

In focusing on political aspects of the Soviet space program, this study has at times emphasized the competitive motives behind the various space missions of the two leading powers. In particular, earlier sections have stressed the political and competitive motives behind the *Vostok* and *Voskhod* programs and the American decision to land and return men from the moon in the decade of the 1970s. In effect, the President of the United States had challenged the Soviet leaders to a race to the moon. In light of this challenge, Soviet interest in *one* space feat, manned lunar landing and return, is of particular interest as regards both the reaction to the President's challenge and the political factors affecting the context of Soviet decisions and actions relating to manned flights to the moon. Consideration of these questions will begin with a brief review of the American commitment to manned lunar landing and return.

On March 29, 1961, the month after the Gagarin flight and Bay of Pigs invasion, John F. Kennedy informed the public:

> I believe that this nation should commit itself to achieving the goal, before the decade is out, of landing a man on the Moon and returning him safely to Earth. No single space project in this period will be more exciting or more impressive to mankind, or more important for the long-range exploration of space; and none will be so difficult or expensive to accomplish . . . In a very real sense, it will not just be one man going to the Moon . . . it will be an entire nation . . . I am asking Congress and the country to accept a firm commitment to a new course of action—a course which will last for many years and carry very heavy costs . . . If we were only to go half way, or reduce our sights in the face of difficulty, it would be better not to go at all.[22]

In fact an American manned lunar landing had been planned as early as 1957 under the auspices of the US Air Force.[23] The Air Force Ballistic Missile Division planned to land a man on the moon by the end of 1965. After the creation of NASA in 1958, the Space Agency took over the Air

Force plans and chose Apollo as the planned successor to Project Mercury.[24] The lunar mission met with a lack of support and positive objections by the Eisenhower Administration, which planned to cancel all manned flights after Project Mercury.[25]

As described in Chapter 4, Kennedy asked his highest advisors if and how the US could "beat" the Russians in space.[26] The race to the moon was chosen because it was winnable. Kennedy's statement setting the deadline as within the decade, i.e., December 31, 1969, at the latest, included a sizable margin of safety in that NASA had told him a manned lunar landing and return would be possible by 1967.[27] All the major technical problems involved had already been solved, at least theoretically, and NASA was in fact able to meet the goal of "before this decade is out," despite a two-year delay occasioned by the Apollo 204 disaster, without budgeted or "cost of living" increments—even though it had always given Congress and the public to believe any delays or cutbacks would squelch the 1969 deadline.[28] NASA's real target date of 1967 was chosen in anticipation of a Soviet moon flight commemorating the fiftieth anniversary of the Bolshevik Revolution on November 7 of that year.[29] In the meantime 60 per cent of NASA's direct spending and about 75 per cent of all NASA's spending were funneled into the Apollo Project.[30]

The political and economic organization of the United States was of a kind that favored broad competition rather than specific competition with the Soviet Union such as a moon race. The American organization for such a race could not be so easily stopped and started as that of the Soviets. For these reasons, a number of differences between the two nations emerged in the conduct of the moon race. First, in deciding to go to the moon, the US required a total public commitment on the part of the top leaders to "get the ball rolling." Otherwise, lack of public and congressional support would have made such an expensive mission as Apollo impossible. The Soviets by contrast, although having less money to spend, could make huge commitments in the chambered meetings of top party and government officials without the necessity of procuring broader support. Moreover, NASA could not cancel the moon mission if the Soviets did not accept or eventually reneged on President Kennedy's challenge, and would instead be inclined to insist that the USSR was still very much in the race no matter what the Soviets did or said. The Soviet leaders, however, could quit, cut back, or reaccelerate their lunar program easily within the limits occasioned by the state of the national economy and the pressures of other commitments.[31]

The preliminary planning for a Soviet manned lunar program probably began as early as 1953 without a firm timetable or competitive goal of achieving this feat ahead of the United States. Buoyed up and probably surprised by the international impact of the first Sputniks and *Vostoks*, the Soviet leaders were likely more earnest about the early scheduling of a manned flight to the moon in the first years of the 1960s. Although no highly placed Soviet political or space official ever announced a deadline for manned lunar flight, Kennedy's public challenge must have encouraged hopes and plans to match or beat the American schedule.[32]

The long leadtime required to mount a manned mission to the moon would have necessitated that the preliminary technical developments begin at about the same time as they did in the United States, that is, around 1961–1962. It is probably the first stages of these developments that led Soviet cosmonauts in the early sixties to predict a Soviet flight to the moon during the decade, despite the lack of any such announcements by the top leadership.[33]

Such an attempt to be first in the moon race was fraught with complications for the Soviet leaders, Khrushchev in particular. The first seven-year plan of 1959–1965 committed the USSR to sizable investments in those sectors of the economy that would return more tangible and immediate returns to the economy than would a manned landing on the moon. As proclaimed in the New Program of the CPSU, the Party considered the seven- and twenty-year (1960–1980) plans as part of building "the material and technical basis" of the Communist stage of social and economic development. Through these plans, the goal was set to overcome the United States in total and per capita industrial production by 1970.[34]

At the same time Soviet commitment to a moon race was further complicated by a perceived need to increase military expenditures. Khrushchev's policy of substituting threats based on nonexistent or exaggerated capabilities had helped to produce an acceleration of strategic weapons development in the United States. The first indication of the bankruptcy of this policy was a substantial increase of the 1961 and 1962 Soviet defense budgets over previously planned figures.[35] The *coup de grâce* befell Khrushchev's policy of exaggerating weapons capability during the Cuban missile crisis in October, 1962.

It seems possible that the economic managers and planners as well as the military leaders who played a role in Khrushchev's 1964 dismissal may have pressured for a diversion of funds from the space program to their own sectors in light of economic plans and military needs. This would push the top leaders, including Khrushchev, toward adopting policies which would hamper the development of a manned lunar program and perhaps into attempts to discourage the United States from meeting its goals for an early lunar landing.

One of the first indications that the Soviet commitment to the moon race was not wholehearted came in March, 1962, when, in a letter to Kennedy, Khrushchev cited the tremendous expenses involved in space exploration in calling for cooperation between the US and USSR in this field.[36] In an interview with an American correspondent in April, 1962, Khrushchev himself suggested that funds formerly allocated to the lunar program would be diverted to the sagging agricultural sector of the economy.[37]

Proposals made by Presidents Kennedy and Johnson and by UN Ambassador Stevenson for a joint US-Soviet lunar program,[38] although probably put forth largely for propaganda purposes, offered the Soviet leaders a golden opportunity for quitting an expensive moon race. Even if the US had no intention of cooperating with the Soviets in this venture, the Kremlin could at least have scored a propaganda victory by accepting the offer and revealing its disingenuousness. If the proposal was even partly sincere, a strong Soviet response could have pressured the American leaders into going along and thereby eroding support for the unilateral American moon program. If the USSR had no hope to beat the US to the moon, a joint lunar program would save it a prestige defeat and give it an equal share in the prestige afforded both for the manned landing and for a dramatic act of international cooperation.

There are probably two principal motives behind the Soviet failure to respond to this American initiative despite the advantages it would offer if the Soviets wanted to opt out of the moon race. First, the Soviet leaders may not have believed in 1962 and 1963 that the US would really land a man on the moon before the USSR; and that with a little determination the USSR would accomplish this first. They may have doubted that the US could maintain the total commitment required to land on the moon in the 1960s. At the time the total cost was estimated to be $40 billion. During this period the USSR enjoyed a comfortable lead in the accomplishments of its space technology, which persisted into the mid-1960s. Not only might this lead have caused Soviet leaders to consider their chances as better than those of the Americans to get a man to the moon first, but it also meant that at that stage the USA would probably benefit more from technical cooperation than would the USSR.

Second, the Soviet leaders have always indulged an overriding passion for secrecy, which has complicated their foreign relations in every area from cultural exchange to disarmament. As detailed in previous chapters, this passion for secrecy has had an especially strong impact on the Soviet space program, which has in some ways been more tightly wrapped in a cloak of secrecy than has Soviet weapons technology.

In setting the landing by both nations of instrumented payloads on the moon as a prerequisite of lunar cooperation,[39] the Soviet Union was probably acting both to make a plausible excuse to defer discussion of a joint lunar program and to make sure that if cooperation were to take place, the Americans would have as much to contribute as the Soviets. When both nations accomplished this feat in early 1966 after three years of failures by the USSR, the independent development of the technical capabilities of both sides had greatly increased technical incompatibility and reduced the chances for and benefits of cooperation. Also by 1966 both countries had in development the hardware required for a unilateral manned lunar program. By attempting to accomplish an instrumented landing first, the Soviets may have hoped to strengthen their bargaining position to the point where they would get their way on issues which would emerge in the operation of a joint program. As mentioned, their first attempt to soft-land on the moon occurred a full three years before the first (and successful) American attempt. By the time the instrumented landings took place, cooperation was further inhibited by the international tensions emerging from the Vietnam War and by the war in 1967 in the Mideast.

Although it is impossible to assess the relative impact of the adduced Soviet motives for, in effect, rejecting American proposals for a joint lunar mission, some disentangling may make them a little clearer. In retrospect the USSR continued the development of technologies that offered *a chance* of getting a man to the moon or at least into lunar orbit before the United States. However, the USSR did not institute a crash program to beat the US to the moon, probably hoping it would be unnecessary. Such a crash program would have precluded both the Soviet strategic weapons buildup in the middle and late 1960s and the investments put into the *Soiuz-Saliut* space station program, which had no direct connection with lunar flights. Looking back, it would have benefited the chances for success of this policy to have done more to stall or delay the American lunar program by positive, if vague and limited, responses to American initiatives for lunar cooperation. Here the passion for secrecy probably precluded a wiser choice of policies, and the two principal considerations, hope for victory and secrecy, combined with financial constraints, led to a strategy of "muddling through," which has been increasingly characteristic of Soviet policy, foreign and domestic. By contrast, the US, which is supposed to epitomize the administrative strategy or lack of strategy of "muddling through," made bold initiatives in both space cooperation and unilateral development of space technology. Some description of how the Soviets "muddled through" follows.

The year 1963 followed on the Soviets' embarrassment in Cuba and was marked by a disastrous harvest in the Soviet Union and the emergence of open and extremely vituperative polemics between the CPSU and the

Chinese Communists abroad. These factors and perhaps an attempt to promote criticism of NASA's moon mission in the United States produced a note of pessimism about Soviet flights to the moon. Academician Sedov took note of the difficulties confronting manned lunar flight, saying that they could not likely be solved in the next few years.[40]

A great flap arose over the visit of British astronomer Sir Bernard Lovell to the USSR in June and July of 1963. Lovell's observations, gleaned from his conversations with Academy President Keldysh and with other scientists, were relayed by him to NASA's Dr. Hugh Dryden[41] and widely publicized in the American press. According to Lovell, Academician Keldysh was very concerned about the expenses involved in safely sending men to the moon and back, particularly because of the peak of solar radiation that would occur in the years 1967–1970. As a result Keldysh felt that (1) a lunar landing would be rejected for the time being, (2) the Soviet Union would emphasize creation of manned stations in earth orbit and unmanned vehicles in deeper space, and (3) the USSR was interested in cooperating with scientists abroad in discussing why and if men should go to the moon. Lovell concluded that the Soviets did not then put any budgetary priority on getting men to the moon and back, that Soviet scientists disagreed about the need for such a mission, and that no "deadline" existed as to when they should accomplish it.[42]

Lovell reported in his letter that he had been chosen by Keldysh to communicate to "the appropriate authorities," i.e., NASA, that the Soviet Union desired that the manned exploration of the moon be turned into a cooperative international enterprise.[43] Lovell's letter and the publicity surrounding his comments had the effect of complicating NASA's budget struggle in Congress.[44] NASA was inclined to minimize the significance of the letter and emphasize the incorrectness of Lovell's conclusions. For example, NASA's Assistant Administrator for International Programs, Arnold Frutkin, considered the Soviet proposal for a discussion of the lunar landing by world scientists an attempt to muster support for the view that an early manned lunar landing as planned by the United States was a bad idea. Frutkin said that the incorrect (to him) conclusion that the Soviets had abandoned their manned lunar program increased debate of the Apollo Project in the United States and probably contributed to cuts in NASA's budget.[45]

In October, 1963, Khrushchev stated that he wished the United States luck with Apollo and its goal of lunar landing before the 1970s. The USSR, he said, was not competing with the US to get to the moon, would require thorough preparation to do so, and did not have any deadline on when to go there.[46] When visiting American businessmen interpreted this to mean that the USSR had abandoned its manned lunar program, Khrushchev denied it.[47] Even Lovell's contention that Keldysh informed him of Soviet plans to cancel an effort at early lunar landing was denied by President Keldysh and by Lovell himself.[48]

Just at the time of Khrushchev's dismissal, Soviet plans about the moon were not clear. Academician Sedov had indicated a manned landing by 1971, but Academician Blagonravov pointed to space stations as the Soviet goal in space, saying the USSR had *no plans to land a man on the moon in this decade.*[49]

By the time of Khrushchev's retirement in October, 1964, the USSR had launched four *Lunas*, one unannounced moon attempt that failed, and four *Elektron* satellites, which cosmonaut Nikolaev claimed were intended to "study possible routes to the moon," in this case by probing for zones of radioactivity that could endanger men going to the moon.[50] For reasons already described, the new leadership team of Brezhnev-Kosygin-Podgornii was at best not a positive gain for the Soviet space program. Unlike Khrushchev, they had not been closely associated with Soviet space triumphs in the late fifties and early sixties. Neither were they given to the dramatic gestures such as space coups which so delighted their flamboyant predecessor. Moreover, the new leaders evidently favored a more "rational" division of technical talents and research funds throughout the whole economy, particularly to applied research in the industrial sector for immediate benefits. Responsible as they were for a rapid buildup of Soviet strategic and conventional forces, the new Soviet leaders may have felt, as did some in the US, that the space program was absorbing resources required for national defense.[51] Any breakthroughs in space cooperation with the USA were unlikely during the first years of rule of this cautious triumvirate and quite possibly also less likely after the succession of Lyndon B. Johnson to the US Presidency.[52]

Whatever the effects of changes in leadership, Soviet cosmonauts and others were still optimistic enough to foresee a manned Soviet moon flight. Before the change, Ambassador to the United States Dobrynin and cosmonaut Gagarin told a UN audience that the Soviet Union would land a man on the moon in the sixties.[53] Cosmonauts Egorov and Feoktistov foresaw a lunar landing based on construction of an orbital launch facility.[54] They probably had in mind the *Proton* launch vehicle, roughly the equivalent of the Saturn I, which could in eight to ten launches assemble a manned lunar lander or put men in orbit around the moon, but which was too small and inefficient for a manned landing and return based on a single launch. The first successful test of the *Proton* vehicle came in July, 1965. A *Proton*-based mission similar to Apollo would have been extremely complex and liable to break down at many points.[55] In describing a manned lunar mission based on *three Proton* launches, cosmonauts Leonov and Beliaev probably had in mind manned orbiting of the moon and return.[56] This is an early indication that either the USSR planned to precede its own manned landing with one or more manned missions in lunar orbit or that it hoped to at least precede Apollo's manned landing with such missions.

In 1965 an added number of Soviet cosmonauts and scientists linked lunar flight to the construction of orbital launch facilities, but they were vague about just when this would happen.[57] Administrator Webb, probably anxious lest the American public get the idea that the Soviets had dropped out of the moon race, wound up contradicting himself in testimony that the USSR was not developing the technology required for manned lunar landing and return (as was the US), but that the Soviets could nonetheless "do anything they want," including explore the moon.[58] At about that time there began a hiatus of over two years in Soviet manned flights, while Project Gemini was going ahead full steam in the United States. This might seem to indicate that the Soviets were concentrating their efforts on the development of a new booster sufficient for a mission of manned lunar landing and return. Some evidence, encountered later, suggests that at least in part this was the case. However, one is pressed to conclude that budget stringencies were also affecting the Soviet space program at this time. Ever since 1962 military flights were consuming a greater proportion of Soviet space launchings, amounting by 1965 to some 45 of 65 launchings and by 1966 to some 28 of 44 launchings.[59] Between *Voskhod* 2 launched in March, 1965, and *Soiuz* 1 launched in April, 1967, the entire Gemini program of ten orbital flights was completed, including several dockings of one spacecraft with another, an operation required in any manned lunar-landing mission employing a booster or boosters within the range of feasible technologies. After the first docking of Gemini 8 in March, 1966, over a year passed before the USSR accomplished unmanned docking of two automatic space vehicles, and nearly three years before a manned Soviet vehicle docked with another.[60] Even if the USSR were banking on development of a super-booster for an early flight to the moon, it seems very likely that it would have wished to gain experience in docking spacecraft if it intended to beat or match American efforts with Apollo.[61]

Regardless of what was by 1965 a clear lead by the US in the race to the moon, Soviet spokesmen still maintained a vague optimism about their own lunar plans. In August, 1965, Academician Sedov claimed that the success of the *Proton* vehicles demonstrated that a man could be sent around or even landed on the moon with existing technology, saying that the time for that was "not far off."[62] In 1966 cosmonauts Gagarin, Beliaev and Leonov predicted a manned Soviet landing on the moon by 1970, and cosmonaut Komarov said that the USSR "will not be beaten by the United States in a race for a human being to go to the moon." Notably, the remarks of highly placed academicians like Keldysh and Blagonravov were a good deal more cautious.[63] In the meantime both space powers had paved the way toward a manned lunar landing by orbiting and soft-landing unmanned vehicles on the moon (*Lunas* 9-13, Surveyor 1, and Lunar Orbiters 1 and 2).[64]

In retrospect it seems clear that by 1967 the USSR had largely abandoned whatever hopes it might have had of getting a man to the moon ahead of or shortly after the United States. The Apollo 204 fire gave General Kamanin an opportunity to accuse the United States of risking the lives of its astronauts in its haste to get to the moon, and perhaps gave some hope of resultant delay in the American schedule. Such opportunity and hope were dashed on April 24, when cosmonaut Komarov perished in the reentry of *Soiuz* 1. Academician Sedov now said that manned flight to the moon was "not in the forefront of Soviet astronautics" and that the problem of returning a man from the moon had not yet been solved.[65] Still, it was announced that work was now under way on a superbooster capable of mounting manned landings on the moon, and cosmonette Tereshkova claimed that a team led by Gagarin and including herself had been assembled for lunar landing.[66] Nonetheless, it is now apparent that the USSR would concentrate its efforts on discerning and solving the difficulties of the *Soiuz* spacecraft, which was designed not for lunar work but for the assembly of a permanent manned space station in earth orbit.[67] Any manned operations to the moon would have to be those of orbit and not landing, as indicated by the unmanned series of *Zonds* evidently aimed at this mission, which flew between 1968 and 1969. Instead of repeating earlier cosmonaut predictions that manned lunar landing would come in five years, cosmonaut-engineer Feoktistov now extended this to "from five to ten years."[68]

In 1967 the US flew eight spacecraft to the moon and the Soviets none. The year's delay in Soviet flights can be attributed in retrospect to several factors. First and most important to the theme of this section, the USSR had evidently backed away from any hope for a manned lunar landing. The Soviets could have continued to use smaller launch vehicles to map the lunar terrain and probe the lunar surface in preparation for a manned landing as the United States did that year. Instead they spent the year preparing a switch to use of the large *Proton* booster as a launcher for *unmanned* missions of lunar orbit and landing that were simply unnecessary as part of a manned program. To the extent that beating the United States to the moon was a consideration, the Soviets evidently felt that the only realistic way to accomplish this was to soft-land an unmanned soil retriever or lunar rover before Apollo 11. If *Zond* was not already a dead-end or at least a severely underfunded program, presumably *Zond* flights would have also taken place in 1967 in hopes of using one or more *Proton* boosters to send a manned *Zond* around the moon in 1968 or 1969. No *Zond* flights related to the moon took place until March, 1968, and the last *Zond* to date, number 8, was launched in October, 1970, still without human cargo. Between the time Apollo 8 orbited the moon in December, 1968, and Apollo 11 landed in July, 1969, further efforts with *Zond* beyond those already scheduled must have been canceled, if they had not been earlier.[69]

Commentary by Soviet cosmonauts, scientists and engineers during 1968 reflected the expectation that the USSR might not be the first country to send men to the moon. One engineer who expected a manned landing based on *Proton* launch vehicles was probably poorly informed, because the Soviet policy of tight security extends to all but the highest placed personnel. Other spokesmen chided the United States for the riskiness and unscientific nature of Apollo and stressed interim goals other than manned lunar landing, such as manned circumlunar flight and unmanned landing and return. They also emphasized technical difficulties which could delay manned landing on the moon.[70]

In March of 1968 cosmonaut Gagarin, presumably the leader of the cosmonaut team scheduled to land on the moon, died in an air crash. This probably helped to quash any remaining plans for manned lunar landings.[71] By October, 1968, Academician Sedov said that the USSR had no plans to send a man to the moon "in the near future" and that exploration of the moon was "not a priority" in the Soviet space program.[72] Shortly afterward there were indications that *Zonds* might be used to send men around the moon, but as has been shown, these plans were probably squelched by American accomplishments, revealing that despite frequent disclaimers, competitive motives are involved in Soviet space planning.[73]

The year 1969 presents many paradoxes to one who has to attempt to divine Soviet intentions concerning plans about manned lunar landings. This is the case with the author's own contention that the Soviet leaders were muddling through with diverse strategies and no bold plan to beat the schedule of Apollo. In particular, those who contend that the USSR was *intensively* developing the hardware for a lunar landing in the late sixties or early seventies can draw some support from the statements of two Soviet cosmonauts.[74]

Cosmonaut Shatalov predicted a Soviet manned landing on the moon within "six, seven, and perhaps more months" on April 9, 1969. Cosmonaut Leonov saw such a landing as taking place in late 1969 or early 1970.[75]

According to NASA announcements as early as 1967, the USSR was developing a superbooster more or less in the class of the Saturn V used to launch Apollo.[76] In retrospect initial development of such a vehicle seems likely, despite the vagueness of Soviet public statements. Presumably, any Soviet flight to the moon in 1969 or the early 1970s would have employed this vehicle.

For the sixties and early seventies the only practical uses of such a large launch vehicle would be to send men to the moon and/or to orbit large space stations, possibly including use of these for moon launches from earth orbit. Cosmonauts Shatalov and Leonov evidently had this vehicle in mind, since, as already mentioned, the smaller *Proton* launch vehicle was poorly suited for manned landing operations. There may indeed have been the possibility of such use should unforeseen delays occur in the Apollo program. Since the first indication of the existence of this vehicle came from NASA in 1967, and since its first testing probably occurred in 1969, it seems likely that its design was similar to the standard Soviet launch vehicle, *Vostok*, i.e., that it employed multiple clusters of engines in strap-on configuration rather than the simpler tandem construction of Apollo-Saturn, which has only five engines in the first stage in a single rocket body. Otherwise, it is hard to see how it could have been ready for testing so soon. Moreover, to date not a single flight of this enormous vehicle has occurred. In the summer and fall of 1969 stories reached the West that a preliminary test or tests of this vehicle resulted in explosion.[77] The clustered strap-on design would explain the early development of this launcher, the difficulties in getting it to fly, and the evident abandonment of the design, since no launch of such a vehicle has occurred to date.

It is likely that the USSR does not currently possess an operational equivalent of the F-1 engine employed in the Saturn booster, second and third stages. Therefore, any vehicle of the purported size of the one under discussion would involve something like the twenty main engines in the first stage of the *Vostok* launch vehicle and perhaps even more.[78] If this is the case, the vehicle is obsolete and was obsolete when NASA first revealed it in 1967. First, the configuration is extremely inefficient in terms of thrust/weight ratio—one of the reasons the *Vostok* was never extensively deployed as an operational ICBM. Second, this complex configuration of such a large vehicle dramatically increases the possibilities of launch failure should a single engine malfunction and makes it more likely that the launch vibrations will do fatal damage to the entire structure. If the arguments put forth in the United States for the replacement of Saturn by the space-shuttle vehicle have merit, the Soviets would have no reason to continue development of a launcher which is clearly inferior to the now-obsolescent Saturn V.

Although the argument presented here necessarily becomes more tenuous with every postulation, the alternative would be no argument at all, so the postulates will be pushed one step further. It seems likely that a vehicle such as that described would lack the qualifications for long-term operational use as a space launcher, both because of the postulated characteristics and the lack of any indications of further work on it to date. Let it be assumed, then, that it was intended merely for interim use, rather like the Saturn, which will see no use now that the Apollo and Skylab programs are completed. As an interim vehicle, obsolete before it was developed, this launcher was almost certainly designed to send men to the moon and back and/or to launch a large permanent space station into earth orbit.[79] Other missions could be accomplished sooner, more safely, more cheaply by operational Soviet launch vehicles. The vehicle as described lacked the capability to mount manned missions to the planets, which require qualitatively different technologies, such as nuclear propulsion, earth-orbital assembly and new life-support systems.

As such, this vehicle appears as a halfhearted attempt to capture the "lunar prize" or else mount some similar spectacular mission to offer the Soviet answer to Apollo. After Neil Armstrong set foot on the moon in January, 1969, and after it became evident that it would take months or years to man-rate this vehicle, the whole project could have been canceled, the designers could have returned to the drawing board and funds could have been diverted to the construction of a more modern vehicle as well as to fuller use of existing operational space technologies, especially those associated with the *Soiuz* and *Saliut* orbital vehicles.

If the Soviet leaders were as intent and as committed to a lunar launch as NASA was with Project Apollo, they would not have gone ahead with the *Lunas, Zond* and *Soiuz* missions in 1968–1971. NASA probably had greater funds at its command but delayed or canceled planned missions similar to the later *Lunas* and *Soiuz-Saliut*.[80] The USSR went ahead with these missions—*Lunas* 15-21 into 1973—solved the problems leading to Vladimir Komarov's death aboard *Soiuz* 1 in 1967, launched twelve more *Soiuz* craft through 1973, put into orbit *Saliut* space stations in 1971 and 1973 and, since the fatal accident in *Soiuz* 11 in June of 1971, has evidently been working on solving difficulties with *Soiuz* now scheduled for cooperative use with the US Apollo spacecraft in 1975, after an evident failure occurred in Spring, 1973. In the meantime military applications and Venus flights continued, and the Soviet Mars program was renewed with two launches to the red planet in 1971 (and one failure, *Kosmos* 419)[81] and four more in 1973.[82]

Returning to the statement of cosmonauts Shatalov and Leonov in the spring and summer of 1969, it would appear in retrospect that they were simply overoptimistic about the schedule for testing and man-rating the new superbooster. The *Proton* launcher, in operational use since 1965, has still not been used to put men in space, although it does launch the *Saliut* space station to which men are carried in *Soiuz* vehicles. More highly placed Soviet space spokesmen never gave any unqualified indications of manned landings on the moon in 1967–1972. If the remarks of Shatalov and Leonov had been calculated in the Kremlin to push the USA into "all-out haste" or a "tragic miscalculation," as one book suggests, they surely would have been repeated by higher and more responsible spokesmen. NASA probably chose to publicly believe the indications by cosmonauts and others not so much to fortify its desperation as to pad its budgets.[83]

Examining the statements of other Soviet space spokesmen for the period in question, that is, the months in 1969 leading up to the July 16 launch of Apollo 11, one gets a different impression from that conveyed by the statements of Shatalov and Leonov. Six of the most authoritative spokesmen in the Soviet space program gave the impression that the USSR had no plans to land men on the moon in the near future.

Academician Boris Petrov claimed that the Soviets were in no way competing with the USA in attempts to put a man on the moon, saying that the *Soiuz* vehicle was for use only in terrestrial orbit and emphasizing that the Soviet space program aimed at "practical and scientific objectives" rather than a race to the moon. Academician Blagonravov emphasized the divergence of the Soviet and American space programs, saying that the Soviet emphasis was on orbital stations rather than on moon landing. He set a permanent orbital space station as the next goal of Soviet astronautics. Such space stations were also indicated by Academy President Keldysh as being among the principal tasks of the Soviet space program. Notably, Keldysh said in April that a permanent space station could make "gigantic carrier rockets" (such as the Soviet superbooster) unnecessary. This supports the contention made here that the superbooster became a dead-end project. Ari Shternfeld, space scientist and popularizer of Soviet astronautics, also emphasized the advantages of orbital launch facilities for sending manned missions to the moon and planets. It should be noted that construction of such a facility would require not months but years as well as extensive testing of materials and techniques such as the space welding carried out in the group flight of *Soiuz* 6, 7 and 8 in October, 1969. Stress on the advantages of unmanned deep-space flights were made by a man identified only as the "Chief Designer of Automatic Interplanetary Stations" who saw no use for men outside earth orbit so far. Cosmonaut Feoktistov, who also holds an administrative position in the space program, indicated in May that the current Soviet emphasis was on automated unmanned flights and that such flights should be used extensively before humans were sent to the moon. Two weeks before the launch of Project Apollo the Associated Press reported from Moscow that "Soviet sources" indicated that two unmanned attempts to return soil from the moon had already been made and another would be launched July 10. This must have been *Luna* 15 launched on July 13, which coincided with the Apollo 11 flight and failed to make a soft landing on the moon.[84]

After Apollo 11, Soviet spokesmen were even more clear and insistent, flatly stating that the USSR had no plans for manned moon flight in the coming months. While generally praising the accomplishments of Project Apollo, they defended the advantages, especially the relative cheapness, of unmanned vehicles for lunar exploration and made a few vague criticisms of Apollo and its role in the American space program. According to Boris Petrov, the USSR saw no necessity of sending men to the moon. One may qualify this somewhat in that in the same remarks he denied that *Luna* 15 was intended to land on the moon, and, at least in this, he was very probably dissembling. Academician Sedov said that the USSR had "no immediate plans" to land men on the moon, preferring automatic stations. In October Party Secretary Brezhnev identified earth-orbital stations as the "decisive means to an extensive conquest of space" and the "main route for man to follow into space." Academy President Keldysh said that the USSR would not send men to the moon until "later on," and when it did, it would employ earth-orbital assembly of the spacecraft. This was a further indication that the superbooster was scratched along with orbiting of the moon by manned *Zonds* and that "later on" would be several years in coming. Especially after the successful flights of *Lunas* 16 and 17 in September and November, 1970, Soviet spokesmen emphasized the relative cheapness of such automated flights, claiming they were anywhere from one-fifth to one-fiftieth the cost of Apollo-type craft. They denied that the Soviet Union had completely abandoned the idea of manned lunar exploration, but stressed the importance of other missions and Brezhnev's emphasis on the contribution of space to the national economy. A number of highly placed space specialists suggested that the high costs of manned operations could be overcome in the future by US-Soviet cooperation.[85]

An article appeared in 1971 by the authoritative Boris Petrov emphasizing automated vehicles for deep-space research and manned laboratories and observatories in earth orbit. Unmanned flight, Petrov said, is "much less expensive" and "in the foreseeable future" would be virtually the only means of directly exploring the outer reaches of space.[86] In the long-term future the USSR still maintains its long-standing commitment to establishing permanent outposts on the moon and to making them self-supporting and profitable to the earth.[87] At present, it would be impossible to predict when, if ever, the necessary technology will be available to the USSR.

NOTES

[1]James E. Webb, address of September 27, 1961, NASA News Release No. 61-215, p. 11, as cited in Vernon Van Dyke, *Pride and Power, The Rationale of the Space Program* (Urbana: University of Illinois, 1964), p. 195. Cf. the statement of Hugh L. Dryden, NASA's deputy administrator, in "News Conference on Russian Space Shot," NASA News Release, August 15, 1962, p. 19. See also memo from JFK to LBJ, April 20, 1961, as cited in John Logsdon, *The Decision To Go to the Moon: Project Apollo and the National Interest* (Cambridge, Mass.: MIT Press, 1970), pp. 109-110.

[2]See Warren Weaver, "Dreams and Responsibilities," *Bulletin of the Atomic Scientists, 19* (5), May 1963, pp. 10-11; and Van Dyke, *op. cit.*, p. 118.

[3]Sheldon, "Summary" in *Soviet Space Programs, 1966-1970, op. cit.*, xxviii.

[4]Sheldon, "Program Details of Unmanned Flights," *Soviet Space Programs, 1966-1970, op. cit.*, pp. 201-202; and Amitai Etzioni, *The Moon-Doggle, Domestic and International Implications of the Space Race* (Garden City, N.Y.: Doubleday and Co., Inc., 1964), p. 14.

[5]G. Ivanov, "Space Business and US Policy," *International Affairs, 10* (6), June 1964, p. 65. See also Robert Salkeld, *War and Space* (Englewood Cliffs, N.J.: Prentice-Hall, Inc. 1970), p. 168.

[6]James E. Webb, address of October 1, 1962, NASA News Release, p. 2, as cited in Van Dyke, *op. cit.*, p. 100.

[7]Simon Bourgin, "Impact of US Space Cooperation Abroad," in *International Cooperation in Outer Space, Symposium, op. cit.*, p. 172.

[8]Charles S. Sheldon II, "Soviet Circumlunar Flight Seen," *Aviation Week, 80* (20), May 18, 1964, p. 77; and Erlend A. Kennan and Edmund H. Harvey, *Mission to the Moon, A Critical Examination of NASA and the Space Program* (New York: William Morrow and Co., 1969), pp. 279-280.

[9]See Neal Stanford, "US Space Officials Say 'Race' Still On," *Christian Science Monitor*, May 26, 1964, p. 3; Joseph G. Whelan, "Political Goals and Purposes of the USSR in Space," in *Soviet Space Programs, 1962-1965, op. cit.*, pp. 80, 123; Leonard N. Beck, "Soviet Space Plans and Economic Capabilities, Part I: Soviet Projections of Space Plans," in the same doument, p. 360; Arnold Frutkin in *International Cooperation in Space*, (Englewood Cliffs, N.J.: Prentice-Hall, 1965), pp. 108-111; and *New York Times*, November 7, 1963, p. 1.

[10]Fermin J. Krieger, *Soviet Space Experiments and Astronautics*, Rand Corporation, Paper #P-2261, March 31, 1961, pp. 35-37; and G. V. Petrovich, "Pervii Iskusstvennii Sputnik Solntsa," *Vestnik Academii Nauk SSSR, 29* (3), March 1959, pp. 8-14 as cited in *ibid.*, p. 42.

[11]Charles S. Sheldon II, *Review of the Soviet Space Program, With Comparative US Data* (New York: McGraw-Hill Book Co., Inc., 1968), pp. 22, 26-27; and Krieger, *supra*, p. 15.

[12]Bubnov and Kamanin as cited by F. J. Krieger, *The Space Programs of the Soviet Union*, Rand Corporation, Paper #P-3632, July 1967, pp. 16-17.

[13]Sheldon, "Program Details of Unmanned Flights," *op. cit.*, pp. 195, 206; and Leonid Vladimirov, *The Russian Space Bluff* (New York: Dial Press, 1973), pp. 20-21, 139-140.

[14]Sheldon, "Program Details of Manned Flights," in *Soviet Space Programs, 1966-1970, op. cit.*, pp. 240-243.

[15]*Ibid.*, pp. 243-245; and Sheldon, "Projections of Soviet Space Plans," in *Soviet Space Programs, 1966-1970, op. cit.*, pp. 390-391. In fact the Soviets still had not sent men to the moon when this study went to press.

[16]Sheldon, "Program Details of Unmanned Flights," *op. cit.*, pp. 196-200; and Sheldon, "Projections of Soviet Space Plans," *op. cit.*, p. 374.

[17]Sheldon, "Program Details of Unmanned Flights," *supra*, pp. 201-202.

[18]*Ibid.*, pp. 202-204; and "Postscript" in *Soviet Space Programs, 1966-1970, op. cit.*, pp. 513-515.

[19]See statements by Alimov (p. 375), Blagonravov (p. 376), Adamov and Feoktistov (p. 379), Denisov (p. 380), B. Petrov (pp. 381, 383), Raushenback and Beregovoi (p. 383) in Sheldon, "Projections of Soviet Space Plans," *op. cit.*

[20]*Ibid.*, pp. 375-376.

[21]Fermin J. Krieger, *Behind the Sputniks, A Survey of Soviet Space Science* (Washington, D.C.: The Rand Corporation, Public Affairs Press, 1958), p. 17 and Chapter 3; and Sergei Gouschev and Mikhail Vassiliev, editors, *Russian Science in the 21st Century* (New York: McGraw-Hill Book Co., Inc., 1960), pp. 208, 214.

[22]As cited by Eugene M. Emme, *Historical Perspectives on Apollo*, AIAA Paper No. 67-389, delivered at AIAA Annual Meeting, Anaheim, California, October 23-27, 1967, as cited in Kennan and Harvey, *op. cit.*, p. 82.

[23]Thomas A. Sturm, *The USAF Scientific Advisory Board. Its First Twenty Years, 1944-1964* (Washington, D.C.: Government Printing Office, 1967), pp. 81-83.

[24]Logsdon, *op. cit.*, pp. 40, 47.

[25]*Ibid.*, pp. 35-36; and Kennan and Harvey, *op. cit.*, p. 75.

[26]Memo from JFK to LBJ, April 20, 1961, NASA Historical Archives, as cited in Logsdon, *supra*, pp. 109-110.

[27]James E. Webb, address of September 27, 1961, NASA News Release No. 61-215, p. 11, as cited by Van Dyke, *op. cit.*, p. 147; and Logsdon, *supra*, p. 40.

[28]Logsdon, *supra*, p. 62; Kennan and Harvey, *op. cit.*, pp. 94-95.

[29]*Space: The New Frontier*, NASA (Washington, D.C.: US Government Printing Office, 1967), O-223-636, p. 2, as cited by Kennan and Harvey, *supra*, p. 86.

[30]*Ibid.*, p. 87; and Hugh Odishaw, "Science and Space," in Lincoln P. Bloomfield, editor, *Outer Space, Prospects for Man and Society* (New York: Frederick A. Praeger, Inc., 1968), p. 76.

[31]Etzioni, *op. cit.*, p. 161.

[32]See Sheldon, "Program Details of Manned Flights," *op. cit.*, p. 245.

[33]*Ibid.*, pp. 245-246; and Daniloff, *The Kremlin and the Cosmos* (New York: Alfred A. Knopf, 1972), p. 140.

[34]See Herbert Ritvo, annotator, *The New Soviet Society: Final Text of the Program of the Communist Party of the Soviet Union* (New York: New Leader, 1962), pp. 114-115.

[35]*Ibid.*, p. 61, note 70; p. 103, note 126; and p. 183, note 256.

[36]*Soviet Space Programs, 1962, op. cit.*, pp. 244-245.

[37]*Ibid.*, p. 244; and Gardner Cowles, "An Editor's Report on Europe and Russia," *Look*, 26, June 5, 1962, pp. 32-37.

[38]*International Cooperation and Organization for Outer Space, 1965, op. cit.*, pp. 157-160.

[39]A condition stipulated by Academician A. A. Blagonravov to NASA's Hugh Dryden and House Space Committee Chairman Miller. See Alton Frye, *The Proposal for*

a Joint Lunar Expedition: Background and Prospects, Rand Corporation, Paper #P-2808, January 1964, pp. 8-9.

40Leonid Sedov, "The Outlook for Space Exploration," *New Times,* No. 26, July 3, 1963, pp. 16-17.

41Full text of letter in *Congressional Record, 109* (123), August 9, 1963, p. 13903.

42See John R. Walsh, "NASA: Talk of Togetherness with Soviets Further Complicates Space Politics for the Agency," *Science, 142* (3588), October 4, 1963, pp. 35-38; Joseph G. Whelan, "Western Projections of Soviet Space Plans," *Soviet Space Programs, 1962-1965, op. cit.,* pp. 374-375; Sir Bernard Lovell, "Russian Plans for Space," *Survey, 52* (52), July 1964, pp. 78-82; Sir Bernard Lovell, "Soviet Aims in Astronomy and Space Research," *Bulletin of the Atomic Scientists, 19* (8), October 1963, pp. 36-39; Shelton, *Soviet Space Exploration* (New York: Washington Square Press, Inc., 1968), p. 8; and Frye, *The Proposal for a Joint Lunar Expedition: Background and Prospects, op. cit.,* p. 7.

43Frutkin, *International Cooperation in Space, op. cit.,* p. 105.

44Frye, *The Proposal for a Joint Lunar Expedition: Background and Prospects, op. cit.,* pp. 7-8.

45Frutkin, *International Cooperation in Space, op. cit.,* p. 108.

46*Izvestiia,* October 26, 1963, p. 1.

47Beck, "Soviet Space Plans and Economic Capabilities . . . , " *op. cit.,* p. 360; *New York Times,* October 27, 1963, p. 1 and November 7, 1963, p. 1; and Frutkin, *International Cooperation in Space, op. cit.,* p. 109.

48Frutkin, *supra,* pp. 110-111; and Beck, *supra,* p. 361.

49Beck, *supra,* p. 358; and "Blagonravov Reviews Soviet Space Program," *Trud,* October 4, 1964, p. 2, as cited in *ibid.,* pp. 353-354 (emphasis added).

50Leonard N. Beck, "Recent Developments in the Soviet Space Program: A Survey of Space Activities and Space Science," in *Soviet Space Programs, 1962-1965, op. cit.,* p. 176.

51See the remark of Dr. Phillip Abelson cited by Lillian Levy in Levy, editor, *Space: Its Impact on Man and Society* (New York: W. W. Norton and Co., Inc., 1965), p. 205.

52Etzioni cites a number of reasons why Johnson would be less favorably disposed to space cooperation than Kennedy. See xiv-xv.

53Beck, "Soviet Space Plans and Economic Capabilities . . . ," *op. cit.,* p. 359.

54Cited by Sheldon, "Projections of Soviet Space Plans," *op. cit.,* p. 359.

55Sheldon, "Program Details of Manned Flights," *op. cit.,* pp. 251-252.

56Beck, "Soviets Space Plans and Economic Capabilities . . . ," *op. cit.,* pp. 364-365.

57See the remarks of Keldysh (p. 359); Titov, Sedov (p. 360); Sedov (p. 361) in Sheldon, "Projections of Soviet Space Plans," *op. cit.*

58Testimony of March 8 and 22, 1965, as cited in Whelan, "Political Goals and Purposes of the USSR in Space," *Soviet Space Programs, 1962-1965, op. cit.,* pp. 123, 125.

59See Table VI in Sheldon, "Overview, Supporting Facilities and Launch Vehicles of the Soviet Space Program," in *Soviet Space Programs, 1966-1970, op. cit.,* p. 120.

60Sheldon, "Program Details of Manned Flights," *op. cit.,* p. 232, 234-235.

61See Shelton, *Soviet Space Exploration, op. cit.,* pp. 146, 265-266.

62"The Outlook for Space Exploration," Interview with Academician Leonid Sedov, *New Times,* No. 33, August 18, 1965, p. 5; and Sedov's article of the same title in *New Times,* No. 8, February 23, 1966, pp. 1-3.

63Sheldon, "Projections of Soviet Space Plans," *op. cit.,* pp. 362-363.

[64]See Shelton, *supra*, pp. 231-232; and Sheldon, Table I, in "Program Details of Unmanned Flights," *op. cit.*, p. 205.

[65]Kamanin as cited in Sheldon, "Projections of Soviet Space Plans," *op. cit.*, pp. 364, 365.

[66]Commentator Pacner in *ibid.*, p. 366; and Shelton, *supra*, p. 159.

[67]Shelton, *supra*, p. 268.

[68]Sheldon, "Projections of Soviet Space Plans," *op. cit.*, p. 366.

[69]For dates of flights, see Sheldon, "Program Details of Manned Flights," *op. cit.*, Table I, pp. 205-206.

[70]See the remarks of engineer Pokorni, cosmonauts Beliaev and Bykovskii and Academicians Sedov, Keldysh, B. Petrov and Blagonravov in Sheldon, "Projections of Soviet Space Plans," *op. cit.*, pp. 368-371.

[71]See Shelton, *Soviet Space Exploration, op. cit.*, pp. 276-277.

[72]See Kennan and Harvey, *op. cit.*, p. 276.

[73]*Ibid.*, p. 279. This was the first Soviet indication that the second series of *Zond* flights (4-8) was a test of manned precursors for circumlunar flight. See also Sheldon, "Program Details of Manned Flights," *op. cit.*, p. 243.

[74]As does Sheldon in *supra*, pp. 245-253 and in "Projections of Soviet Space Plans," *op. cit.*, pp. 391-392.

[75]Sheldon, "Projections of Soviet Space Plans," *supra*, pp. 372-374.

[76]Sheldon, "Overview, Supporting Facilities and Launch Vehicles of the Soviet Space Program," *op. cit.*, pp. 146-148; and Sheldon, "Program Details of Manned Flights," *op. cit.*, p. 246.

[77]*Ibid.*, p. 147 and *ibid.*, p. 247. See also "Soviets Suffer Setbacks in Space," *Aviation Week and Space Technology*, 91 (20), November 17, 1969, p. 26.

[78]See the postulated configuration in Sheldon, Appendix B in *Soviet Space Programs, 1966-1970, op. cit.*, pp. 562, 563.

[79]See Daniloff, *op. cit.*, p. 171 for a slightly different interpretation.

[80]See the comment about the US Prospector Program made in Sheldon, "Program Details of Unmanned Flights," *op. cit.*, p. 202. The US Skylab mission of the mid-1970s was broadly similar to *Soiuz-Saliut*.

[81]See Sheldon and DeVoe, Appendix A of *Soviet Space Programs, 1966-1970, op. cit.*, pp. 545-558; and for more recent launches the "Satellite Report" in *Space World* (monthly, 1969-).

[82]Charles S. Sheldon II, *United States and Soviet Progress in Space: Summary Data Through 1973 and a Forward Look*, Library of Congress, Congressional Research Service, January 8, 1974, p. 48.

[83]See Kennan and Harvey, *op. cit.*, pp. 279-280.

[84]See the quotations in the chronology of Sheldon's, "Projections of Soviet Space Plans," *op. cit.*, pp. 371-374. See also Sheldon, "Program Details of Manned Flights," *op. cit.*, pp. 236-238.

[85]See the chronology in *supra*, pp. 375-384. Also *New York Times*, October 26, 1969, p. 1.

[86]B. Petrov, "Space Exploration: Progress and Trends," *World Marxist Review, 14* (4), April 1971, pp. 84-85, 87. Academician Petrov is identified as a "hero of socialist labor" presumably for his work in the space program.

[87]See the remarks of Academician Vararov as cited by Gouschev and Vassiliev, *op. cit.*, p. 214.

Planetary Programs

As described in previous chapters, the USSR takes as its long-range goal in space the exploration and exploitation of the planets. This goal was indicated by early publications on Soviet goals in space.* More recently a 1964 book coauthored by General Kamanin looked forward to manned flights to the planets by 1975–1990.[1] In contrast to NASA's major mission orientation, which focused on the goal of bringing Project Apollo to a successful conclusion rather than on its place in a longer range plan of space exploration, Soviet discussion of the USSR's space program has generally been linked with the long-range goal of manned exploration and colonization of the planets. Both those Soviet missions conducted in earth orbit as well as Soviet accomplishments and plans concerning the moon have consistently been linked to planetary exploration.[2]

The attitude of looking toward future exploration of the planets in appreciating existing programs of earth orbital and lunar research is generally characteristic of those men in the West who can be categorized as space scientists. By contrast, others involved in space planning and decision-making are much more likely to emphasize other goals involving immediate or short-range payoffs. Because American politics does not generally coopt leading members of the scientific community into responsible political roles, because political leaders in the US do not make use of or require science and pseudoscience to legitimize their political roles, and because American leaders are more subject to the influences of myriad political interests of which scientific interests are a tiny fraction, the goals seen by space scien-

*See Chapter 4 and G. V. Petrovich (pseudonym), "Pervii isskustvennii sputnik solntsa," *Vestnik Akademii Nauk SSSR*, 29 (3), March, 1959, pp. 8-14.

tists appear to have had less influence on the American space program than on the Soviet program. The American program has been more preoccupied than the Soviet with national pride and prestige, stimulus to the economy, meeting the demands of nonscientific interests, and acquiring quick benefits through applications of space technology to economic and military needs.

Discounting those efforts of both the US and the USSR to land men on the moon, the most readily observable difference in the American and Soviet space programs as measured by numbers of launchings has been that the former has put a greater relative effort into applications of space research to earthbound needs and the latter has put a greater relative effort into exploration of the planets.[3] In terms of payload weight, the USSR has committed about eleven times as much to exploration of the planets than the USA. Estimates have been made that the Soviet planetary program costs between five and ten times the American.[4] The first Soviet attempts to impact or reach the planets were two unannounced Mars flights in 1960 that failed to reach earth orbit. The United States did not attempt a planetary launch until nearly two years later, in 1962.[5] Through 1973 there have been 19 Soviet launches to Venus, 14 to Mars and no interplanetary flights. By comparison the US has launched five payloads to Mars, two to Venus, one to Mercury, two to the outer planets, and five interplanetary flights.* The total payloads related to the planets are: USSR, 33; USA, 15. Of the total flights of each country, 12 per cent of Soviet flights have been "earth escapes" as compared to eight per cent of American flights.[6]

Soviet eagerness to get to the planets is demonstrated by their determination to do so despite repeated failures. The evidence available indicates that few, if any, Soviet planetary missions have been completely successful in carrying out their intended tasks. Despite this, between 1960 and 1967 the USSR attempted at least two planetary launches during every launch window of the planets Venus and Mars.[7] Launch windows, or minimum energy requirements to reach the planet, occur every 19.2 months for Venus and every 25.6 months for Mars.[8] It is possible and even quite likely that the USSR made additional launches toward Mars and Venus which have not been confirmed by unclassified sources.[9] The only Venus or Mars windows skipped by the USSR appear to have been those for Mars in 1967 and 1969 and for Venus in 1973.[10]† By way of comparison, the United States did not

---

*In January 1974 the Soviets announced cancellation for the time being of all plans for planetary flights except those to Mars and Venus. See *Aviation Week and Space Technology*, 100 (1), January 7, 1974, p. 9.

†The gap in flights to Venus in 1973 probably indicates the intention to introduce new technology, most likely the Proton booster and possibly a roving vehicle or soil retriever. See *Aviation Week and Space Technology*, 100 (5), February 4, 1974, p. 11.

start flights to the planets until 1962 and has skipped windows for Venus in 1964, 1965, 1969, 1970 and 1972 and for Mars in 1962, 1967 and 1973.[11]

Not until 1967, after ten launches at Venus and one Venus systems test, did the USSR succeed with *Venera* 4 in returning data from the atmosphere of Venus. Transmission of data from the surface of Venus by *Veneras* 7 and 8 in 1970 and 1972 lasted for a few minutes only. The US has not yet landed probes on Mars or Venus, nor were the Soviets successful in returning data from the atmosphere or on or near the surface of Mars until Mars 3, which reached the planet in late 1971. The US in March, 1972, became the first and only country to launch a probe toward Jupiter—Pioneer 10—and, likewise, the first toward Saturn (Pioneer 11) and Mercury (Mariner 10) in 1973.[12] Commitments to "grand tours" of the planets made possible by the configuration of the outer planets in the mid- and late 1970s were canceled in the US and never made in the USSR.

American sources contend that Soviet haste to impact and soft-land on the earth's nearest planetary neighbors has prompted the USSR to reject the rigorous methods of sterilization applied to US planetary probes. They conclude that the Soviets have probably transferred terrestrial microorganisms to the surfaces of Mars and Venus and possibly contaminated these planets.[13]

The long-term Soviet goal of man's conquest of the planets continues to be an underlying goal of their space program as suggested by statements already referenced. No detailed plans or timetable for this process has been made public by the Soviet Union. Presumably recent reversals in manned flights connected with the fatalities aboard *Soiuz* 1 and 11 have delayed such plans as exist. Moreover, several missions will probably be accomplished before men are sent to the planets and new techniques developed. Future progress will hinge on the level of commitment to and the successes and failures encountered in the course of these developments. Prerequisites to manned planetary flight would probably include the following:

1. Assembly by docking and/or mechanical assembly of deep-space vehicles in earth orbit.
2. Manned circumflight of the moon.
3. Manned landings and exploration on the moon.
4. More soft-landings and orbitings of the near planets (Mars and Venus).
5. Landing of roving vehicles on the planets similar to *Lunakhods* 1 and 2.
6. Manned flybys and/or orbitings of the near planets.[14]

Although the Soviet interest in the planets may not be urgent, it appears nonetheless to be genuine, so that one may presume the early phases of this program are under development and that planning is taking place concerning the later stages. The first stage listed in this hypothetical program should be conducted soon. In fact, a good deal of the preliminary work and experiments have already been done. The *Soiuz* and *Saliut* craft have been used to carry out orbital docking as well as to test space welding for assembly of vehicles in orbit. Soviet work in bioastronautics has proceeded very rapidly to discover the problems faced by men in lengthy space flights, providing techniques of life support for extended periods and ways of protecting men from the hazardous effects of radiation and long-term weightlessness.[15] A closed-cycle life-support system of the type to be used for manned voyages to the planets was displayed at the space museum in Moscow in 1970.[16]

Construction of an orbital launch facility has been a consistent prediction of well-placed figures in Soviet space research. Soviet cosmonauts and space leaders such as Academicians Keldysh, Pokrovskii, Boris Petrov, cosmonaut Feoktistov, Party Secretary Brezhnev, Ari Shternfeld and others have since at least 1964 linked Soviet flights to the moon to construction of launch facilities in earth orbit.[17] Academician Blagonravov has also predicted the launch of planetary flights from bases on the moon.[18]

Such manned flights to deep space of necessity involve technological developments beyond the Apollo-Saturn V, which represents the operational limit of current space technology. Specifically manned and advanced unmanned flights to the planets will be preceded by the development and testing of reusable shuttle boosters, electric plasma rockets and nuclear propulsion.

Shuttle vehicles such as that proposed as NASA's major technological goal after the completion of the Skylab missions offer significant advantages over current booster technology in cases of multiple payloads and manned tasks as well as for the orbiting of heavy payloads which may involve assembly in earth orbit. Earth-orbital assembly is generally required in turn for the creation of large space stations and manned flights to the planets.[19] In light of these technological conditions, it is noteworthy that Soviet spokesmen have for a long time expressed an interest in and intention of constructing reusable shuttle and boost-glide space vehicles.[20] This interest and intention continue to be expressed and are often linked with plans for flights to the planets.[21]

Evidence of Soviet interest in nuclear and/or plasma propulsion goes back to Colonel Penkovsky's report of the early 1960s that a Soviet marshal was killed by the explosion of a nuclear rocket during a test.[22] Although Soviet research and development in this area have since come under a blanket of strict security, recent years have seen several expressions of interest in nuclear and rocket propulsion as a key interplanetary travel by Soviet spokesmen.[23] The American NERVA Project aimed at developing an operational nuclear rocket has been canceled.[24]

Although the possibility of manned Soviet flights to the planets in the 1980s or 1990s has been hinted at, no hard date or timetable has been announced by Soviet space leaders. In fact, the tenor of Party Secretary Brezhnev's 1969 speeches on space indicated that for the immediate future the USSR would concentrate on operations in earth orbit which promised direct and immediate benefits to the national economy.[25] Academicians Keldysh, B. Petrov, Sedov and Blagonravov have all followed the Brezhnev "line" in emphasizing practical benefits of space exploration and a concentration on work in near-earth orbits.[26] Nonetheless, in contrast to the United States, which has dropped any commitment to manned space operations after Apollo-*Soiuz* and has no plans for future manned operations outside earth orbit,[27] the USSR has continued not only to express its interest in manned flight to and ultimate settling of the planets but also to view these goals as the logical end of current space research.[28]

For the immediate future, Soviet planetary exploration will continue to rely on unmanned vehicles. The large *Proton* vehicle has already been used for flights to Mars, having soft-landed the Mars 3 probe on the surface. A US landing on Mars has through repeated delays and cancellations been put off until the three planned Viking missions in 1975–1976 and 1979. By that time the Soviets are likely to have already completed most, if not all, of the work planned for Viking. American Mars probes to date have not used the maximum of existing booster technology and therefore have been smaller than those launched by the old Soviet *Vostok* and much smaller than the *Proton*-launched probes. This gap in size and achievement is, until Viking, likely to become greater as Soviet techniques are perfected and, possibly, as greater use is made of the *Proton* and even larger Soviet boosters.[29] In the absence of any American "competition," Britain's Bernard Lovell is convinced that the USSR will go on in the exploration and colonization of the planets, very likely making unilateral and permanent alterations of their environments.[30] Academician Boris Petrov predicted in late 1971 that the USSR would send soil returners and roving vehicles to the planets similar to those already employed by the Soviets on the moon. He did not, however, envision a manned landing within the next ten years.[31]

Both the Soviet Union and the United States have expressed an interest in sending spacecraft to Mercury.[32] The US accomplished this with a spacecraft also sent to Venus in 1973.[33]

FUTURE SPACE ACTIVITIES

Because the planetary programs of both the superpowers, especially the USSR, are generally aimed at the discovery of information and techniques which will only find practical application in the distant future, the previous section placed some emphasis on the future of the planetary programs of both countries. Future space missions are not, however, necessarily limited to investigation of the planets. This section, therefore, will examine existing trends pointing at future space activities not connected with planetary research and will generally speculate about the future space activities of both the USSR and the USA.

In terms of overall funding, available resources and degree of commitment comparisons of present and future levels are difficult to pin down. Since peaking in 1965 at around $5.5 billion, the NASA budget has fallen off to around $3.2 billion. With the end of the Vietnam War, the completion of the Apollo program and the formulation of new mission goals and the space shuttle, NASA's budget will probably show a modest increase for the first time in nearly a decade. However, interest in space in the United States has evidently decreased, as reflected by a falling-off in university enrollments in space-oriented curricula.[34] Recent years have also seen cutbacks in the various programs associated with the National Defense Education Act prompted by a perceived "science gap" in the wake of the Soviet Sputnik.

It has been claimed that while NASA budgets were decreasing in recent years, Soviet space spending has been increasing ever since the Sputnik.[35] It is impossible to substantiate these claims from published materials. In estimating expenditures, some indication can be seen in the number and type of space launchings. Although the USSR has in recent years launched more payloads to orbit, it has not yet mounted an extensive program of manned lunar research. Development of the much-touted Soviet Saturn V class superbooster has evidently been delayed, frozen or canceled, as witnessed by the lack of any successful test flights to date. NASA continues to maintain that the Soviets are near completion of work on such a superbooster.[36] It would also appear that there has been some delay in the *Soiuz-Saliut* missions. A *Soiuz*-type vehicle was tested in orbit in 1972 after the tragedy of *Soiuz* 11 in June, 1971. Two attempts at a *Soiuz* 12-*Saliut* 2 mission in 1973 failed to materialize, evidently owing to difficulties with the *Saliut* 2 station and with *Kosmos* 577, also evidently a *Saliut* station.[37]*

*Successful docking of a *Soiuz* vehicle with a *Saliut* station took place in May 1975 while this study was in proof.

Not only has there been no sudden spurt or rapid acceleration of Soviet space activities as with Apollo in 1961–1965, but the trend appears, if anything, to be the reverse. It is only in the area of military flights and flights to Mars that there has been any marked increase in Soviet activities extending through 1973.

Whatever the Soviet commitment to space spending, the relative position of the two superpowers in space is also a function of American commitment. In spite of signs of minor improvement in NASA's budgetary position indicated above, the long-term trend appears to favor the Soviet Union if past Soviet commitments are extended into the future. The US operates in space very much by fits and starts according to approval of major mission goals and Soviet "competition." Having won first prize in the "moon" olympics and possessing a wide technological advantage in terms of the Apollo-Saturn system, the US is not likely to be propelled by the same motives produced by the Sputnik embarrassment and the drama of the race to the moon. Recent recommendations for several new space vehicles by an American space task force were generally rejected, and the plans for the proposed space shuttle were scaled down.[38] No manned missions are contemplated past the Apollo-*Soiuz* hookup in 1975. This could be altered considerably if the space shuttle is developed as scheduled by 1980.

Although Skylab and previous programs, which have been canceled and funneled into Skylab, were probably intended in part to maintain American public interest in space after the drama of Apollo and before some new space goal could be sprung on the public, public and congressional reaction to Skylab and its successors has been tepid at best. NASA seems to feel that at this stage it would be impossible to marshal the support required for embarking on more advanced missions of the distant future such as manned flight to the planets or further manned activities on the moon.[39]

At present the key to the American future in space appears to be the space-shuttle vehicle. NASA has not identified advanced manned missions for the shuttle, evidently feeling that the Congress and the public will not now approve the shuttle if it is to be built primarily for missions which are too remote and ambitious. Instead it draws attention to the cost effectiveness of the shuttle for missions in existing operation and hints about the strategic importance of the vehicle. With both Skylab and the shuttle, NASA has tended to find strategic justifications because of the erosion of the broader support on which Apollo was built.[40] If NASA gets the shuttle, it will be possible to greatly reduce the cost of manned missions beyond Skylab, and Soviet-American cooperation may also offset some of the cost of more ambitious programs. Then, presumably, proposals for such missions would enjoy a greater prospect of congressional and public approval than they would now.[41] Of course, such a "game plan," if it exists, is based on getting the shuttle. The prospects for full funding of the shuttle now face determined congressional opposition and considerable public apathy.[42]

Even with the shuttle, NASA will find it difficult to get support for its programs. Some help may come from the influential, but currently hard-pressed, aerospace industries and trade unions; from the Air Force, which will use the shuttle for some of its own missions, and from the several geographic areas impacted by the space program. As with Apollo, spending for the shuttle itself will peak in the developmental rather than the operational stage. Actually *using* the shuttle for advanced space missions would not have the diverse and large payoffs and, presumably, supports that could be expected from *building* it. This seems to have been the case with the Saturn V, developed for the Apollo Project, which since Apollo has been used only to launch the Skylab station. There are no plans for the future use of this remarkable booster owing to its high cost and the hope of replacing it with the space shuttle.

It has been argued that the research, development and operations spending of civilian space activities provides a politically viable alternative to similar military spending.[43] Although some differences between the political viability of the two were indicated in Chapter 3, this is at least partly correct. One must point out, however, that, at least at the current time, there is no clearly discernible indication of forthcoming reductions in military spending by the two superpowers and, therefore, no discernible "slack" to be taken up.

The leaders of the USSR do not have to marshal the support of diverse interests, an institution like the American Congress, or the general public to the extent that American leaders must in order to secure approval of their policies. Thus, the future of the USSR in space is much more directly a function of the will of the highest leaders. Previous chapters have argued that the motives behind the Soviet space program, and the privileged and influential position enjoyed by the scientists who are supporters of planetary programs in both countries, favor to some degree a long-term and scientific orientation in Soviet space activities. In addition to this, soon the USSR will have for the first time a larger pool of scientific and technically trained manpower than the United States.[44] These people will provide political support and human resources for a continuing space effort. The "science gap" may then be genuine in contrast to the one declaimed after Sputnik, and it will grow if the relative proportion of technical graduates continues to favor the USSR. If the USSR does forge ahead of the US in space, this is likely to create support for American space activities, but foreseeable advances beyond the shuttle do not pose the element of immediate strategic threat as did the Sputnik. Indeed one must recall that even the reaction compounded of the strategic threat perceived in the case of Sputnik and the "strategic bluster" by Khrushchev did not produce a dramatic change in American space activities until a new administration took office.

Paradoxically, the Soviet Union, the land of the crash program and "campaigns" in every area from philosophy to fertilizer, appears in the "space race" to be playing the role of the tortoise to the American hare. The next few years will indicate the extent to which the tortoise is still plodding on while the hare, having won the laurels in a breathtaking sprint, loses interest in the race.[45]

In terms of current technical capabilities, the US clearly enjoys an advantage over the Soviets in the case of the Apollo-Saturn system. With the addition of the Skylab launched by Saturn V, the US has, in the absence of rapid Soviet advances, a much larger and more complex system than *Soiuz-Saliut* system now stalled by the *Soiuz* 11 tragedy and subsequent difficulties but likely to be employed further very soon. Despite its tremendous capabilities and flexibility,[46] the only manned use now planned for Saturn technology beyond Skylab is the 1975 Apollo-*Soiuz* mission with the Soviet Union. If NASA gets what it has asked for in funding for the shuttle, the vehicle will not be in operational use until the 1980s. That it will get the shuttle is still in doubt, although development is now moving ahead.

The most advanced Soviet space technology in the sense of putting into orbit the heavy payloads required for complex manned and planetary research are the heavy launch vehicle, *Proton*, and the secret superbooster or very heavy launch vehicle. *Proton* has been operational since 1965, and its recent use in *Zond*, *Luna*, Mars and *Saliut* missions indicates that it is likely to replace the original ICBM-based vehicle as the mainstay of Soviet manned and planetary missions.[47] Lack of evidence of any recent work on the very heavy launch vehicle in the Saturn V class is quite possibly a reflection of the Soviet intention to use a shuttle vehicle and/or construction of an orbital launch facility as the technological basis for advanced missions to the moon and planets.

As regards the moon, NASA currently has no plans for further manned missions in the development stage.* The Soviets have on numerous occa-

---

*There are plans for a lunar orbiter in polar orbit for 1979, two more lunar orbiters and two lunar rovers in the 1980s, a communications satellite on the far side of the moon and two advanced rover/retrievers in 1990 and 1991. See *Aviation Week and Space Technology, 100* (8), February 25, 1974, pp. 12-13.

sions expressed their intent to send men around the moon and afterward land on its surface. They identified the *Zond* circumlunar vehicle as capable of carrying men, so that one might have expected a manned circumlunar flight by *Zond* in 1971 or 1972.[48] That it did not occur is an indication that the USSR, like the United States, intends to conduct manned circumlunar flight with the same technology employed for lunar landing, thereby testing the associated equipment in deep space before the landing itself takes place. Because the Soviets have not moved dramatically ahead in the development of a Saturn V-type booster in the past few years, it is quite possible that they intend to mount manned missions to the moon on the basis of such post-Saturn technology as a reusable shuttle and/or orbital launch facility. Academy President Keldysh himself indicated in 1969 that orbital launch facilities make it possible to "avoid the building of gigantic carrier rockets."[49]

Use of such new technology would delay manned exploration of the moon by the Soviet Union until the late 1970s at the earliest. By waiting, however, they should be able to accomplish at least as much as Apollo, and to have in addition acquired experience with the equipment and techniques required for manned flights to the planets which are beyond the capability of the Apollo-Saturn-Skylab system. Moreover, they will be able to move directly to establishment of manned bases on the moon, which has been one of their goals from the start.[50]

Before the development of such advanced systems, the USSR has announced its intention to accomplish other lunar missions which are within the capabilities of existing Soviet technology. These are the creation of a moon-based automatic observatory of earth and of astronomical observatories on the moon, very likely on its far side, which is shielded from the light reflections and radio emissions of earth.[51] Although the US also possesses the capability for such missions, it has no current plans to carry them out. The US Explorer 49 space probe launched in 1973 did do radio astronomy from the far side of the moon while in lunar orbit.[52]

In the area of manned operations in space, the Soviet Union and the United States will be operating along similar lines in the immediate future. The creation of manned space stations has been a long-term Soviet goal.[53] The *Soiuz* spacecraft which first flew in 1967 is evidently the first operational facet of a program aimed at creating a permanent or semipermanent manned orbital station, which is now the main goal of Soviet astronautics in the current period. In 1971 the *Saliut* orbiting station was added and joined in orbit by *Soiuz* 10 and, later, by *Soiuz* 11.[54] A similar American program, Skylab, was planned to succeed Apollo in 1972.[55] It was delayed until May, 1973, and was carried out over a period of eight months ending in early

1974. Apollo will be briefly revived in 1975 when a Soviet *Soiuz* and American Apollo spacecraft will dock in earth orbit. Although the two countries have chosen similar missions, the Soviet manned program, insofar as it is connected with the eventual construction of orbiting space platforms, looks outward toward the planets, while the exclusive focus of the American manned program is directed inward to operations in earth orbit.[56] This could well change with the development of the space shuttle.

Military programs now account for the bulk of the launchings of vehicles to orbit by both space powers. Soviet military launchings, mostly surveillance satellites, have grown in number in recent years. Although US military missions have decreased in number in the same period, this is largely the result of improving the coverage and useful life of American military satellites. Unlike the NASA budget, which has fallen off since 1965, US spending for military space activities has increased in the same period. Moreover, as the Apollo program has wound down and general public interest in space declined, NASA has more and more leaned toward strategic justifications of its programs and the absorption of military goals into its mission planning. So, for example, Skylab has emerged as the successor to a series of programs such as MODS (Military Orbital Development System), MOL (Manned Orbiting Laboratory) and AAP (Apollo Applications Program), which were not funded past the preliminary stages. Both MODS and MOL were to be carried out by the US Air Force.[57] It seems likely that if Skylab included aspects of military interest, the similar Soviet program involving the *Soiuz* and *Saliut* space vehicles do as well.

As described previously, various Soviet space spokesmen have for years expressed an interest in the development of reusable shuttle vehicles on the order of the space shuttle now planned in the United States. It is probably true that in the current period the justifications for the development of a shuttle vehicle are largely strategic.[58] The cost of developing shuttle vehicles can only be compensated by their use in orbiting large numbers of heavy satellites which would otherwise involve the building and consumption of many nonreusable launchers. At the present time only military launches are made in numbers sufficient to promise large and early savings from the use of shuttle vehicles. Moreover, the shuttle will make it more likely that certain military missions can be carried out at a cost acceptable to political decisionmakers, enabling both the Soviet and American military to get approval for missions that would not otherwise be funded. Lovell expects that the shuttle will enable the US military to develop a system of satellite interception and attack.[59] The Air Force has planned and requested authorization for such a system for a long time, and the USSR has already developed one.

Paradoxically, further advances in arms control agreements are likely to increase funding and operational levels of military activities in space.[60] The more extensive the coverage of such agreements, the greater will be the felt need of both sides to develop unilateral means of policing them, such as surveillance satellites. Ground-based inspection faces determined Soviet opposition, and even if it could be agreed to, both sides would probably wish to supplement it with space-based inspection as a check and guarantee and even a necessity in the case of arms control in space itself. Moreover, recent progress toward limitation of strategic arms has been made at the same time that competition in conventional armaments has increased. Satellites are likely to play an increasing role in this field, especially in the area of naval surveillance.

Looking further into the future, no technological developments now foreseeable are likely to give either power a clear-cut strategic superiority or military "control" of space.[61] The Space Treaty and existing and future arms control agreements, as well as the unwillingness of either side to yield to the other a strategic advantage, combine to preclude such a development.

After lagging behind the United States for many years in conduct of space activities that find useful application to the domestic economy, the USSR has begun to take a much stronger interest in this area in recent years. Basically, applications activities include communications, observation and mapping of various phenomena on the surface of the earth and in its atmosphere, and navigation. Programs embodying such activities have been carried out since the early sixties by the United States and since the mid- and late sixties by the USSR. The US also now has plans for the use of satellites in the control of air traffic.

New directions in applications programs include the American Earth Resources Technology Satellite, which aids in the discovery, mapping and monitoring of earth resources.[62] The USSR has also announced its intention to carry out such work and to include it among the functions of manned space stations.[63] Soviet space spokesmen have also announced the intention to set up facilities for direct television broadcasts from space, which raises complicated questions in the areas of international law and sovereignty.[64] Applications activities involving manned space stations were also part of the American Skylab program.

Recent statements of Soviet spokesmen stress that earth applications will be the main goal of various manned and unmanned Soviet space activities in the immediate period.[65] Since most earth applications activities can be accomplished with existing technology and promise to return greater economic benefits than their costs of operation, they will unquestionably be continued and expanded by both space powers.

Among the long-term possibilities that could alter the scope and political significance of the application of space activities to the economy would be the discovery of extremely valuable resources in space. Although such an occurrence could have a profoundly disturbing impact on international space politics and international politics in general,[66] it appears highly unlikely in the foreseeable future. The cost of finding and returning such resources would be so high that it is hard to imagine what could be found that would make the effort of returning it worthwhile. Moreover, for some time to come, human technology will be based on materials readily available on earth, so that even if something exists "out there" of immense potential value to men, they are not likely to even be aware of the uses for it. Only when earthbound technology itself moves out into space, to the moon and the planets, is it likely that such discoveries will be made, if at all. Such possibilities appear to be beyond the scope of presently tenable expectations and guesses.

NOTES

[1]Bubnov and Kamanin, *Obitaemye Kosmicheskie Stanstii* as cited by Fermin J. Krieger, *The Space Programs of the Soviet Union*, Rand Corporation, Paper #P-3632, July 1967, pp. 16-17. See also Sir Bernard Lovell, "The Great Competition in Space," *Foreign Affairs*, 51 (1), October 1972, p. 138.

[2]See, for example, the statements of Feoktistov, Egorov and Keldysh (p. 359); Petrovich (p. 360); Petrovich, Nicolaevich and Pokrovskii (p. 361); Keldysh (p. 367); Sedov (p. 369); Shatalov (p. 372); Keldysh (p. 373); Blagonravov (p. 371); Sedov (p. 378); B. Petrov (p. 381); G. Petrov (p. 382); Martynov and Beregovoi (p. 383); Dmitrev and the "Deputy Chief Designer" (p. 384) in Sheldon, "Projections of Soviet Space Plans," in *Soviet Space Programs, 1966-1970, op. cit.*

[3]See B. C. Murray and M. E. Davies, *A Comparison of US and Soviet Efforts to Explore Mars*, Rand Corporation, Paper #P-3285, January 1966, pp. 3, 34; and Alton Frye, "Soviet Space Activities: A Decade of Pyrrhic Politics," in Lincoln P. Bloomfield, editor, *Outer Space: Prospects for Man and Society* (New York: Frederick A. Praeger, Inc., 1968), p. 181.

[4]Sheldon, "Summary" in *Soviet Space Programs, 1966-1970, op. cit.*, xxviii; Frye, *supra*, p. 193; Murray and Davies, *supra*, pp. 29-30. This probably is an overstatement.

[5]Sheldon, *Review of the Soviet Space Program, With Comparative US Data* (New York: McGraw-Hill Book Co., Inc., 1968), p. 29.

[6]See Charles S. Sheldon II, *United States and Soviet Progress in Space; Summary Data Through 1973 and a Forward Look*, Library of Congress, Congressional Research Service, January 8, 1974, pp. 42, 43, 56-59. The number of launchings in this case is probably a fairly reliable indication of costs. Using it one may estimate the total expenditures of the USSR to reach the planets as roughly two to three times that of the US.

[7]William Shelton, *Soviet Space Exploration* (New York: Washington Square Press, Inc., 1968), p. 217.

[8]Davies and Murray, *op. cit.*, p. 2.

[9]See Sheldon, "Program Details of Unmanned Flights," in *Soviet Space Programs, 1966-1970, op. cit.,* pp. 163-165, 212-214.

[10]*Ibid.,* p. 218; and Sheldon, *United States and Soviet Progress in Space* (1973), *op. cit.,* p. 47.

[11]See Sheldon, *United States and Soviet Progress in Space* (1973) *op. cit.,* pp. 56-59; and Frye, "Soviet Space Activities: A Decade of Pyrrhic Politics," *op. cit.,* p. 185.

[12]Sheldon, *supra,* pp. 49, 57-58.

[13]See B. C. Murray, M. E. Davies and P. K. Eckman, *Planetary Contamination II: Soviet and US Practices and Politics,* Rand Corporation, Paper #P-3517, March 1967, pp. 1-2.

[14]A not dissimilar program was described by the "Chief Designer" Korolev in 1961. See *Izvestiia,* December 31, 1961, p. 3. Also see Sergei Gouschev and Mikhail Vassiliev, editors, *Russian Science in the 21st Century* (New York: McGraw-Hill Book Co., Inc., 1960), pp. 208-209, 214; *USSR Probes Space* (Novosti Press Agency, Moscow), no date, no pagination; and Boris Petrov cited in *Christian Science Monitor,* June 9, 1971, p. 13.

[15]See James M. McCullough, "Soviet Bioastronautics: Biological, Behavioral and Medical Problems," in *Soviet Space Programs, 1966-1970, op. cit.,* pp. 265-295; Sheldon, "Projections of Soviet Space Plans," in the same volume, p. 392; and Shelton, *Soviet Space Exploration, op. cit.,* pp. 258-259.

[16]*Aviatsiia i Kosmonavtika,* no. 12, 1970, pp. 33-34 as cited in *Soviet Space Programs, 1971, A Supplement to the Corresponding Report Covering the Period 1966-1970,* US Senate, Staff Report Prepared for the Use of the Committee on Aeronautics and Space Sciences, Library of Congress, Congressional Research Service, 92nd Congress, 2d session, April 1972 (hereinafter referred to as *Soviet Space Programs, 1971*), p. 53.

[17]See the statements of Keldysh (pp. 359, 373, 378, 378-379, 381); Pokrovskii (p. 361); B. Petrov (p. 372); Feoktistov (pp. 359, 365); Brezhnev (p. 378); Shternfeld (p. 472); cosmonauts Egorov, Nikolaev and Popovich (pp. 359, 361, 367); "Scientific Commentator," (p. 361); "unnamed writer" (p. 369); scientist Chembrovskii (p. 380); and Professor Dmitrev (p. 384) in Sheldon, "Projections of Soviet Space Plans," *op. cit.*

[18]*Ibid.,* p. 367.

[19]See Lovell, "The Great Competition in Space," *op. cit.,* p. 136. For cost savings promised by shuttle vehicles, see "Go for the Space Shuttle," *Newsweek,* 80 (6), August 7, 1972, p. 43; and Sheldon, *United States and Soviet Progress in Space* (1973), *op. cit.,* p. 70, who says the currently planned US shuttle vehicle would reduce the cost of putting one kilogram of mass in orbit from $1,500 to $330.

[20]See V. Aleksandrov, "Rocket-plane, Aircraft of the Future," and G. I. Pokrovskii, "Intercontinental Ballistic Missiles and Aviation," in *Soviet Writings on Earth Satellites and Space Travel* (New York, Citadel Press, 1958), pp. 196, 199.

[21]See the comments on shuttle vehicles by Petrovich (pseudonym) (p. 360); cosmonaut Titov (pp. 361, 362); cosmonaut Popovich (p. 367); scientist Shternfeld (p. 371); and engineers Andanov and Maksimov (p. 377) in Sheldon, "Projections of Soviet Space Plans," *op. cit.* See also Sheldon's prediction for a Soviet shuttle in *ibid.,* p. 390, and his "Summary" of *Soviet Space Programs, 1966-1970, op. cit.,* xxxv. Also see the comments by cosmonaut Beregovoi (p. 55), Academy President Keldysh (p. 56) and correspondent Denisov (p. 57) in *Soviet Space Programs, 1971, op. cit.*

[22]See Shelton, *Soviet Space Exploration, op. cit.,* pp. 260-261.

[23]See the comments of Petrovich (pseudonym), (pp. 360, 373); Sedov and Feok-

tistov (pp. 360-361, 365); and Eliseev (p. 380) in Sheldon, "Projections of Soviet Space Plans," *op. cit.*; and Sheldon's comments in "Summary," *op. cit.*, xxxv.

[24]Sheldon, *United States and Soviet Progress in Space* (1973), *op. cit.*, pp. 22, 69. On page 22 of Sheldon's earlier report of the same title covering 1972, the text read correctly that American nuclear rocket development "has *now* been canceled." In the work referenced here this is misprinted as "has *not* been canceled" (emphasis added).

[25]See Joseph G. Whelan, "Political Goals and Purposes of the USSR in Space," *Soviet Space Programs, 1966-1970, op. cit.*, pp. 35-37; and Sheldon, "Projections of the Soviet Space Plans," *op. cit.*, p. 378.

[26]Whelan, *supra*, pp. 36-37; and Sheldon, *supra*, pp. 378, 379, 380, 381.

[27]Lovell, "The Great Competition in Space," *op. cit.*, pp. 135, 138; and Erlend A. Kennan and Edmund H. Harvey, *Mission to the Moon, A Critical Examination of NASA and the Space Program* (New York: William Morrow and Co., 1969), p. 261.

[28]Soviet flights to and colonization of the moon have been predicted by Keldysh (pp. 367, 368); Sedov (pp. 369, 378); B. Petrov (pp. 381-382); General Kamanin (p. 364); Feoktistov (p. 366); Academician G. Petrov (p. 366); engineer Pokorni (p. 368); cosmonaut Leonov (p. 382); and Professor Victorov (p. 366) as cited in Sheldon, "Projections of the Soviet Space Plans," *op. cit.*

[29]See Murray and Davies, *op. cit.*, pp. 2-3, 13-14, 15-17, 29-30, 34-35; Lovell, "The Great Competition in Space," *op. cit.*, p. 135; and Sheldon, *supra*, pp. 355, 388.

[30]Lovell, *supra*, p. 138.

[31]Cited in *Soviet Space Programs, 1971, op. cit.*, p. 57. A similar prediction about sending roving vehicles to Mars was made by Academician Blagonravov in *Pravda*, December 23, 1971, p. 3.

[32]See the remarks of engineer Vasilev in *Soviet Space Programs, 1971, supra*, p. 57.

[33]Sheldon, *US and Soviet Progress in Space* (1973), *op. cit.*, p. 67.

[34]B. W. Augenstein, *Policy Analysis in the National Space Program*, Rand Corporation, Paper #P-4137, July 1969, pp. 5-6.

[35]Mose L. Harvey, "The Lunar Landing and the US Soviet Equation," in Eugene Rabinowitch and Richard S. Lewis, editors, *Man on the Moon—The Impact of Science, Technology and International Cooperation* (New York: Basic Books, Inc., 1969), pp. 72-73.

[36]Sheldon, *US and Soviet Progress in Space* (1973), *op. cit.*, p. 23.

[37]*Ibid.*, pp. 9-10.

[38]*Ibid.*, pp. 68-71.

[39]*Ibid.*, p. 69. See also Kennan and Harvey, *op. cit.*, p. 261.

[40]Kennan and Harvey, *supra*, p. 310; Robert Salkeld, *War and Space* (Englewood Cliffs, N.J.: Prentice-Hall, Inc., 1970), p. 104; Lovell, "The Great Competition in Space," *op. cit.*, p. 136.

[41]James C. Fletcher, current NASA administrator, has indicated that all advanced missions past Skylab hinge on the space shuttle. See *Christian Science Monitor*, June 9, 1971, p. 13.

[42]Sheldon, *US and Soviet Progress in Space* (1973), *op. cit.*, pp. 68-69.

[43]See Lillian Levy in Levy, editor, *Space: Its Impact on Man and Society* (New York: W. W. Norton and Co., Inc., 1965), p. 210; John M. Logsdon, *The Decision To Go to the Moon: Project Apollo and the National Interest* (Cambridge, Mass.: MIT Press, 1970), p. 169; Christopher Wright, "The United Nations and Outer Space," in Hugh Odishaw, editor, *The Challenges of Space* (Chicago: University of Chicago Press, 1963), p. 289; and John K. Galbraith, *The New Industrial State* (Boston: Houghton Mifflin Co., 1967), p. 342.

[44]See Sheldon, "Projections of Soviet Space Plans," *op. cit.*, p. 354.

[45]Mose L. Harvey, *op. cit.*, thinks it "almost certain" that the USSR will soon register "firsts" with a manned space station, truly maneuverable spacecraft and reusable vehicles, p. 72.

[46]Augenstein, *op. cit.*, pp. 87-89; and Salkeld, *op. cit.*, p. 168.

[47]Sheldon, "Projections of Soviet Space Plans," *op. cit.*, p. 355.

[48]See the comments of B. Petrov (p. 376) and Sheldon (p. 390) in *ibid.*

[49]*Ibid.*, p. 373.

[50]See Varvarov, president of Astronomical Institute, quoted in Gouschev and Vassiliev, *op. cit.*, p. 214.

[51]See the comments of B. Petrov, Andreanov and Kondratev in Sheldon, "Projections of Soviet Space Plans,"*op. cit.*, pp. 383, 384.

[52]Sheldon, *US and Soviet Progress in Space* (1973), *op. cit.*, p. 54.

[53]See Bubnov and Kamanin, *Obitaemye Kosmicheskie Stantsii* as cited in Krieger, *The Space Programs of the Soviet Union*, *op. cit.*, pp. 16-17.

[54]See Sir Bernard Lovell, "Russian Plan for Space," *Survey 52* (52), July 1964, p. 82; Sheldon, "Program Details of Manned Flights," "Projections of Soviet Space Plans," and "Postscript," in *Soviet Space Programs, 1966-1970, op. cit.*, pp. 253, 372-384, 390, 516-519, 527-529.

[55]See Salkeld, *op. cit.*, p. 98.

[56]See Lovell, "The Great Competition in Space," *op. cit.*, pp. 135, 138; and *Christian Science Monitor*, "Soviets To Follow Own Path in Space," June 9, 1971, p. 13.

[57]See Salkeld, *op. cit.*, pp. 149-150; and Kennan and Harvey, *op. cit.*, p. 310.

[58]Salkeld, *supra*, p. 104; and Lovell, "The Great Competition in Space," *op. cit.*, p. 136.

[59]Lovell, *loc. cit.*

[60]See Augenstein, *op. cit.*, p. 78.

[61]Karl W. Deutsch, "Outer Space and International Politics," in Joseph M. Goldsen, editor, *Outer Space in World Politics* (New York: Frederick A. Praeger, 1963), p. 146.

[62]"New Eye in Space," *Newsweek, 80* (6), August 7, 1972, p. 43.

[63]Barbara M. DeVoe, "Soviet Application of Space to the Economy," in *Soviet Space Programs, 1966-1970, op. cit.*, pp. 315-321.

[64]*Ibid.*, pp. 318-319; and the remarks of Boris Petrov in Sheldon, "Projections of Soviet Space Plans," *op. cit.*, p. 381.

[65]See Keldysh (pp. 372, 373); Feoktistov (p. 379); and Volkov (p. 381) as quoted in Sheldon, *supra*. Also see the comments of G. Petrov (p. 54); B. Petrov (pp. 55, 58); Blagonravov (p. 55); and Fiodor Chukhrov (p. 55) in *Soviet Space Programs, 1971, op. cit.*

[66]See Sidney Hyman, "Man on the Moon. The Columbian Dilemma," in Rabinowitch and Lewis, *op. cit.*, p. 50.

8 International Cooperation in the Exploration and Use of Outer Space

Perhaps the one factor which more than any other establishes a great *potential* for space cooperation is that only two nations, the US and the USSR, have developed extensive programs of space technology and science. An indication of this is seen in the number of payloads successfully launched to earth orbit and beyond by the world's nation states. Through 1973 the figures are: USA, 831; USSR, 900; France, 8; Japan, 4; Italy, 2; China, 2; United Kingdom, 1; Australia, 1. The International European Launcher Development Organization has recorded four failures in its attempts to launch satellites.[1]

To the extent that the rest of the world's countries are interested in space, they have a great deal to gain by cooperation or association with one or both of the two space powers. This would be the case if the world's countries wished to take part themselves in space activities or merely to reap the benefits of such activities as carried on by the space powers.

NASA's director of international programs, Arnold Frutkin, argues that third countries are interested in space because space activity is "an effective stimulus" to more general and diverse interest, study and work in technological fields of direct significance to economic development.[2] Moreover, earth applications of space technology such as those in communications, meteorology, navigation and earth resources, which are made possible by the achievements of the two space powers, can in many cases be employed to the benefit of third countries and to the profit of individuals and organizations in them.[3]

Certain aspects of space and the scientific research devoted to it make this field especially well suited to international and cooperative efforts.

Science itself is international; no nation has a monopoly on it. Hugh Odishaw, chairman of the Space Science Board of the National Academy of Sciences, contends that "there has been international cooperation in science even when nations have been unable to cooperate on anything else."[4] Moreover, cooperation, when deliberately undertaken, promises to accelerate the development of space science.[5]

Generally speaking, the work of scientists is carried on internationally. Scientific method is applied to natural rather than national phenomena* and requires that scientific work be carried out in the open to permit replicability and testing by many scientists. Scientists themselves meet through publication of their work and through numerous international conferences and personal contacts. The space age was born in and as a part of the International Geophysical Year, a coordinated effort by the world's scientists through the International Council of Scientific Unions and the Special Committee of the IGY (CSAGI).

Scientists are aided by learning about the work of their colleagues and by discussing their own work with them. They are usually eager to do both. The control of science and scientists in the USSR is, according to one British scientist, "without parallel" in the US or Great Britain.[6] The Soviet Government is reluctant to disseminate details of *technologies* applied in scientific work even among Soviet scientists. Nonetheless, Soviet scientists are still eager to discuss their own work and that of their colleagues abroad, and they are "sincere" in doing so.[7]

Exchange of information among scientists is *one level* of international cooperation in space. It does not as a rule involve national governments except negatively insofar as they may choose to prohibit or control the work, travel and openness of their national scientific community. The IGY was a collaborative effort among scientists rather than governments in which governments were involved only in that their positive support was solicited by *national* groups of scientists to facilitate work agreed to in the international context. Agreements *between* governments played no significant role.[8]

The international activities of scientists do form a basis for and an impetus toward other kinds of international cooperation, including cooperation directly involving political authorities. Science cannot, however, serve effectively as a substitute for political cooperation or by itself overcome the political barriers to cooperation. Scientists are as likely as other citizens to be loyal to their national governments and as subject to the authority and

---

*The distinction does not hold in every case but is most evident of the *natural* sciences.

policies of their governments which may impede international cooperation.[9]

Space exploration has characteristics which make it the subject of *international* interest and concern. Orbital space flights cross over the frontiers of many, and often all, countries. They can be observed from all countries and can serve as a means of observing all countries with or without their political consent. Like the high seas, space is of interest to various governments in that useful activities can and are being carried on in it, but it has not yet been appropriated to the exclusive use or ownership of a single country or group of countries. Today, and to a potentially much greater extent in the future, activities carried on in space may result in significant benefits and/or deprivations to the people of each and every nation in the world.

Space is a "new threshold" of mankind which, like the discovery of atomic energy, can be the potential source of great benefit or great disaster to mankind. As such, there is a special urgency attached to international cooperation and international regulation of space activities. Near space is a limited resource which may require international cooperation and regulation to be used to maximum benefit.[10] At the same time deep space is so vast that it can appear to dwarf the "petty" differences of nations and blocs of nations.[11] The perspective of earth when viewed from space is one in which the "boundaries of the system" are the surface of the earth and its atmosphere, not the political boundaries which seem so important from an earthbound perspective. The ecological perspective of the world system is more compelling when one knows that men have looked at earth from deep space and when one has seen representations of what they saw. This breakdown of the national perspective can lead to greater international cooperation.

The space powers themselves have compelling reasons to foster space cooperation with other countries. Pride and prestige have already been described as important motivations for their space programs. International cooperation has a certain prestige and propaganda value in itself. NASA's Arnold Frutkin claims that comparisons of Soviet space cooperation with the broader efforts of American cooperation are "inescapable" and boasts that in such international forums as COSPAR (the Committee on Space Research of the International Council of Scientific Unions), the United States has been able to call attention to its cooperative programs while the USSR has not. This is part of what Frutkin, as NASA's chief officer for international programs, has referred to as part of "the struggle for minds" aimed at a "scientific, technical, and official elite."[12] Former Administrator Webb spoke of the "effort to project the *image* of the US as a nation wanting to work with other nations . . . a nation leading in this field and willing to share this knowledge with other nations."[13] From its earliest days NASA wished that the world would be in the same relation to the American space program as the American people, sharing in its successes and failures. Cooperation itself was one means of recouping prestige lost to the Soviets through Sputnik and Gagarin.[14] Political considerations also entered into US policy on space cooperation when locations were sought for tracking stations at the same general location as the existing one in Johannesburg, which is considered dispensible in light of the international stigma attached to the white regime in South Africa.[15] US tracking stations for manned flight and scientific satellites are now operating in Nigeria and the Malagasy Republic.[16]

Space cooperation can serve other policy goals as well. Frutkin says that it builds support at the UN for US positions and has "lent credibility to our posture" which contrasts sharply "with our competitor's performance."[17] It is further suggested that space cooperation can help solidify America's "tenuous alliances" and "nudge the Soviet Union into a more forthcoming posture in the world."[18] Soviet spokesmen have not been so forthright in advancing these political goals for their cooperative space programs. As detailed below, they have not mounted such extensive cooperative projects as the United States. American initiatives therefore give them an incentive to be more cooperative and to attempt to discredit the cooperative efforts of the United States.

In addition to these political goals, both superpowers are also impelled by more mundane, instrumental goals in seeking to cooperate with other countries in various aspects of space exploration. For example, early in the American program it was hoped that foreign skills could help the US to overcome the Soviet space lead. Access to foreign scientific manpower is still a benefit conferred by international cooperation.[19] The United States in particular favors cooperation with Western Europe, which Mr. Frutkin sees as "far stronger scientifically and technologically than the USSR."[20] Tapping foreign manpower also appears to be one motive for Soviet space cooperation with other countries.[21]

Space cooperation holds out the promise of saving money, and this is one advantage that the highest Soviet spokesmen have given more attention to in their public remarks than their American counterparts, perhaps because one function of American space programs is to find politically acceptable and rewarding ways of spending money, not saving it.[22] American financial goals in cooperative space programs are by comparison aimed more at strengthening markets for American products, ensuring "a fair return to the American people," improving the balance of trade[23] and securing a rich new market for the American aerospace industry.[24]

Certain space activities require facilities which can be provided only through international cooperation. So, for example, both space powers have reached agreements with other countries establishing ground-based tracking stations and communications relays at longitudes and latitudes outside their own frontiers.[25] To the extent that the Soviet Union has been unwilling and/or unable to conclude such agreements with as many of these countries as has the United States, it has had to fall back on a relatively expensive system of ship-borne tracking stations and communications relays. This tracking fleet is composed of 16 ships operated by the Academy of Sciences and Strategic Rocket Forces ranging in size from 5,000 to 45,000 metric tons displacement.[26]

Both space powers have also claimed that they support space cooperation because it encourages international cooperation in general and thereby promotes the cause of peace. Those in the United States who envisioned the *whole* of space exploration being carried on through an international effort expected that this would be the way to prevent an arms race in space.[27] Others saw space cooperation as an important area for installing cooperative habits—the "functional" route to peace.[28] Still others hoped that cooperative space exploration would provide a nonlethal way of working out human energies and frustrations.[29] President Eisenhower evidently shared the view already described that, as a new threshold of human activity, space represented both a new opportunity and a new critical danger that required international cooperation and control. Public Soviet pronouncements have also taken this view. Academician Sedov has claimed that space cooperation between scientists can safeguard peace.[30] Space research, says Sedov, promotes international cooperation, which in turn "has a positive influence on relations between states."[31]

Just as many factors promote space cooperation, others serve as barriers and limits to it. The very fact, as mentioned before, that two countries are far in front of all the others in the exploration of space, while giving other countries incentives to cooperate, limits the competence of third countries to contribute to and employ the space technology of the superpowers.[32] The military hardware employed in space research and the strategic significance of some of the nonmilitary technologies involved inhibit space cooperation both because the space powers want to keep these details secret and because third countries often do not want to involve themselves in the military or quasimilitary activities of either of the superpowers. Although both space powers have been anxious not to share their secrets, the Soviet Union is particularly marked by attitudes of xenophobia and compulsive secrecy.[33] It has been mentioned above that these characteristics limit the contributions of individual Soviet scientists to space cooperation. So far as space competition prevails between the USA and the USSR, both space powers are sure to be opposed to such cooperative space activities as will make them reliant on the capabilities of others to maintain a competitive posture. Cooperation is one form of competing, but one which is not permitted to get in the way of other forms. The complex sequence of events which forms any major space mission can be indefinitely complicated and delayed by a breakdown at thousands of places in the sequence. Neither space power is likely to rely on its friends or enemies in order to meet the schedules of such complex missions as manned and planetary exploration of space.[34] In this regard the cooperative Apollo-*Soiuz* mission planned for 1975 is an unprecedented breakthrough in space cooperation.

The greatest barriers to space cooperation are unquestionably political differences between states and the various factors which contribute to them. Soviet spokesmen have identified "international tensions" as barriers to space cooperation,[35] and the record of space cooperation generally indicates that *ceteris paribus*, cooperation is most likely between those nations with the fewest basic policy disagreements, i.e., among members of existing alliances, states with similar social systems, etc. To these policy differences must be added factors which make for *preconceived* hostilities between states and in particular the Communists' definition of capitalism and capitalist states as the inevitable enemies of socialist states.

Both space powers have publicly expressed their commitments to the principle of international cooperation in space, the only difference being not the degree of professed commitment but that the United States has been more frank and explicit in describing its motives. United States cooperation is mandated in the Space Act of 1958.[36] One of the immediate deputies of NASA's administrator is the assistant administrator for international programs, the post held by Arnold Frutkin. The expansion of the American space program in 1961–1962 was in part the result of the belief that "space . . . offers exciting prospects for international cooperation."[37] A representative of the State Department has more recently listed encouraging international space cooperation as the first on a list of "six major policies . . . which interact . . . to shape our space policy."[38]

Similarly, the Soviet Union has for years maintained an "Intercosmos" Council, or Council for International Cooperation in the Exploration of Outer Space, under the Academy of Sciences and currently chaired by Academician Boris N. Petrov.[39] Like the USA, the USSR takes part in the activities of UN agencies interested in space, makes bilateral agreements with other states affecting space activities, and sends its scientists to nongovernmental world organizations, meetings and conferences connected with space research. Party Secretary Brezhnev has contended on various occasions that the USSR supports international cooperation in space and that space should be the "arena . . . of international cooperation . . . and not . . . hostile clashes." Similar sentiments have been expressed by Soviet space leaders such as Academicians Keldysh, Blagonravov and Sedov and by such governmental spokesmen as D. M. Gvishiani, vice-chairman of the State Committee on Science and Technology. Space cooperation is claimed to be one aspect of the Soviet commitment to "peaceful coexistence" and devotion to peace and humanity.[40]

Cooperative Programs with Various Countries: USA

The United States has gone much further than the Soviet Union in establishing cooperative programs with other countries. This section will not detail the cooperative programs of the USA but will review the main features of American international space programs as a basis for comparison with those of the USSR.

From the inception of the space age in the IGY, the US adopted a policy giving out "advance details" of experiments, "extensive information" of satellite instrumentation and design, and other information, while the Soviet Union provided "virtually no advance information" and released only routine information after its space experiments had been carried out.[41]

In the deliberation of the National Space Act of 1958 which created NASA, the bulk and variety of witnesses supported an active stance of space cooperation.[42] The creation of NASA as a civilian space agency was intended to provide the greatest possibilities for international cooperation.[43]

Both inside and outside NASA exists an organizational framework to facilitate international cooperation in space. NASA maintains an Office of International Programs headed by an assistant administrator, and the Department of State has Offices for International Scientific Affairs and International Organization Affairs. NASA and the State Department maintain close liaison with the semigovernmental National Academy of Sciences and its Space Science Board.[44] The Space Science Board advises NASA on scientific aspects of its programs, and NASA advises the Academy concerning its participation in the space activities of the international scientific community.[45] The Academy serves as a channel of communication and implementation for policies of international space cooperation adopted by NASA. Completely outside the governmental sphere are agencies like the American Astronautical Society, which participates with other American voluntary groups in the activities of the International Astronautics Federation (IAF), founded in 1951.[46] NASA propagandizes its own activities at home and abroad and is greatly aided in doing so by the United State Information Agency (USIA).[47] Although all US organizations involved with the various aspects of space work take a more active approach to space cooperation than their Soviet counterparts, the governmental agencies of NASA, the State Department and the USIA are the most moved by political considerations and the least "innovative" and open with respect to international cooperation. Nongovernmental agencies such as those affiliated with IAF are least likely to see the necessity of hedging cooperation about with restrictions and requirements for quid pro quos. The semiofficial National Academy of Sciences (NAS) and its Space Science Board fall somewhere in between on this continuum.[48] When the organizational structure of the IGY was superceded in 1959 by the Committee on Space Research (COSPAR) of the International Council of Scientific Unions, NASA, in conjunction with NAS, decided to extend to COSPAR an offer whereby NASA would launch "worthy experiments proposed by scientists of other countries." British, Canadian and, later, other scientists took advantage of this offer.[49]

Although there has been a reorientation of NASA's policies toward the USSR since Apollo, cooperative efforts were originally aimed at the "friends" of the United States rather than at its "principal adversary." To do otherwise, Arnold Frutkin contended, would be "pernicious." This seems a clear indication of the extent to which political considerations shaped NASA's cooperative policies. Frutkin even considered that "to neglect our friends" until we become friendly with the Soviet Union through space cooperation was the *reductio ad absurdum* (?!) which revealed the perniciousness of a cooperative strategy aimed at the Soviet Union.[50] It would seem that Frutkin must have believed that to concentrate on Soviet-American cooperation was immoral neglect of "our friends" or that it would be more feasible to use space cooperation to cement friendships than to end emnities. In fact, his subsequent remarks suggest that he had in mind a kind of "position of strength" policy whereby the success of NASA's cooperative program with nations other than the Soviet Union would somehow *force* ("nudge" is the word he uses) the Soviets into a more obliging stance.

Before embarking on its cooperative activities NASA established a list of criteria, in effect requirements, for cooperative programs. It was evidently assumed that the initiatives for cooperation would usually come from abroad rather than from NASA itself and would be the result of the general statement made in COSPAR and elsewhere of NASA's receptiveness to cooperative programs. Such programs would require:

(1) literal cooperation without the passing of dollars,

(2) solid rather than token program content,

(3) a project-by-project procedure,

(4) negotiation with central and authorized civil agencies on a direct

technical, rather than indirect diplomatic, basis,

(5) encouragement of foreign scientific interests, rather than imposition of domestic concepts, and

(6) stress upon the less sensitive, purely experimental areas, at least to begin with.[51]

In addition, cooperative projects were to be of *mutual* scientific interest, and scientific results were to be generally published.[52] These requirements determined the general direction of NASA's cooperative policies. First, they established definite limits on the extent and kinds of cooperation, limiting it largely to scientific cooperation, excluding technology and economic applications. Second, the cooperating country required a high minimal competence in science and technology, so that most of NASA's cooperation would be with the advanced countries rather than the underdeveloped ones. Last, these requirements limited most cooperation to activities carried on *bilaterally* rather than multilaterally.[53]

Within these requirements NASA has strived to keep its cooperative programs free of any military taint and has shied away from projects which might have dramatic political effects but would be relatively meaningless scientifically. So, for example, NASA turned down a hastily drawn Egyptian proposal to launch a sounding rocket, which was evidently timed to accomplish this feat ahead of a similar project mounted by Israel.

NASA has authority to classify anything of military significance, and it has applied this authority to technological rather than scientific information. NASA was willing to share technological information extensively only when it hoped thereby to capture influence in the multilateral European Launcher Development Organization (ELDO) and the European Satellite Research Organization (ESRO). Testimony of NASA officials before Congress on agreements involving space cooperation inevitably reveals the concern of legislators lest American technological secrets get into the wrong hands or be given away without compensation.[54]

NASA requirements for quid pro quos in space cooperation probably look suspicious to the Soviet Union and may account for a certain sincerity in Soviet charges that American cooperation aims at control of foreign science and/or the extraction of large profits for aerospace firms. This reinforced the impact of NASA's early intent to seek cooperation with nations other than the USSR.[55]

NASA space facilities are generally open to the public, and NASA is anxious to show foreign visitors through them. According to Kash, "everyone recognizes" that these gestures are in part attempts to point out the openness and cooperativeness of the American space program as compared with Soviet secrecy. Although Soviet cosmonauts have recently visited the Johnson Spacecraft Center in Houston and American astronauts have visited "Star Town," the cosmonauts training center near Moscow, all invitations by NASA to Soviet representatives for visits to American launch sites have been turned down, apparently because the Soviets are unwilling to reciprocate. Representatives from other socialist states, which have no launch facilities of their own, have made visits to Cape Kennedy. The Soviet Military Attache has visited Cape Kennedy. Since the Military Attache may be considered a spy, the USSR may have felt it had no obligation to reciprocate.[56]

Cooperative American programs can generally be divided into four categories: information exchange, personnel exchange, cooperative projects and operations support.[57] The last category includes the array of 29 stations throughout the non-Communist world, some manned by the United States, some by foreign nationals and some jointly. Each is operated on the basis of a bilateral agreement between the United States and the 19 other political entities in which the stations are located.[58] In contrast with other cooperative programs, the stations do not involve equal mutuality of benefit in that they are a necessity for NASA and a convenience for the other political entities. Moreover, the initiative for their construction or use in the American space program has generally come from NASA itself. Maintaining the image of a civilian and cooperative space program was important to NASA in securing some of these stations.[59] Of all of NASA's international programs, operations support has the highest priority. Former Administrator Webb claimed that the system of stations gives the United States a great advantage over the Soviet Union and is closely related to American security policies, indicating that despite the civilian image NASA seeks to project, the possible significance of the stations is not completely innocent.[60]

Other international programs of NASA have included:

(1) the launching of satellites prepared by other nations,
(2) the inclusion of foreign experiments in large US satellites,
(3) joint sounding rocket projects,
(4) ground-based activities in connection with orbiting experiments in meteorology, ionospheric research, earth resources and other fields,
(5) joint research with and training of foreign personnel, and
(6) limited commercial exchange of technology.[61]

In the 1960s NASA concluded 250 international agreements with 34 countries and ESRO involving an expenditure of around $500 million, about a third of which was spent by NASA. NASA worked with scientists from 70 countries, largely in the form of personnel training and exchange.

As of July, 1971, NASA had launched 14 different satellites for foreign countries, agreed to launch 11 more at its own expense and taken part in over 600 cooperative launchings of sounding rockets.[62] By 1972 NASA had cooperated in one form or another with 89 countries, including ESRO.[63] Cooperation has also aided NASA's claims to be running a civilian and non-military program. In making available to several nations the data from its Tiros weather satellites, NASA is able to claim international support against the Soviet charge that NASA's Tiros was an observation satellite serving military purposes.[64] The US Navy has also entered the field of international space cooperation in offering to all nations the use of its Transit system of navigation satellites.[65] The unequalled excellence of American space technology has enabled the US to profit substantially from its space programs. About 75 per cent of the space hardware used in European countries is purchased in the US, and the United States exports between $4 and $5 billion of aerospace equipment annually. Moreover, of the 26 satellites to be launched by NASA in 1974, five were to be paid for by foreign governments.[66]

For the future NASA has commitments to a large variety of cooperative activities including the following: two probes of the sun in cooperation with the Federal Republic of Germany; mapping of global wind circulation by one satellite and 500 balloons in cooperation with France; involvement of 58 countries in NASA's Earth Resources and Technology Satellite program; involvement of British, French and West German firms in support of design studies for the space shuttle; direct broadcast to India from a synchronous satellite to modified televisions of Indian programs on birth control, basic agriculture, "national integration," etc.[67] Cooperation with Western Europe should reach an unprecedented scale with the development of the US space shuttle. ESRO has agreed to develop, at a cost of $300-400 million, a variable configuration multimodular Spacelab satellite which, beginning in 1980, will be carried to orbit by the shuttle and which can be adapted to a wide variety of missions, manned and unmanned. Spacelab is expected to make up some 40 per cent of NASA's payloads in the 1980s and to involve the participation of both American and European scientist-astronauts.[68] Pressed by budget cutbacks, NASA is looking for West European support of unmanned planetary programs scheduled for the 1980s.[69]

Among the most significant and controversial of American space programs involving other countries have been those of Comsat (Communications Satellite Corporation) and Intelsat (International Telecommunications Satellite Consortium). Comsat was established under Public Law 87-624, the Communications Satellite Act of August, 1962. Comsat is a privately-owned, profit-making, government-regulated monopoly of satellite communications. Half of its equity is owned by US common carriers, principally the telephone monopolies and their holding companies, and half was sold on the open market subject to the requirement that individual holdings be of limited size. The President appoints three directors of Comsat, the common carriers six and the public stockholders six.[70] Comsat buys satellites in the US and pays NASA to launch them into synchronous orbits.

Establishment of Comsat occasioned a lengthy filibuster in Congress, was the subject of several presidential policy pronouncements and involved participation by representatives of the Departments of State, Defense and Justice, the Federal Communications Commission and the New York Stock Exchange. It now has thousands of stockholders and involves some 45 countries.[71]

Over 80 nations are members of Intelsat, the private international telecommunications satellite consortium formed in 1964, of which Comsat is the American member. Comsat owns about 52.6 per cent of the equity and had the same per cent of the votes in Intelsat's directing body. Until it was reorganized in 1971, Intelsat was in form and in fact more or less exclusively run by Comsat, which managed it, owned most of its stations and carried out all its space activities. Now Intelsat has its own "space segment" and is jointly run by three different agencies, one of which is based on representation proportional to equity and the others on equal representation. Comsat still owns an absolute majority of the equity and is the dominant partner. Most of Intelsat's money is spent in the US with Comsat contracting for the launch of Intelsat satellites by NASA. Comsat maintains its own relations with foreign governments and firms and has offices in Geneva to "assist European publications in understanding Comsat's viewpoint and activities." Intelsat profits are set by agreement at 14 per cent.[72]

Comsat makes fine grist for the attacks of Soviet propaganda mills. In addition to its more mundane aspects, one Soviet writer accuses it of having leased 30 satellite communications channels to the Pentagon for use in communication with military personnel in Southeast Asia.[73] A detailed study of Comsat and its establishment is outside the framework of this study. Suffice it to say that whatever the Soviet perspective on Comsat, the US Government in many spheres has found private monopolies preferable to public monopolies. It has also been able to accomplish important projects by making the necessary activities profitable to others rather than by undertaking them through its own initiative or through the operation of its negative sanctions.

Summing up, the cooperative activities of the United States have been very diverse and very extensive. Cooperation, however, up until the more

recent Soviet-American breakthroughs, has taken a back seat to competition. Even cooperative projects have reflected the political goals of the
United States rather than merely scientific or disinterested cooperativeness
with other countries. Cooperation as implemented has not been directed at
the goals of cooperation set forth in the 1958 Space Act or routinely
repeated in most NASA literature. American cooperation with various
nations was in the past either part of the competition with the Soviet Union
or in the background of that competition. Cooperation in space as an
alternative to competition is only being realized in the cooperation of the
two space powers.

COOPERATIVE PROGRAMS WITH VARIOUS COUNTRIES: USSR

At the start of the space age during the IGY, Soviet cooperation with other
countries was markedly less than that of the USA. Soviet representatives
restricted IGY agreements for exchange of space information, provided
little or no advance information on Soviet satellite and sounding rocket
programs, and refused to catalogue the data and findings from their experiments so that the scientists of other countries would not even know what to
ask for in proposing exchange or sharing of information. Such findings as
were released by the USSR as part of the IGY were of the type that would
normally be published in Soviet scientific journals, and information on
satellite design and instrumentation, telemetry patterns, orbital elements
and raw data return were either delayed or never announced.[74] These
policies formed the basis for relative uncooperativeness of the Soviet Union
in the first years of its space program.

The organizational framework for implementing Soviet space cooperation with other nations is centered in the Academy of Sciences of the USSR.
Negotiation and ratification of agreements on space cooperation often
involve the Academy's president, Academician M. V. Keldysh, and its
vice-president, A. P. Vinogradov. The Academy is directly subordinated to
the All-Union Council of Ministers of the USSR. Various space activities,
including cooperative activities, which involve participation of certain enterprises in such fields as communications, medicine, meteorology, etc.,
include the involvement of ministries responsible for those enterprises
(Ministry of Communications, Ministry of Health, Ministry of Hydro-
meteorology, etc.). Cooperative agreements may also involve the Ministry
of Foreign Affairs. The Academy of Sciences includes an Intercosmos
Council with direct responsibility for cooperative space ventures, chaired
by Academician Boris N. Petrov and with Academician V. S. Vereschetin
as vice-chairman. The Intercosmos Council, founded in 1967, includes

representatives from Bulgaria, Hungary, East Germany, Cuba, Outer Mongolia, Poland, Rumania, the USSR and Czechoslovakia and meets in full session once a year. Each country finances its own contribution separately with the rockets and other "means of space technology" provided by the USSR.[75] Academicians Leonid I. Sedov and Anatolii A. Blagonravov also hold high positions and are often representatives of the USSR in negotiations involving space cooperation.

Soviet reticence to cooperate with other countries in the exploration of space even extends to other Communist states, including the USSR's fellow members of the Council for Mutual Economic Assistance. Between 1957 and 1962 international cooperation among the "socialist" states did not go beyond visual and photographic observation of Soviet satellites.[76] While Soviet space cooperation has remained quite limited, it has increased gradually since 1962 and involves as its most active partners the other socialist countries, France and the United States. Cooperation with other countries of the "bloc" provides the USSR with added technical competence (especially in the case of cooperation with East Germany, Czechoslovakia and Poland), allows the other socialist countries to share in Soviet prestige, strengthens loyalty to the USSR, and now that China has launched two satellites, may even make it possible to build one small bridge over the hostility separating China and the USSR. China might choose to cooperate with the US instead.*

Activities involving Soviet space cooperation with socialist countries other than those already mentioned include the provision of stations for tracking Soviet satellites. Such a station is operating in Cuba, which also serves as a base for Soviet tracking ships.[77] An *Orbita* ground station for relay of signals from *Molniia* communications satellites has been operating since 1970 in Ulan Bator, the capital of Outer Mongolia. The USSR indicated that it would build such a station in Cuba in 1965 and concluded an agreement to that purpose with Cuba in 1967. Evidently, construction is now complete after a delay of five years, making possible the reception of Soviet television in Cuba.[78]

Beginning in late 1968 the Soviet Union began launching a series of satellites in cooperation with other socialist states under the name of *Interkosmos*. The first launch in this series was the *Interkosmos* precursor, *Kosmos* 261. *Interkosmos* 1, launched nearly a year later, duplicated the mission of *Kosmos* 261 in studying solar radiation. It had on board instruments supplied by East Germany, the Soviet Union and Czechoslovakia.

---

*Recent initiatives toward China suggest this possibility, but this author knows of no actual steps to implement such a policy.

Scientists of these countries, along with some from Bulgaria, Hungary, Poland and Rumania, participated in the experiments conducted by *Interkosmos* 2-5. Through 1973 the USSR launched ten *Interkosmos* satellites.[79] An International Center for Scientific and Technical Information formed in 1969 within the Council for Mutual Economic Assistance may contribute now or in the future for such space cooperation among socialist states.[80]

Socialist states comprise the great majority of the members of *Intersputnik*, the Soviet answer to Intelsat. Soviet expression of an interest in international cooperation in space communications was expressed as early as 1962.[81] One Western specialist, F. J. Krieger of the Rand Corporation, doubted that the USSR really intended to take part in a world satellite communications system.[82] Considering the way in which Comsat and Intelsat are organized and compete with one another, the absence of Soviet participation is no surprise. *Intersputnik* is claimed by Soviet spokesmen as "fair and democratic," based on the equality of its members. Intelsat, they claim, is discriminatory and US-dominated.[83] The draft treaty for setting up *Intersputnik* was originally submitted to the UN Committee on the Peaceful Uses of Outer Space (COPUOS) in 1968 by the USSR, the pro-Soviet East European states represented at the UN, Cuba and Outer Mongolia.[84] Then NASA Administrator Dr. Thomas O. Paine thought that *Intersputnik* was designed "to rival and embarrass" Intelsat, but he concluded that it was not notably successful in this intent since only the UAR, Cuba and the East European Communist states had joined it by the end of 1970.[85] Whether or not Intelsat is embarrassing to Comsat or to the United States, its customers are largely businesses which are not likely to be disturbed that Intelsat is a money-making venture. Unless third countries wish to make great investments in space technology (which they have not), most of the monies involved will be spent in the USA or the USSR. Since the USSR lags behind the United States in communications satellite technology and has not yet launched 24-hour synchronous satellites, *Intersputnik* cannot provide communications channels so cheaply as Intersat even if it collects no "surplus value." Space cooperation between the USSR and its allies through *Interkosmos* and *Intersputnik* programs seems likely to grow, and it is possible that new programs will be added. Other socialist states have not participated in the deep-space programs of the USSR, nor are there any plans to include "bloc" cosmonauts in manned Soviet missions.[86] The USSR has made available to scientists of the Czechoslovak Socialist Republic and other socialist countries samples of lunar soil from *Luna* 16 and launched two *Vertikal* probe rockets in cooperation with Bulgaria, East Germany, Hungary, Poland and Czechoslovakia.[87]

The closest cooperation between the USSR and a nation outside the socialist camp has been with France. Indeed, Soviet cooperation with France has been greater than with any other single country. Soviet-French cooperation began modestly in November, 1965, with an agreement for the transmission between the two countries of color television via the Soviet communications satellite, *Molniia* 1. While most of the world uses an American process for the transmission of color television, the Soviet Union and its allies employ the French SECAM-3 system.[88] Regular broadcast of color television takes place between the USSR and France employing SECAM-3.[89]

Soviet-French cooperation began in earnest in 1966, when the two countries signed a comprehensive agreement which set forth the following goals:

(1) launch of a French satellite by a Soviet launcher in the USSR,
(2) joint meteorological studies,
(3) joint studies of satellite communications,
(4) joint work in several areas, especially television,
(5) exchange of scientific delegations, information, trainees and personnel,
(6) organization of conferences and symposia, and
(7) creation of a standing Soviet-French commission for space cooperation.[90]

The occasion for the signing of this agreement was the visit of French President Charles DeGaulle to the USSR in June, 1966. DeGaulle witnessed the launching of a Soviet weather satellite from Tiuratam ("Baikonur") by a *Vostok*-type launcher.[91] As noted previously, Tiuratam is the main Soviet cosmodrome, and the *Vostok* was not put on public display until a year later at the Paris Air Show. According to one source, scientists and leaders from the other socialist countries did not get a chance to visit Soviet cosmodromes until *after* the visit by DeGaulle.[92] All of the work envisioned by the 1966 agreement went ahead to the satisfaction of both countries, except for item #1 above. The French satellite *Roseau*, which was to be launched from a Soviet cosmodrome, encountered financial difficulties, and its development and launch were canceled in 1969.[93]

In October, 1970, DeGaulle's successor, Georges Pompidou, visited the Soviet Union and witnessed the launch of a Soviet reconnaissance satellite from Tiuratam.[94] Pompidou and DeGaulle are the only Westerners known to have visited a Soviet cosmodrome.[95]

French laser reflectors on the *Lunakhod* roving moon vehicle permitted exact calibration of its position from earth. A French experiment ("Stereo") to measure the solar wind was carried aboard the Soviet spacecraft Mars 3, launched in 1971. The USSR has provided France, Czechslovakia and the US with a sample of the lunar soil collected by *Luna* 16. A wide variety of other projects are now in the planning or execution stage, including the launch of French satellites by a Soviet launcher. Two such satellites, *Oreol* 1 and 2, were launched in 1971 and 1973.[96] French equipment was aboard some of the four Soviet Mars probes launched in 1973.[97]

Besides France, the other socialist states and the United States (which will be examined separately), the Soviet Union does not have extensive programs of international space cooperation, and such programs as exist involve fewer countries and less cooperation than those of the United States. In the area of tracking and relay stations, the USSR, probably because it does not want to share information with other countries, has a much less extensive land-based system than the United States which the Soviets supplement with tracking ships. Between 1967 and 1972 tracking stations of one kind or another have been built with Soviet help in the UAR (two), Mali, Guinea, Cuba and Chad. Reports also indicate Soviet interest in setting up stations in Indonesia, Australia and Chile.[98] Earlier the USSR had unsuccessfully solicited the support of tracking stations operated by the United States and other countries.[99]

The USSR has cooperated with a number of countries in less dramatic ways. So, for example, Soviet equipment and technical assistance to India helped create the rocket range at Tumba which is now an *international* range for sounding rockets.[100] Sounding rocket experiments had previously been conducted with India and Indonesia.[101] Observation of Soviet satellites has entailed cooperation with Holland, Greece, the UAR, Finland, West Germany, Sweden and other states.[102] The Soviet Union has suggested that its *Molniia* communications satellites be used to establish communications links with Japan, the United States and other nations.[103] In 1971 only one orbital launch took place from the Soviet cosmodrome at Kapustin Iar, that of the bloc science satellite, *Interkosmos* 5. All the *Interkosmos* satellites have been launched from Kapustin Iar, and no clearly identifiable military flights have been launched from there in years. At the 23rd Annual Congress of the IAF in Vienna in October, 1972, USSR representatives said that delegates would be allowed to visit a Soviet launch site when the 24th Congress met in October, 1973, in Baku in Soviet Azerbaijan. It seemed likely that at that time Kapustin Iar would be the first Soviet cosmodrome open to the international public.[104] Evidently the visit never took place.

FACTORS AFFECTING SOVIET-AMERICAN SPACE COOPERATION

Most of the general factors described previously as affecting space cooperation generally have the same or similar impact on Soviet-American space cooperation. This section focuses on the particular factors which serve as inducements or barriers to space cooperation between the USA and the USSR, including those of a scientific, technological, economic and political nature.

As stated previously, science is by nature an international human endeavor; it is more rapidly advanced by the widest circulation of scientific methods, techniques and findings. Unlike space technology, space science is of little strategic significance and in most aspects is remote from the security concerns of states. Political restrictions on scientific cooperation therefore tend merely to restrict the progress of science rather than to safeguard national security. Scientific exchange and cooperation between the Soviet Union and the United States can do a great deal to advance space science.[105]

Insofar as the experiments of the two countries are different, cooperation and exchange can fill in the information gaps in the space science of both. This is particularly true of the study of phenomena which are interrelated but widely spread. Simultaneous scientific measurements of the earth's atmosphere, magnetosphere, radiation belts, meteorological phenomena, the oceans, etc., require many widely scattered satellites, sounding rockets and earth stations. Cooperation between the two major space powers greatly enhances the range and number of such measurements. Cooperation and exchange even contribute to science in the conduct of identical or near-identical experiments by the US and the USSR in that replication is a standard requirement of science to confirm results, the usefulness of techniques and the accuracy of scientific apparatus.

In both the US and the USSR the scientific communities and especially those concerned with space are larger and growing faster than in other countries. This means that the influence of scientists is relatively marked and increasing with time in both countries.[106] Having a natural interest in cooperation and frequently interacting through scientific meetings, conferences and symposia, both Soviet and American scientists pressure their governments for greater space cooperation between the two countries.[107] The United States from the outset has been quite willing to share scientific information on space. The USSR, so secretive in other fields, has published the scientific findings from the first Sputniks and become increasingly open about providing scientific information on space in recent years. Moreover, a "special" cooperative relationship developed between Academician Blagonravov and NASA's late Dr. Hugh Dryden, facilitating scientific cooperation between the two countries.[108] Science by itself appears to offer only inducements and no barriers to Soviet-American space cooperation.

In cooperating with one another in space the USSR and USA are, of course, jealous of whatever prestige their counterpart might win or take away from themselves. Moreover, each is sensitive to whatever strategic implications might flow from its competitor's development of space technology. However, aside from prestige and security, both the US and the USSR are the two countries least likely to circumscribe the exploitation of space with international restrictions or to complain, as many third countries have, that the monies spent on space should be diverted to areas of more immediate and direct benefit to themselves. Much more than other countries, the US and USSR are in a position to usefully exploit each other's techniques, knowledge, space spending and commitment to space exploration.[109]

But technology creates not only inducements but barriers to Soviet-American space cooperation. Integrating the space technologies, as in the case of the compatible docking apparatus under development for the Apollo-*Soiuz* mission in 1975, requires significant outlays of money. In general neither power likes to be dependent on the other for launch vehicles, spacecraft, or flight schedules when it can provide these with greater certainty for itself. Originally developed in an atmosphere of competition, which is far from over, the space technologies of the two powers remain largely incompatible.[110] The political leaders of both countries are concerned lest, in sharing technology, the other side is able to acquire information of strategic significance or get "something for nothing." The Soviets in addition are probably concerned about revealing the extent to which their own space technology lags behind that of the United States. One indication of this was their insistence that their recent offer to NASA to purchase ten Apollo lunar space suits be kept strictly secret.[111]

In terms of economic motivations, it is probably true that integration of space technologies would provide savings in the long run, even if initial costs were high. In the meantime savings could be realized by dividing or sharing missions which are currently duplicative or redundant. This is especially true in the case of planetary missions, where sharing information of experiments both checks and supplements the efforts of the two nations taken separately and obviates the need for each to undertake missions aimed at finding what the other has already discovered. The high cost of manned and deep-space missions is another inducement to cooperation, especially in a period in which the space scientists of both countries are chafing under budgetary constraints and may be unable to undertake certain advanced missions without finding another country willing to help bear the costs. Because of the even greater reluctance of third countries to undertake expensive space research, this factor applies with special force to the US and the USSR. Economic reasons are generally ascribed to the shift in 1962 to a more favorable position on Soviet-American space cooperation.[112] Certain programs of earth applications such as use of satellites for weather prediction, earth resources location and study, and communications create additional economic inducements to space cooperation. Some, like communications, however, may cause difficulties as each country is reluctant to pass funds to the other, and one side does not like to reward capitalist enterprises and the other to reward socialist enterprises. Because in the United States the space program is a politically important source of employment and profit-making, the American Government is not disposed to sacrifice these benefits as a part of a scheme of cooperation with the Soviet Union. However, the profit motive is likely to overcome much of the resistance to the *sale* of American space technology to the USSR. It is noteworthy in this regard that in the past few years the USSR has become eager to purchase all kinds of Western technology.

Pressing political considerations are probably the greatest inducements and barriers to space cooperation for both the Soviet Union and the United States. Cooperation can save either side from falling too far behind the other in overall space capability and prestige. As positions shifted, it was first the United States, then the USSR, which fell behind and may fall farther behind if the space programs of the two countries are not integrated. The shift from a bipolar political world to a looser and more complicated configuration of power has sapped much of the rivalry of earlier periods, and the USSR feels particularly pressed to improve relations with the West in consequence of its rivalry with China and its need for Western technology.

Space cooperation in itself, presumably including US-USSR cooperation, is a proclaimed goal of the national policies of both countries as specified in the 1958 National Space Act and voiced by political leaders and space spokesmen in both countries.[113] Such cooperation can serve numerous political goals. For the United States, its continued expressions of willingness and more or less substantive offers of cooperation with the USSR have served as a propaganda policy to demonstrate the openness and cooperativeness of American society while pointing out the secrecy and sinister aspects of the USSR. It behooves the Soviets to counteract this policy by responding to these American offers or at least offering publicly defensible programs of Soviet-American cooperation on its own. If this leads to real cooperation, it will further the American political goal of "opening up" the closed Soviet society and "driving a wedge in the iron curtain."[114] Finally, both space powers have proclaimed in favor of Soviet-American space cooperation in that it will serve the interests of cooperation in other areas and also the cause of peace.[115]

The most formidable barrier to US-USSR space cooperation has always been the political rivalry and hostility that prevail between the two countries. The effect of this factor is compounded by the concern of both sides with advancing their prestige and maintaining the security of militarily relevent technologies employed in their space programs. In the USSR this concern extends to an obsessive preoccupation with secrecy which seems to have blocked other political goals of the Soviet space program, most notably those linked with international cooperation. It is characteristic of Soviet space leaders to cite "international tensions" as the most significant factor impeding Soviet-American cooperation.[116] General disarmament is often mentioned by various Soviet leaders, including those with a special responsibility for space research as a precondition to ambitious programs of cooperation.

Given the above inducements and barriers, Soviet-American space cooperation is likely to take rather definite forms. Space cooperation, much like arms control, is likely to occur at times and in areas where neither space power holds a commanding lead. So, for example, it is likely that the USSR and the other socialist countries did not participate in the 1962 Washington conference of the World Meteorological Organization on space meteorology largely because they did not want to reveal how far ahead of the Soviet Union the United States was in the field of satellite meteorology.[117] In the politically more sensitive areas which involve the exchange of technology, the space power which perceives itself as technologically ahead will oppose cooperation which would benefit its partner more than itself. This is probably one reason why Khrushchev's 1962 initiative for Soviet-American cooperation did not come until after the flights of astronauts Glenn and Sheppard.

Beyond these factors, likely areas of US-USSR cooperation include:

(1) exchange of scientific data, findings and samples rather than technology;

(2) activities like space meteorology and magnetic survey, which require simultaneous measurements in separate areas;

(3) activities in which the information uncovered in one nation's space research can be profitably applied by the other, such as meteorology, earth resources and navigation;

(4) activities with little or no military or strategic significance such as geodesy, deep space and planetary research, research into the earth's magnetosphere, selenology, cosmic radiation, space medicine and biology;

(5) very expensive activities such as manned and planetary missions, stations in space, on the moon, etc.;

(6) tracking and observation of space vehicles;

(7) communications, especially if conducted outside the normal channels of Comsat, Intelsat, and *Intersputnik*, which are commercially and ideologically incompatible;

(8) activities relating to the safety, emergencies of, aid to, and rescue of astronauts and cosmonauts;

(9) activities reflecting *joint concerns* for the regulation of space activities such as contamination and pollution of space and celestial bodies, use of precious space resources and the application of international law in space.[118]

Although almost all space activities may involve at least one of the preceding nine factors to some degree, it will be the degree of applicability or nonapplicability of several factors which is likely to determine whether a particular activity can be brought within the scope of Soviet-American cooperation.

PROPOSALS FOR A SOVIET-AMERICAN MISSION TO THE MOON

Probably the most interesting sidelight (or sideshow) in the history of Soviet-American cooperation centered around proposals made in the early 1960s for a joint mission of manned lunar landing and return. Although nothing substantial ever emerged from these proposals, they did create a brief sensation and helped to illuminate some of the political considerations which enter into the space planning of the two countries, especially as these affect US-USSR cooperation in space.

President John F. Kennedy first invoked the prospect of Soviet-American space cooperation in his inaugural address saying, "Together let us explore the stars."[119] It should be noted that in his campaign Kennedy had chided the Eisenhower Administration for "losing" the space race with the USSR. Kennedy repeated the theme of US-USSR space cooperation in his first State of the Union message. He was later reported to have first broached the topic of a joint Soviet-American lunar mission to Nikita Khrushchev at their summit meeting in Vienna in the summer of 1961.[120] In the meantime, as described previously, the US had suffered a severe setback in the Bay of Pigs invasion and embarked on an ambitious manned lunar program with the expressed aim (Kennedy's) of "beating the Soviets" to the moon.

Etzioni contends that, shortly before his death, Kennedy "realized . . . that the space race was getting out of hand," soaking up too much of the nation's wealth and exacerbating international tensions.[121] In September of

1963 Kennedy suggested in a formal address before the United Nations General Assembly that his goal of landing and returning a man from the moon in the decade of the 1960s be accomplished as a joint Soviet-American enterprise.[122] After Kennedy's death in November this proposal was repeated by UN Ambassador Stevenson and President Johnson in his State of the Union message, both in December, 1963.[123] What lay behind this rather extraordinary suggestion, and what was the Soviet reaction to it?

The original Kennedy suggestion in Vienna of a joint moonflight occurred in an atmosphere of alternative somberness and banter. The serious topics for discussion were Laos and Berlin, and the failure to reach any agreement on the latter presaged the third Berlin crisis in 1961. At the same time less serious issues were interjected. So, for example, Khrushchev's chitchat with Mrs. Kennedy prompted him to send her a puppy from the litter of a Soviet space dog. Kennedy's suggestion of a joint lunar effort was probably neither offered nor taken in complete seriousness. In turning it down, Khrushchev cited the absence of disarmament and the military nature of space hardware as his reasons.[124]

Kennedy's reiteration of his moon offer before the UN in September, 1963, was more significant in that it was made in public and coincided with the release of the fact of the 1961 offer to the American press. Including the offer in the address to the General Assembly was apparently Arthur Schlesinger, Jr.'s, idea, and it was concurred in by the President and his chief advisors. The purposes behind the offer, according to Schlesinger, were to make "an effective political gesture at home and abroad" to aid ratification of the 1963 Test Ban Treaty in the Senate and to lend credibility to America's expressed commitment to detente. Kennedy was also reacting to a change in the Soviet stance on disarmament and trying to (1) demonstrate that the US was as flexible or more so than the USSR in its pursuit of detente, and (2) put some pressure on the Soviets to compromise with American initiatives in arms control, including the prohibition of nuclear weapons in space. In so doing Kennedy was attempting to build a momentum for detente, which was reversed by the arrest of the American professor Barghoorn on charges of espionage in Moscow in November, 1963.[125]

Alternative explanations of Kennedy's initiative have been made by others, who, if not so close to the White House as Professor Schlesinger, are not necessarily less accurate. Etzioni's contentions, mentioned above, take an interpretation which reflects even better on the President than those of Kennedy's official apologist, Schlesinger. Etzioni's view is shared by Marion Levy.[126] Levy and Etzioni take the President's offer at face value. Some support for their contention can be taken, assuming that Kennedy intended to respond to and amplify vague initiatives from the Soviets for US-USSR space cooperation passed on to NASA by the British scientist Bernard Lovell in the summer of 1963.[127] Frye suggests that Etzioni may be correct in thinking that the President hoped to dilute the American commitment to a manned lunar landing that he had initiated over two years previously. Unlike Etzioni, Frye contends that Kennedy may have sought to be disencumbered of his personal commitment to a manned lunar landing in the 1960s because he feared that the necessary public and congressional support would not be forthcoming.[128]

Some of the evidence suggests that a joint Soviet-American lunar mission was at least possible, so that Kennedy's initiative in the UN speech was not an empty gesture. In private conversations between NASA's Deputy Administrator Hugh Dryden and Academician Blagonravov shortly before the President's speech, the latter had allegedly agreed that discussion of such a mission could begin after both nations had soft-landed instrumented payloads on the moon.[129] Such an agreement marked a change in the Soviet stance, enhanced the possibility of Soviet-American cooperation without putting pressure on the American side to respond in the manner that a *public* Soviet initiative would have done.[130]

Those who take a more skeptical view of Kennedy's offer argue that the President was confident that the USSR would not make a positive response. The technological barriers to a joint lunar mission were formidable.[131] They were sure to be greater after both countries had met Blagonravov's precondition of soft landings on the moon before *discussions* could begin. A year earlier Kennedy himself had said that political barriers of distrust and hostility would probably make such a mission impossible.[132] Van Dyke persuasively argues that if the President seriously entertained the possibility of a joint lunar mission as the main basis for his proposal, he would have more thoroughly explored the possibilities behind the scenes before making it. In Van Dyke's view the *dramatic* quality of the Kennedy proposal indicates that it was a "competitive move . . . designed mainly for propaganda purposes."[133] Belying the contention by Etzioni and Frye that Kennedy hoped to dilute the nation's commitment to the moon race or to erase his personal identification with it is the fact that, shortly after making the proposal, Kennedy wrote to the House subcommittee chairman responsible for NASA appropriations, saying that his proposal did not alter the "full-blast" pace of the US lunar program and that "a good many barriers of suspicion and fear" stood in the way of major progress in US-USSR space cooperation.[134]

Although Kennedy's proposal met with near-unanimous support at the UN and in the press, congressional response was distinctly negative. The House attached an amendment prohibiting the use of NASA appropriations

in a joint lunar program with any Communist country. The final appropriation for 1964 incorporated the Senate's amendment, which required joint
congressional approval for such use of NASA funds.[135] The reaction of the
Congress and that of the Soviets, who did not choose to respond to his
proposal, were probably anticipated by the President. These factors support
a middle position between that of the strongest Kennedy adherents like
Etzioni and the most skeptical observers like Van Dyke. Kash concludes
that the Kennedy proposal was a "double-edged sword" that would further
detente and the US positions on arms control. Its major purpose, however,
was as a grand gesture for propaganda effect.[136] As such, it "worked" in
that the Soviets did not respond, but it also complicated NASA's budget
struggle with the Congress. In retrospect, it would seem to have been in
the Soviets' interest to call Kennedy's bluff. Accomplishing a joint lunar
program or at least increasing Soviet-American space cooperation could
further Khrushchev's strategy of detente and strengthen the USSR against
China. It could also keep the US from scoring a *unilateral* coup with Project
Apollo or, at the least, delay that program, allowing the Soviets to beat
Apollo's schedule or mount some other spectacular of their own.[137]

This author thinks it probable that such a move was made impossible by
Soviet opposition to Khrushchev's detente strategy and to Khrushchev himself. That Khrushchev lacked the authority for bold initiatives in foreign
policy after the Cuban missile crisis seems plausible in light of his removal
in October, 1964. The Barghoorn incident was not in line with Khru-
shchevian strategy and tactics and may have been pulled off without his
knowledge in an attempt (successful at that) to wreck them. If Khrushchev
had been responsible for the decision to arrest Barghoorn on trumped-up
charges, why did he arrange for the professor's release after Kennedy's
intervention on his behalf?[138]

Whatever the reasons, Kennedy's proposals for a joint lunar mission were
not realized, and after modest beginnings in 1962, Soviet-American space
cooperation remained quite limited until major breakthroughs occurred in
1970–1972.

OTHER AREAS OF SOVIET-AMERICAN COOPERATION

Until 1962 Soviet-American space cooperation was largely limited to the
nongovernmental and multilateral cooperation of scientists in the IGY and
the organizations that grew out of the IGY, CSAGI and COSPAR.[139] Both
countries also participated in such nongovernmental international scientific
organizations as the International Astronautics Federation and the International Council of Scientific Unions. Space cooperation involving govern-

ments, as in the United Nations, was blocked by disagreements centering around the relative representation of socialist and nonsocialist countries.

The United States took a number of initiatives to begin cooperation with the Soviets. In late 1959 NASA Administrator T. Keith Glennan verbally offered the USSR use of the tracking network developed for Project Mercury and arranged to have this conveyed in writing to the Soviet Academy of Sciences through the American National Academy of Sciences. The offer was acknowledged with thanks, but when the Soviets began their program of manned flights, they did not ask for American help in tracking despite the inadequacy of their own tracking system.[140] The Soviets evidently did not see the need to cooperate with a "second-rate" space power like the United States. Moreover, use of American tracking facilities would require prelaunch announcement of Soviet flights, which not only would depart from the Soviet policy of secrecy but would have the effect of guaranteeing easy detection and wide publicity of delays and failures of Soviet launches. As such, NASA could expect the Soviet response, and this expectation probably entered into the decision to offer the use of American facilities.[141]

In 1961 Mr. Frutkin suggested to Soviet representatives that the two countries communicate via the passive reflector satellite Echo 1. Initial acceptance of the proposal by Academician Blagonravov was withdrawn owing to "unspecified difficulties."[142] In general, no significant US-USSR space cooperation took place in this early period despite the efforts of NASA and US scientists.[143]

The first sign of a change occurred in November, 1961, when the Soviet delegate to COPUOS, the UN's Committee on the Peaceful Uses of Outer Space, took his seat after a lengthy dispute over the Committee's membership.[144] Shortly afterward, the achievement of a limited consensus between the two space powers resulted in the passage of UN Resolution 1721 (XVI), the first international agreement on the peaceful uses of outer space. This change was probably in part the result of American successes in two manned suborbital Mercury flights in the spring of 1962. Following the first of these two flights, Nikita Khrushchev, in sending his congratulations to President Kennedy, broached the possibility of US-USSR space cooperation. An exchange of letters between the two was followed up by negotiations between NASA's Hugh Dryden and Academician Blagonravov.[145] These talks led to the three-part bilateral space agreement of June, 1962. The agreement called for (1) coordinated launchings by both countries of meteorological satellites and exchange of data over a direct communications link between Moscow and Washington (the "cold line"), (2) launching of satellites by both countries to measure earth's magnetic field and exchange of data, and (3) joint communications experiments employing the passive reflecting satellite, Echo II.[146]

Although this agreement marked the first major milestone in Soviet-American space cooperation, its implementation was relatively unsuccessful. Exchange of meteorological data began in September, 1964. Generally on the American side, the Environmental Science Services Administration (ESSA) was much less than satisfied with the promptness, coverage, quality and regularity of the data supplied by the Soviets. No coordinated launching of weather satellites by the two countries has taken place.[147] Implementation of the agreement for exchange of data on the earth's magnetic field began well enough in 1966 but remained limited, owing in NASA's view to the unwillingness of the USSR to contribute extensively and to irregularities of their participation in data exchange. Joint experiments with Echo II in 1966 were rather more successful, but some difficulties were encountered here as well.[148] Generally speaking, the areas marked off for cooperation in 1962 were very limited. No extensive coordination of integration of activities was planned, and none was achieved. No exchanges or transfers of personnel or technology were contemplated, and none took place. Early in this period the USSR, Poland and Czechoslovakia reneged on their acceptance of invitations to an International Meteorological Satellite Workshop in Washington, and the USSR also refused to accept a display of the American Friendship 7 spacecraft.[149]

In the exchange of letters between Kennedy and Khrushchev, both leaders had suggested sharing information on space medicine, but the Soviets begged out a few months later.[150] Also, in the summer of 1964 the Soviets repudiated in principle any cooperation with the United States in establishing a global satellite communications network, charging that Comsat would be US-dominated and would extend monopoly capitalism into space.[151] Had the United States made Comsat a public rather than a private venture, the substantial US lead in satellite communications would probably have precluded any meaningful cooperation anyway, at least until the USSR had a respectable satellite communications capability of its own and could take part in the production and launching of Comsat vehicles.

By the fall of 1965 a lasting agreement was reached for the exchange of information on space medicine and bioastronautics. This was to take the form of volumes based on the contribution of monographs by both sides selected and compiled by a joint editorial review board. The agreement marked the second major milestone in Soviet-American space cooperation. The Soviets delayed its implementation for *four years*. That is, it was not until the spring of 1969 that the Soviet side identified its compilers, confirmed an outline of the material to be included and agreed to a schedule for exchange of materials and selection of manuscripts for the first volume. The first exchange of chapters did not take place until January, 1970.[152] Repeated initiatives by NASA for shared use of tracking facilities in the 1960s met with no positive response from the Soviets. In international scientific forums, however, Soviet space scientists were quite willing and eager to share scientific information and discuss mutual problems with their American counterparts.[153] Soviet secrecy, international tensions heightened by the Vietnam War, and bureaucratic caution and inertia were the major factors that stood in the way of making and implementing space agreements with the United States on space cooperation in the 1960s. For its part, NASA's continued insistence on an equal quid pro quo in every exchange probably contributed to Soviet suspiciousness and caution.

Many initiatives made in the 1960s by NASA, the National Academy of Sciences and President Johnson for Soviet-American space cooperation in various fields received perfunctory replies or no replies at all from the Soviets. In the view of the US Department of State, it was largely political considerations on the Soviet side rather than technical difficulties which blocked space cooperation between the two countries.[154]

Just as the success of the first American attempts to put men into earth orbit evidently served to trigger the first agreements on bilateral space cooperation between the Soviet Union and the United States in 1962, the success of the Apollo program as marked by Apollo 11 in July, 1969, evidently had a similar effect. In both cases dramatic American successes seemed to have convinced Soviet leaders that cooperation with the United States would be beneficial to Soviet space research and that the USSR could not automatically win any competition in space by way of American default. The first sign of a major improvement in cooperation was the exchange of visits by the American astronaut Frank Borman and the Soviet cosmonauts Beregovoi and Feoktistov in the summer and fall of 1969. Borman toured the USSR as an official guest and conferred with top governmental and space leaders.[155]

In a series of letters to Academicians Keldysh and Blagonravov in 1969, NASA Administrator Thomas O. Paine solicited increased cooperation in various areas and, in particular, in discussion of compatible systems for rendezvous and docking in space as a means of providing mutual capabilities for the aid and rescue of cosmonauts and astronauts. Keldysh responded favorably in December, 1969. By the summer of 1970 the two sides had agreed to hold a conference in Moscow during October on compatible systems for rendezvous and docking of spacecraft and other problems. In December a summary of the results of the talks was released. It defined the technical problems to be overcome and provided for (1) exchange of technical information on radio guidance and rendezvous systems, (2) definition and exchange of technical requirements by each side, and (3) formation of three joint working groups to assure compatibility of the systems developed and to begin meeting in March, 1971.[156]

Shortly afterward, in January, 1971, the talks of October, 1969, were supplemented by an equally significant agreement between NASA and the Soviet Academy of Sciences to extend the cooperation planned for rendezvous and docking into other fields. The two sides agreed to undertake and further consider:

(1) improvement in the exchange of data from meteorological satellite systems;
(2) coordination of meteorological sounding rockets, including cooperation with other countries;
(3) coordination in study of the natural environment from space;
(4) exchange of space plans and objectives with a view toward coordinating programs near earth, to the moon and planets;
(5) exchange of lunar soil samples, and
(6) improving exchange of information in space biology and medicine, including biomedical data from manned flights.

Four attachments to the agreement spelled out details of implementation and specific objectives. Again, joint working groups were set up to meet regularly and carry out the agreement.[157]

In May of 1972 the plans laid out in the summary of October, 1970, and the agreement of 1971 were raised to the level of an executive agreement signed by Alexei Kosygin and Richard Nixon during the visit of the latter to Moscow. In the meantime discussion of the possibility of a joint US-Soviet mission leading to the docking of Soviet and American craft in space had made it possible to include a commitment to such a mission in 1975 in the executive agreement.[158] In the months before Nixon's trip to Moscow the various joint working groups had had several productive meetings; lunar samples from Apollos 11, 12, 14 and 15 and *Lunas* 16 and 20 were exchanged; and "real time" exchange of significant findings from the Mars probes, Mariner 9 and Mars 2 and 3, was accomplished.[159] Work is now going ahead in both countries and jointly toward the rendezvous and docking of an Apollo and *Soiuz* spacecraft in 1975. Implementation has gone ahead except for a brief delay at the start of the joint effort, when the USSR was about two weeks late in providing complete technical details of its existing docking equipment.[160] Recently, the Soviets have released technical details and depictions of the *Soiuz* space vehicle which will be visited by American astronauts in orbit in 1975.

The projects which were included in the May, 1972, executive agreement are the biggest and most dramatic step toward Soviet-American space cooperation. Whether they will lead to further advances is not yet clear and will depend in large measure on a relatively equal commitment to space research in the two countries and the continuance or improvement of the current atmosphere in which tensions between them have been notably eased.[161] The Apollo-*Soiuz* mission promises to be the first major space event in which direct *technological* cooperation has and will take place between the two countries. More importantly, it directly links the relative prestige of the two countries, which previously have been inversely related to one another in this respect. Should this become characteristic of the two national space programs generally, it will reverse the relationship of those space programs to the political relations between the US and the USSR, and to the extent that space research remains a dramatic endeavor for general publics, will exert a powerful force weakening the divisions and strengthening the political bonds between the two countries. The political significance of Apollo-*Soiuz* is considerably greater than the meager scientific results promised by the mission. However, it remains questionable whether the two countries will support an ambitious space program outside the context of political competition. Space has already lost some of its dramatic appeal to the general public, and, of course, unforeseen political clashes could topple the structure of direct space cooperation that has so recently been built.

Preliminary work on Apollo-*Soiuz* has already resulted in a number of advances in Soviet-American cooperation. The *Soiuz* vehicle will carry an American transceiver for communications with Apollo and the ground. Representatives from both countries will also be permitted to be present as observers and advisors at the manned control centers *and* at the launch sites, where Americans will get their first view of a Soviet launch.[162]

As a follow-on to Apollo-*Soiuz*, it has already been agreed that future spacecraft of both nations, such as the proposed American space shuttle, will have docking and crew transfer apparatus fully compatible with one another. The United States will in the future use the same space cabin atmosphere, equivalent to earth atmosphere at sea-level pressure, that has been employed in Soviet spacecraft all along.[163] The principal aim of compatible docking equipment is to make possible aid and rescue of one nation's spacemen by those of another. Although Alexei Kosygin in 1970 had already offered Soviet assistance to the crew of the endangered Apollo 13,[164] no nation has yet provided direct aid or rescue to the astronauts of another. As long as both the USSR and the USA pursue active manned space programs, this event becomes increasingly likely. It seems certain that such an occurrence would dramatize space cooperation and have a powerful impact on public opinion.

Among other suggestions by NASA to which the Soviets have not responded are those for coordination of the planetary programs of the two countries and the placement of one nation's experiments in the other's spacecraft.[165] These are real possibilities if US-USSR space cooperation continues on the upswing. The budgetary constraints currently felt by the space programs of both countries give space researchers and administrators a strong impetus to cooperation if the alternative would be that certain programs, such as manned visits to the planets, could not take place at all within the foreseeable future. Since NASA has not been given the go-ahead for planning manned planetary missions on its own, it has shown some interest in cooperating with the USSR, which has made manned colonization of the planets the proclaimed ultimate goal of its space program.[166]

Besides the Apollo-*Soiuz* mission, Soviet astronauts and scientists have already taken part in the Apollo 12 Lunar Science Conference in Houston in 1970.[167] Although there has as yet been no discussion of an astronaut-cosmonaut exchange,[168] a trend has been established toward closer work between the space researchers of the two countries and this can, of course, continue and grow. Applications activities of both countries, which offer direct and immediate benefits to the other, such as NASA's Earth Resources Technology Satellite program, also offer strong inducements to further cooperation. For example, the USSR has agreed to permit ERTS photographs of its territory and has received the imagery from the United States. The same is true in the field of space communications, although, as noted previously, the structures established for accomplishing this by the two countries are competitive and ideologically "loaded." Should the Soviets continue to lag behind the United States in space communications technology, they may overcome their ideological scruples and make some use of Comsat and/or Intelsat.

INTERNATIONAL ORGANIZATION FOR SPACE COOPERATION

The dominant pattern of both American and Soviet space cooperation has been bilateral. This section will focus on the involvement of the two space powers in multilateral, international organizations. The one aspect of space exploration in which international organizations have played the leading role to date is legal regulation of space activities. This will be discussed in the next chapter.

International space organizations may be divided into four types: (1) those that are almost entirely nongovernmental and nonpolitical, (2) those that are quasigovernmental, (3) those that directly involve states and arise out of agreements between states, and (4) those that may fall into any of the three foregoing categories but are regional or bloc-oriented in scope.

The International Astronautics Federation (IAF) is the one space-related organization that falls into the first category. The IAF is composed of scientists and scientific organizations without regard to national origin. So, for example, amateur organizations like the American Rocket Society and professional organizations like the American Astronautical Society take part in IAF. Space scientists working for governments take part in IAF only as individuals. American scientists have participated in the activities of the IAF since its founding in London in 1951.[169] The Soviet Academy of Sciences, officially a "public" rather than a "state" institution, applied for membership in IAF in 1956 and joined the Federation at its Seventh Congress in September of that year at which occasion the lone Soviet delegate, Academician Leonid Sedov, was selected as the Federation's vice-president.[170] Private organizations of the type of the American Rocket Society are comparatively rare in the Soviet Union, and their members are neither sufficiently prominent nor trustworthy to be allowed to go on junkets to exotic foreign cities. Private citizens would lack both permission and the necessary funds. The Soviet or other socialist member of IAF is likely to be more "official," and hence cautious, than most of his counterparts.[171]

The IAF has a certain political significance apart from bringing together in a congenial atmosphere the scientists of different countries in that it maintains an International Institute of Space Law for study and discussion of legal questions involving space.[172] The USSR has also made use of the IAF for political purposes by using its delegation to refute allegations that the Soviet manned program incorporated a disregard for safety and human life.[173] Soviet participation in IAF has increased since Academician Sedov forayed alone to Rome in 1956, but at a recent Congress, many of the scheduled Soviet speakers did not show. The continuing Soviet interest in the organization is indicated by the fact that the 1973 Congress (the 24th) was held in Baku in October.[147]

A multiplicy of international scientific unions exists for cooperation in the study of various scientific problems. These represent quasiofficial institutions such as national scientific unions as well as professional organizations in certain fields. The parent organization of these various unions is the International Council of Scientific Unions (ICSU); among them are the International Astronomical Union, Scientific Radio Union, Union of Geodesy and Geophysics, Pure and Applied Physics, Theoretical and Applied Mechanics, Biochemistry, Physiological Sciences, Biological Sciences and Mathematics.[175]

Although all of these unions make possible contact between scientists interested in space, including Soviet scientists, the first noteworthy contribution of ICSU to space cooperation was its sponsorship of the IGY. Although IGY did not set out to accomplish closely coordinated international cooperation, it did successfully implement an intensive international scientific effort and sharing of information in geophysics. The "year" lasted for 30 months from July, 1957, to December, 1959, and involved 30,000 scientists and observers from over 70 nations. As a part of IGY the United States launched eight satellites and three space probes; the Soviet Union Sputniks 1-4 and two space probes.[176] ICSU has sponsored programs similar to the IGY in recent years. These include the International Years of the Quiet Sun (1964–1965), the World Magnetic Survey, and the International Years of the Active Sun (1968–1970).[177] The IGY was successful in avoiding political controversy with the exception of its attempt to involve the "two Chinas." When the Nationalist Chinese accepted a belated invitation to join, scientists from the People's Republic quit.[178]

ICSU's greatest contribution to space cooperation was the formation in 1959 of the International Committee on Space Research (COSPAR). COSPAR was the result of the conviction of those who took part in organizing IGY that space cooperation ought not to end when the IGY came to a close.[179] As originally composed by ICSU, COSPAR consisted of (a) representatives of the scientific unions of seven countries actively involved in space research: Australia, Canada, France, Japan, USSR, United Kingdom and USA; (b) additional representatives on a rotating basis for the scientific unions of other countries otherwise involved in space research, and (c) representatives from nine different functional international scientific unions.[180]

The Soviet delegate expressed his dissatisfaction with the composition of the committee. He suggested that the Ukraine and Belorussia be added as permanent members and that the representatives of the scientific unions of six East European socialist states be added to the category of rotating members. The proposal was rejected, and the Soviet Academy withdrew from COSPAR until the committee was reformed by compromise in 1960.[181] The Soviet behavior was consistent with other claims to superpower status elsewhere, such as for the sovereign equality of the 15 Soviet Socialist Republics (or at least two of them) and for the political equality of the "socialist camp" with the "imperialist" and nonalligned states.

Eleven members were added to COSPAR as a part of the compromise. The inclusion of the socialist states, Czechoslovakia and Poland, stretched the rule which was retained, that full members were to be scientific unions whose home country was actively involved in space research. Other new members operated tracking stations in cooperation with NASA or were participants in one of the space organizations of Western Europe. One more functional international scientific union was added as well.[182] A bureau was formed for executive functions consisting of a president (H. C. Van de Hulst of the Netherlands), two vice-presidents (Richard Porter from the United States and Academician Blagonravov) and four additional members, two to be chosen from the slates of nominees compiled by the two vice-presidents. These turned out to be men from Poland, Czechslovakia, the United Kingdom and a Frenchman from the International Union of Theoretical and Applied Mechanics, a subtle touch by Mr. Porter, one suspects. The organization also maintains an executive council, four working groups and a secretariat. The Frenchman, Maurice Roy, has succeeded to the presidency.[183]

According to its charter, COSPAR is nonpolitical and does not recommend specific actions to national governments. It is interested, however, in learning about and providing a forum for discussions and proposals of international space cooperation.[184] COSPAR, of course, provides a setting for personal exchanges among scientists of various countries and indirectly involves national scientific institutions and governments.[185] Each year COSPAR representatives report on the space programs of their respective countries. In addition the organization sponsors symposia, compiles technical manuals and reference books, forms working groups on such problems as space law and space contamination, and relays announcements of launches and other space events to other scientific space organizations.[186] COSPAR cooperates with the UN, especially UNESCO, and with individual national governments interested in space cooperation.[187]

An example of this last kind of activity was the offer made to COSPAR members by NASA through the representative in COSPAR of the American National Academy of Sciences. NASA offered in 1959 to "support COSPAR" by launching the "suitable and worthy experiments proposed by scientists of other countries." Britain and Canada quickly took action to accept the American offer. Soviet delegates did not respond to this American action or put forward a similar proposal of their own.[188] The United States' delegates also took the initiative for the listing of all world space tracking stations with COSPAR, and their proposal was adopted and implemented by other nations, including the USSR.[189]

After initial difficulties over representation, Soviet representatives participated regularly in COSPAR. At first they were often unwilling or unable to provide much information about their national space program. They now contribute freely, but according to Dr. Lovell, COSPAR does not exhibit the "uninhibited" exchange characteristic of other scientific unions.[190] Publicly, Soviet space scientists have expressed the highest regard for IAF and COSPAR and their activities. In 1970 the Soviet Union hosted the thirteenth annual meeting of COSPAR in Leningrad.[191]

The third type of international space organization is that which involves direct participation of governments. This type includes the International Telecommunications Union (ITU), the World Meteorological Organization, the International Civil Aviation Organization, the World Health Organization, the International Atomic Energy Agency, the UN, UNESCO and the UN's Committee on the Peaceful Uses of Outer Space (COPUOS).[192]

Organized under the UN since 1949, the ITU has accepted the responsibility for negotiating radio frequency allocations for space and earth-space services. Previously it exercised this responsibility only for earth communications in accordance with international treaties involving nearly every nation on earth. The United States obtained ITU permission for use of the band on which its first satellites broadcast their signals; the Soviet Union did not do so for the first Sputniks and employed frequencies which occasionally interfered with earth radio traffic (and were readily receivable by most home shortwave sets). Despite American charges, the Sputniks did not violate ITU regulations. By 1959, when treaties extended the ITU regulations to space communications, the USSR complied.[193]

The Soviet Union was the first country to propose in March, 1958, the creation of an international space program under UN control which would be the exclusive agency through which national space programs would be carried out. Part of the proposal was the formation of an agency similar to what became COPUOS, and the proposal was part of a disarmament package which included the familiar Soviet demand for the liquidation of foreign military bases. The United States in responding wished to separate the issues of peaceful use of outer space and disarmament, but in fact disagreed in principle with *both* aspects of the Soviet proposal.[194]

The US in November, 1958, secured passage for its own proposal which set up a UN ad hoc committee to make recommendations concerning the peaceful uses of outer space. The ad hoc committee was to consist of the United States and eight of its allies, the Soviet Union and two of its allies, and six neutrals. The Soviet Union was not satisfied with the composition of the committee and the national origins of its officers. The US proposal was cautious, envisioning limited and exploratory functions for the committee.[195] The USSR had attempted to block the American proposal by dropping its demands for the elimination of bases and a UN space agency and substituting a proposal similar in scope to that of the United States, but with ad hoc membership of four socialist states, three NATO states and four neutrals. (They probably considered two of the latter, Sweden and Argentina, sufficiently "western.") Upon passage of the American proposal, the USSR, Czechoslovakia and Poland announced they would not participate in the work of the committee. Later India and the UAR dropped out as well.[196]

Although President Eisenhower had on one occasion spoken in favor of the conduct of all space activities under UN auspices,[197] the ad hoc committee rejected such an idea, suggesting instead merely that a permanent committee be established with an office in the Secretariat to facilitate international cooperation. On the basis of these recommendations, the US and USSR were able to reach a compromise on the membership of a reconstituted ad hoc committee, which was to confine itself to an examination of legal questions and to the review and study of international space cooperation, generally including the possibility of programs undertaken by UN agencies. The proposal for a permanent office within the Secretariat was dropped. The compromise proposal which passed the General Assembly unanimously allotted 12 committee seats to the West, seven to the Soviet bloc, and five to neutrals.[198]

For two years the new ad hoc committee was prevented from accomplishing much because it was split over voting procedure, the USSR and its allies insisting on the chairmanship and a two-thirds or unanimity rule, the Western states favoring a simple majority.[199] Academician Blagonravov demanded "parity" on the committee, reminding interested persons that the lack of it had forced the USSR to quit COSPAR that same year.[200] In spite of these difficulties the committee agreed to principles governing the registration, launch announcement and identification of spacecraft.[201] In late 1961 COPUOS was established as a permanent UN committee over further Soviet objections about parity. It included the members of the revised ad hoc committee plus three additional neutrals. Of the eight neutrals, only two (UAR and India) could then be considered as leaning toward the USSR. The three Latin American countries, two European neutrals, and Lebanon might then be viewed as pro-West. After a show of reluctance, the USSR took its seat on the committee.[202] This was about the time when the USSR shifted to a more cooperative stance in COPUOS, COSPAR and bilaterally with the USA.[203]

The activities of COPUOS have led to international agreements on the banning of "weapons of mass destruction" from outer space and the rescue of astronauts.[204] Soviet-American disagreement and rivalry, which has characterized almost every UN agency, is just as noticeable in COPUOS. The USSR for a time sought to impose various legal restrictions on space activities and opted for general rather than specific agreements. The United States by contrast has resisted international control of space "as a free society" and concentrated on arms control and narrow legal issues.[205]

Original hopes for a truly international space agency[206] have met with almost no success, the only exception being the establishment of an international range for sounding rockets in Tumba, India.[207]

The fourth and last type of international space organization is limited in scope to a single region or military political bloc. Examples of this type of organization include ESRO, the European Satellite Research Organization; ELDO, the European Launcher Development Organization, and Europace, a consortium of European industrial firms interested in manufacture and application of space technologies.[208] Technical and political considerations tie Comsat-Intelsat and *Intersputnik* to the political blocs led by the two space powers, although both organizations purport to be universal in scope.

ESRO and ELDO are both intergovernmental organizations aimed at ensuring that Europe will not be totally eclipsed in space techniques and applications by the two space powers or forced into relying entirely on the US and/or the USSR to dole out such benefits of space research to Europeans as they see fit. Both organizations were founded in 1964; ELDO with seven members, the original six of the Common Market less Luxemburg plus the United Kingdom and Australia; ESRO was founded by ten nations, the five EEC members of ELDO plus Denmark, Spain, Sweden, Switzerland and the United Kingdom. ESRO has concentrated on building satellites, ELDO on launch vehicles.[209] The Soviet Union, consistent with its other policies, has objected to West German membership in ELDO on the grounds that the Federal Republic might use space technology for military purposes.[210] Plans were made in 1973 to merge ESRO and ELDO into a common organization which would be a sort of NASA for the Common Market countries.[211]

As mentioned above, the USSR has also voiced objections to Intelsat in which the US through Comsat plays the leading role for largely technical and economic reasons. Intelsat was founded by 11 members of the Western alliance system, Ireland, Switzerland, and the Vatican City State, which the Soviets used to call the ideological capital of imperialism.[212] *Intersputnik* and the Intercosmos Council, of course, have their political and ideological centers in the mecca of "anti-imperialism," Moscow. A draft agreement for the establishment of *Intersputnik* was presented at the UN in 1968 by the six Warsaw Pact members of the UN, Cuba and Mongolia.[213] The UAR has since adhered to *Intersputnik*.[214] The organization has developed very slowly in comparison with Intelsat, which has been in active operation for several years. The USSR has yet to launch any 24-hour synchronous satellites (*Statsionar*) as promised for 1970 and 1971, while the United States successfully launched the first 24-hour synchronous satellite in 1963. Completion of a ground station in Cuba for *Intersputnik* was delayed until 1973.[215]

The Intercosmos Council of the Soviet Academy of Sciences is chaired by Academician Boris N. Petrov with Vladen Vereshchetin as vice-chairman. It was founded in 1967. Delegates from nine socialist states (Warsaw Pact members, Mongolia and Cuba) meet annually.[216] At least in form, the Intercosmos Council more clearly establishes multilateral principles than comparable American programs outside the field of space communications, but it is nonetheless dominated by the USSR.

In summary, one can conclude that neither the United States nor the Soviet Union to date has made any large-scale contributions to the development of a meaningful role in space exploration and research for international organizations of any kind, especially universal organizations like the UN. The United States has been fearful of having any kind of controls imposed on its national space activities, and the Soviet Union has maintained a passion for secrecy and a proclivity for troublesome political maneuvering over questions of parity for itself and the socialist camp. Both space powers have relied largely on *bilateral* mechanisms to implement space cooperation, and the most dramatic events affecting space cooperation in recent years have been agreements designed to implement bilateral space cooperation between the two space powers themselves.

NOTES

[1]Calculated from Table I figures in Charles S. Sheldon II, *United States and Soviet Progress in Space: Summary Data Through 1973 and a Forward Look*, Library of Congress, Congressional Research Service, January 8, 1974, p. 11.

[2]Arnold W. Frutkin, *International Cooperation in Space* (Englewood Cliffs, N.J.: Prentice-Hall, 1965), p. 60.

[3]Don E. Kash, *The Politics of Space Cooperation* (Purdue University Studies, Purdue Research Foundation, 1967), p. 20; and G. Ivanov, "Eurospace: Rocket Race," *International Affairs*, No. 12, December 1964, pp. 101-102.

[4]Quoted in Kash, *supra*, pp. 32-33.

[5]*Ibid.*, p. 19.

[6]Sir Bernard Lovell, "The Great Competition in Space," *Foreign Affairs*, 51 (1), October 1972, pp. 124-125.

[7]Frutkin, *International Cooperation in Space, op. cit.*, pp. 102-103; and Charles S. Sheldon II, "Summary" in *Soviet Space Programs, 1966-1970, op. cit.*, xxxvi.

[8]Frutkin, *supra*, p. 18.

[9]*Ibid.*, pp. 14-15.

[10]*Ibid.*, pp. 11-12, 15.

[11]See the statement attributed to Senator Lyndon B. Johnson by Kash, *op. cit.*, p. 11.

[12]Frutkin, *International Cooperation in Space, op. cit.*, pp. 74, 73.

[13]US Congress, House, Committee on Appropriations, *Independent Offices Appropriations for 1963, Hearings Before a Subcommittee*, 87th Congress, 2d Session, 1962,

Part 3, pp. 418-419. Emphasis added.

[14]According to Kash, *op. cit.*, pp. 16, 17.

[15]US Congress, House, Committee on Science and Astronautics, Subcommittee on Applications and Tracking and Data Acquisition, *1964 NASA Authorization*, Hearings, 88th Congress, 1st session, 1963, Part 4, p. 2850; and Vernon Van Dyke, *Pride and Power: The Rationale of the Space Program* (Urbana: University of Illinois, 1964), p. 240.

[16]Frutkin, *International Cooperation in Space, op. cit.*, p. 103.

[17]*Ibid.*, p. 78.

[18]*Ibid.*, p. 73; and Leon Sloss, "Unilateral Space Observation and Atlantic Alliance," in Frederick J. Ossenbeck and Patricia C. Kroeck, editors, *Open Space and Peace, A Symposium on Effects of Observation* (Stanford, California: Stanford University, The Hoover Institute, 1964), p. 161.

[19]Kash, *op. cit.*, pp. 18, 19; and Van Dyke, *op. cit.*, p. 234.

[20]Frutkin, *International Cooperation in Space, op. cit.*, p. 73.

[21]See Joseph G. Whelan, "Soviet Attitude Toward International Cooperation in Space," in *Soviet Space Programs, 1966-1970, op. cit.*, pp. 443, 448; also D. Gvishiani, "Soviet Scientific and Technical Cooperation with Other Countries," *International Affairs*, Nos. 2-3, February-March, 1970, pp. 46-52. Gvishiani is a vice-chairman on the State Committee of Science and Technology.

[22]Kash, *op. cit.*, p. 20. See also the remarks of Leonid Sedov and S. Petrov in Sheldon, "Projections of Soviet Space Plans," in *Soviet Space Programs, 1966-1970, op. cit.*, pp. 365, 382.

[23]Herman Pollack, Director, Bureau of International Science and Technical Affairs, Department of State, "Impact of the Space Program on America's Foreign Relations," in *International Cooperation in Outer Space, Symposium, op. cit.*, p. 601.

[24]Kash, *op. cit.*, p. 21.

[25]See *ibid.*, p. 20; and T. Keith Glennan, "The Task for Government," in American Assembly, Columbia University, *Outer Space Prospects for Man and Society* (Englewood Cliffs, N.J.: Prentice-Hall, Inc., 1962), p. 102.

[26]Charles S. Sheldon II, "Overview, Supporting Facilities and Launch Vehicles of the Soviet Space Program," in *Soviet Space Programs, 1966-1970, op. cit.*, pp. 150-152; and *Soviet Space Programs, 1971, op. cit.*, pp. 58-60, 72-75.

[27]Kash, *op. cit.*, p. 12; and Pollack, *op. cit.*, p. 601.

[28]Such as Lyndon B. Johnson in his introduction, "The Politics of the Space Age," to Lillian Levy, editor, *Space: Its Impact on Man and Society* (New York: W. W. Norton and Co., Inc., 1965), p. 7; "New Stage in Space Research," an interview with Academician L. I. Sedov, *New Times*, No. 9, March 3, 1971, pp. 23-24.

[29]Kash, *op. cit.*, p. 13; and Lovell, "The Great Competition in Space," *op. cit.*, p. 138.

[30]Hessian Television Service, March 20, 1968, as cited in Whelan, "Soviet Attitude Toward International Cooperation in Space," *Soviet Space Programs, 1966-1970, op. cit.*, p. 404.

[31]TASS International Service, Moscow, January 22, 1969, as cited in *ibid.*, 404n.

[32]Arnold W. Frutkin, "Patterns of International Space Applications and Their Extension," American Astronautical Society, Reprint 67-146, May 2, 1967, p. 385.

[33]Whelan, "Soviet Attitude Toward International Cooperation in Space," in *Soviet Space Programs, 1966-1970, op. cit.*, p. 451.

[34]See Sheldon, *United States and Soviet Progress in Space* (1972), *op. cit.*, p. 71.

[35]See, for example, the remarks of Academician Blagonravov cited by Whelan, "Soviet Attitude Toward International Cooperation in Space," in *Soviet Space Pro-*

grams, 1966-1970, op. cit., p. 404.

[36]See *ibid.,* p. 405.

[37]Jerome B. Wiesner in a report to President John F. Kennedy as cited by Kash, *op. cit.,* p. 10.

[38]Pollack, *op. cit.,* p. 601.

[39]TASS dispatch of October 14, 1970, as cited by Nicholas Daniloff, *The Kremlin and the Cosmos* (New York: Alfred A. Knopf, 1972), p. 184.

[40]See Whelan, "Soviet Attitude Toward International Cooperation in Space," in *Soviet Space Programs, 1966-1970, op. cit.,* pp. 403, 443-444, 447, 448; and also D. M. Gvishiani, *op. cit.,* pp. 46-52; "International Cooperation in Science and Technology," an interview with D. M. Gvishiani, *New Times,* No. 3, January 20, 1970, p. 7.

[41]Frutkin, *International Cooperation in Space, op. cit.,* p. 20.

[42]*Ibid.,* pp. 28-30.

[43]US Congress, House, Select Committee on Astronautics and Space Exploration, *Hearings on HR 11881, Astronautics and Space Explorations,* 85th Congress, 2d session, 1958, p. 135.

[44]Van Dyke, *op. cit.,* p. 236.

[45]Frutkin, *International Cooperation in Space, op. cit.,* p. 37.

[46]See Kash, *op. cit.,* p. 29.

[47]See Simon Bourgin, "Impact of US Space Cooperation Abroad," in *International Cooperation in Outer Space, Symposium, op. cit.,* pp. 166-172.

[48]Kash, *op. cit.,* p. 30.

[49]Frutkin, *International Cooperation in Space, op. cit.,* p. 38.

[50]*Ibid.,* p. 73.

[51]*Ibid.,* p. 35.

[52]*International Programs,* Pamphlet prepared by the Office of International Programs, NASA, January 1963, p. 1.

[53]Kash, *op. cit.,* p. 43; and Lincoln P. Bloomfield, *The Peaceful Uses of Outer Space,* Public Affairs Pamphlet No. 331, 1962, p. 19.

[54]Kash, *op. cit.,* pp. 33-34, 40, 52.

[55]*Ibid.,* p. 50; and G. Khozin, "Pentagon Seeking Control of Space Research in Asia and Africa," *International Affairs,* No. 2, February 1968, pp. 30-34.

[56] Kash, *op. cit.,* pp. 60-61. See also Whelan, "Soviet Attitude Toward International Cooperation in Space," in *Soviet Space Programs, 1966-1970, op. cit.,* p. 422. An exchange of visits by the crews of the Apollo-*Soiuz* mission to Soviet and American launch facilities took place for the first time in 1975.

[57]Kash, *supra,* pp. 50-51.

[58]See Frutkin, *International Cooperation in Space, op. cit.,* p. 72; and Van Dyke, *op. cit.,* p. 233.

[59]Glennan, *op. cit.,* p. 102.

[60]Kash, *op. cit.,* p. 66.

[61]Frutkin, *International Cooperation in Space, op. cit.,* p. 41; and Frutkin, "International Progress of NASA," *Bulletin of the Atomic Scientists,* 17 (5-6), May-June 1961, pp. 229-232.

[62]Frutkin, "NASA's International Space Activities," *International Cooperation in Outer Space, Symposium, op. cit.,* pp. 13-15. *International Programs,* Pamphlet prepared by the Office of International Programs, NASA, July 1967, p. 37.

[63]See *Space Agreements with the Soviet Union,* Hearing Before the Committee on Aeronautics and Space Sciences, US Senate, 92nd Congress, 2d session, June 23, 1972, pp. 82-97.

[64]Kash, *op. cit.,* pp. 40-41; and Bloomfield, *The Peaceful Uses of Space, op. cit.,* p. 19.

[65]Seth T. Payne and Leonard S. Silk, "The Impact on the American Economy," in Bloomfield, editor, *Outer Space: Prospects for Man and Society* (New York: Frederick A. Praeger, Inc., 1968), p. 96; and Sheldon, *United States and Soviet Progress in Space, op. cit.,* pp. 39-40.

[66]*Aviation Week and Space Technology, 99* (21), November 19, 1973, pp. 15-16; the editorial by Robert Hotz, "A Better Aerospace Budget," *Aviation Week and Space Technology, 100* (6), February 11, 1974, p. 7; and *Space World,* K-4-124, April 1974, pp. 12-13.

[67]Frutkin, "NASA's International Space Activities," *op. cit.,* pp. 15-20.

[68]See Donald E. Fink, "Europeans See Wise Use of Spacelab," *Aviation Week and Space Technology, 99* (19), November 5, 1973, pp. 42-47. So far, the bulk of the development of Spacelab has been done by West Germany. Italy, France, Britain, Belgium, Spain, the Netherlands, Denmark and Switzerland are participating to a much smaller extent.

[69]See Craig Covault, "Planet Flight Cooperation Sought," *Aviation Week and Space Technology, 100* (8), February 25, 1974, pp. 12-13.

[70]Erin B. Jones, *Earth Satellite Telecommunications Systems and International Law* (Austin: University of Texas, 1970), pp. 105-106.

[71]Joseph M. Goldsen, *Research on Social Consequences of Space Activities,* Rand Corporation, Paper #P-3220, August 1965, pp. 5-6.

[72]John A. Johnson (Vice-President, Comsat), "The International Activities of the Communications Satellite Corporation," *International Cooperation in Outer Space, Symposium, op. cit.,* pp. 196-214. The reorganization of Intelsat is described on pp. 609-637 of the same Senate document. See also Jones, *op. cit.,* pp. 114-115.

[73]Khozin, *op. cit.,* p. 32.

[74]Frutkin, *International Cooperation in Space, op. cit.,* pp. 19-21.

[75]Vladen Vereshchetin, "Intercosmos: Results and Prospects," *Space World,* I-1-97, January 1972, pp. 38-39.

[76]*USSR Probes Space* (Moscow: Novosti Press Agency Publishing House), no date, no pagination; and Hugh Odishaw, "Science and Space," in Bloomfield, editor, *Outer Space: Prospects for Man and Society, op. cit.,* p. 91.

[77]Sheldon, "Overview, Supporting Facilities and Launch Vehicles of the Soviet Space Program," *op. cit.,* pp. 150-151.

[78]Barbara M. DeVoe, "Soviet Applications of Space to the Economy," in *Soviet Space Programs, 1966-1970, op. cit.,* pp. 301-302, 304; and Sheldon, *United States and Soviet Progress in Space* (1973), *op. cit.,* p. 68.

[79]Sheldon, "Program Details of Unmanned Flights," and Whelan, "Soviet Attitude Toward International Cooperation in Space," in *Soviet Space Programs, 1966-1970, op. cit.,* pp. 188-190, 439-440; *Soviet Space Programs, 1971, op. cit.,* p. 11; and "Satellite Report," *Space World,* K-6-126, June 1974, p. 30.

[80]See D. Gvishiani, "Soviet Scientific and Technical Cooperation with Other Countries," *op. cit.,* pp. 46-52.

[81]See Georgii Pokrovskii, general and professor, "Profitability of the Peaceful Uses of Outer Space," *International Affairs,* No. 9, September 1962, pp. 117-118.

[82]F. J. Krieger in Jerry and Vivian Grey, eds., *Space Flight Report to the Nation* (New York: Basic Books, Inc., 1962), p. 189.

[83]See V. Aldoshin, "Outer Space Must Be a Peace Zone," *International Affairs,* No. 12, December 1968, pp. 38-39; Joseph G. Whelan, "Soviet Attitude Toward International Cooperation in Space," in *Soviet Space Programs, 1962-1965, op. cit.,* pp. 473-474; and DeVoe, "Soviet Applications of Space to the Economy," *op. cit.,* p. 304.

[84]Text in Appendix C of *Soviet Space Programs, 1966-1970, op. cit.,* pp. 597-602.

[85]US Congress, House, Committee on Science and Astronautics, *Hearings on NASA Authorization, FY 1971,* 91st Congress, 2d session, February 1970, Part 1, p. 14.

[86]Barbara M. DeVoe, "Soviet Applications of Space to the Economy," *op. cit.,* pp. 304-305; Sheldon, "Projections of Soviet Space Plans," *op. cit.,* p. 370 (Keldysh); and Whelan, "Soviet Attitude Toward International Cooperation in Space," in *Soviet Space Programs, 1966-1970, op. cit.,* p. 449.

[87]*Soviet Space Programs, 1971, op. cit.,* pp. 26, 62; Vereshchetin, *op. cit.,* pp. 38-39.

[88]Whelan, "Soviet Attitude Toward International Cooperation in Space," in *Soviet Space Programs, 1962-1965, op. cit.,* p. 475; and DeVoe, "Soviet Applications of Space to the Economy," *op. cit.,* pp. 303-304.

[89]K. L. Plummer, "Soviet-French Cooperation in Space," in Appendix F of *Soviet Space Programs, 1966-1970, op. cit.,* p. 614. The appendix consists of Plummer's article of the same title which appeared in *Spaceflight, 12,* September 1970, pp. 360-361. See also *USSR Probes Space, op. cit.*

[90]Plummer, *supra,* p. 611.

[91]Sheldon, "Overview, Supporting Facilities and Launch Vehicles of the Soviet Space Program," *op. cit.,* p. 126. Alexei Kosygin visited France in December 1966.

[92]Alton Frye, "Soviet Space Activities: A Decade of Pyrrhic Politics," in Bloomfield, editor, *Outer Space: Prospects for Man and Society, op. cit.,* p. 204.

[93]Plummer, *op. cit.,* pp. 611-613.

[94]Sheldon, "Overview, Supporting Facilities and Launch Vehicles of the Soviet Space Program," *op. cit.,* p. 126.

[95]According to NASA Deputy Administrator George Low in *Space Cooperation Between the US and the Soviet Union,* Hearings Before the Committee on Aeronautics and Space Sciences, US Senate, 92nd Congress, 1st session, March 17, 1971, p. 11.

[96]*Soviet Space Programs, 1971, op. cit.,* pp. 14, 19, 21, 26, 62; Plummmer, *op. cit.,* p. 614; and "Satellite Report" in *Space World,* K-6-126, June 1974, p. 30. The French got their sample of .lunar soil three weeks before the Czechs and two months before the US.

[97]Sheldon, *United States and Soviet Progress in Space* (1973), *op. cit.,* p. 48.

[98]*Soviet Space Programs, 1971, op. cit.,* p. 58; *USSR Probes Space, op. cit.*; and Sheldon, "Overview, Supporting Facilities and Launch Vehicles of the Soviet Space Program," *op. cit.,* p. 150.

[99]According to NASA's Homer Newell, US Congress, House Subcommittee of the Committee on Appropriations, *Hearings, Independent Offices for 1963,* 87th Congress, 2d session, 1962, Part III, p. 413.

[100]*USSR Probes Space, op. cit.*; and Frutkin, *International Cooperation in Space, op. cit.,* p. 62.

[101]"Soviets Planning Extensive IQSY Effort," *Aviation Week, 81* (8), August 24, 1964, p. 67.

[102]*USSR Probes Space, op. cit.*

[103]DeVoe, "Soviet Application of Space to the Economy," *op. cit.,* p. 304.

[104]See "Soviets Seen Downgrading Space Congress," *Aviation Week and Space Technology, 97* (16), October 16, 1972, p. 15; and Sheldon, "Overview, Supporting Facilities and Launch Vehicles of the Soviet Space Program," *op. cit.,* p. 128.

[105]Kash, *op. cit.,* pp. 33-34.

[106]*Ibid.,* p. 45.

[107]*Soviet Space Programs, 1962, op. cit.,* p. 245; and Academicians Keldysh and

Blagonravov as cited by Whelan, "Soviet Attitude Toward International Cooperation in Space," *Soviet Space Programs, 1966-1970, op. cit.,* p. 403.

[108]Whelan, *supra,* pp. 417, 426-428.

[109]See Frutkin, "Patterns of International Space Applications and Their Extension," *op. cit.,* pp. 379, 385, 389, who paradoxically had argued earlier that the US had more to gain technologically by cooperation with Western Europe. By 1967 he had changed his tune.

[110]Alton Frye, *The Proposal for a Joint Lunar Expedition: Background and Prospects,* Rand Corporation, Paper #P-2808, January 1964, p. 17; Sheldon, *Review of the Soviet Space Program, With Comparative US Data* (New York: McGraw-Hill Book Co., Inc., 1968), p. 83; and Dr. Thomas O. Paine in US Congress, Senate, Committee on Aeronautics and Space Sciences, *NASA Authorization for Fiscal Year 1970,* Hearings, 91st Congress, 1st session, 1969, Part 1, p. 32.

[111]See Craig Covault, "Soviet Plan To Buy Apollo Suits for Comparative Study Purposes," *Aviation Week and Space Technology, 98* (13), March 26, 1973, pp. 17-18.

[112]Sheldon, *supra,* p. 83; Kash, *op. cit.,* p. 83; *Soviet Space Programs, 1962, op. cit.,* pp. 244-245; Testimony of NASA Deputy Administrator George Low in *Space Cooperation Between the US and USSR, 1971, op. cit.,* p. 5; Remarks of Academician Blagonravov in Whelan, "Soviet Attitude Toward International Cooperation in Space," *Soviet Space Programs, 1966-1970, op. cit.,* p. 404; and Testimony of George Low in *Space Agreements with the Soviet Union, 1972, op. cit.,* p. 22.

[113]See Whelan, "Soviet Attitude Toward International Cooperation in Space," *Soviet Space Programs, 1966-1970, op. cit.,* pp. 405-406.

[114]Testimony of Geogre Low before the Senate Committee on Aeronautics and Space Sciences, *Space Cooperation Between the US and USSR, 1971, op. cit.,* p. 5; Senator Lyndon B. Johnson as cited by Van Dyke, *op. cit.,* p. 77; President Lyndon B. Johnson in Levy, *op. cit.,* p. 7.

[115]Pollack, *op. cit.,* p. 601; and Blagonravov, Brezhnev, Keldysh and Sedov as cited by Whelan, "Soviet Attitude Toward International Cooperation in Space," *Soviet Space Programs, 1966-1970, op. cit.,* pp. 403-404.

[116]See the remarks of Blagonravov in Whelan, *supra,* p. 404.

[117]See Kash, *op. cit.,* p. 89.

[118]See Klaus Knorr, "On the International Implications of Outer Space Activities," in Joseph M. Goldsen, editor, *International Political Implications of Activities in Outer Space; A Report of a Conference,* Rand Corporation, Report #R-362-RC, May 1961, p. 151.

[119]Amitai Etzioni, *The Moon-Doggle, Domestic and International Implications of the Space Race* (Garden City, N.Y.: Doubleday and Co., Inc., 1964), xii-xiii.

[120]*New York Times,* September 22, 1963, p. 1; Arthur Schlesinger, Jr., *A Thousand Days* (Boston, Mass.: Houghton-Mifflin, 1965), pp. 361, 373. Schlesinger reports Khrushchev as replying, "Why not?" to Kennedy's suggestion but changing his mind a day later.

[121]Etzioni, *op. cit.,* xiv.

[122]See *Department of State Bulletin, 49* (1267), October 7, 1963, pp. 532-533.

[123]*Department of State Bulletin, 49* (1279), December 30, 1963, p. 1011; and Lyndon B. Johnson's State of the Union speech in US Congress, House, *House Document No. 251,* 88th Congress, 2d session, January 8, 1964, p. 6.

[124]Schlesinger, *op. cit.,* pp. 361, 367, 373; and Sheldon, *United States and Soviet Progress in Space, op. cit.,* p. 71, who says offers by both sides for a joint lunar mission were largely "rhetorical."

[125]Schlesinger, *supra*, pp. 919-922; Frye, *The Proposal for a Joint Lunar Expedition . . . , op. cit.*, pp. 6, 11-14; and Sir Bernard Lovell, "Soviet Aims in Astronomy and Space Research," *Bulletin of the Atomic Scientists, 19* (8), October 1963, pp. 36-39.

[126]Lillian Levy in Levy, *op. cit.*, p. 200.

[127]See Frye, *The Proposal for a Joint Lunar Expedition . . . , op. cit.*, p. 7; and Frutkin, *International Cooperation in Space, op. cit.*, p. 105.

[128]Frye, *supra*, pp. 3-6.

[129]*Missiles and Rockets, 13*, September 23, 1964, p. 14.

[130]See Frye, *The Proposal for a Joint Lunar Expedition . . . , op. cit.*, pp. 8-9.

[131]Hugh Odishaw, "Science and Space," in Bloomfield, *Outer Space, Prospects for Man and Society, op. cit.*, p. 91; and Hugh Dryden in US Congress, House, Committee on Appropriations, *Independent Offices Appropriations for 1964*, Hearings Before a Subcommittee on Appropriations, 88th Congress, 1st session, 1963, Part 3, pp. 104-105.

[132]*International Cooperation and Organization for Outer Space, 1965, op. cit.*, p. 158.

[133]Van Dyke, *op. cit.*, p. 160.

[134]John R. Walsh, "NASA: Talk of Togetherness with Soviets Further Complicates Space Politics for the Agency," *Science, 142* (3588), October 4, 1963, p. 37.

[135]Frutkin, *International Cooperation in Space, op. cit.*, pp. 1-2; Van Dyke, *op. cit.*, p. 246.

[136]Kash, *op. cit.*, pp. 85-86.

[137]See Frye, *The Proposal for a Joint Lunar Expedition . . . , op. cit.*, pp. 18-19.

[138]See Frederick C. Barghoorn, *Politics in the USSR* (Boston: Little, Brown and Co., 1972), pp. 291-292.

[139]Frutkin, *International Cooperation in Space, op. cit.*, p. 85.

[140]*Ibid.*, p. 89; and *Department of State Bulletin, 42*, January 11, 1960, p. 62.

[141]See Kash, *op. cit.*, pp. 68-69.

[142]Frutkin, *International Cooperation in Space, op. cit.*, p. 90.

[143]*Ibid.*, pp. 90-91; and NASA Deputy Administrator Hugh Dryden in Grey, *op. cit.*, p. 191.

[144]Whelan, "Soviet Attitude Toward International Cooperation in Space," in *Soviet Space Programs, 1962-1965, op. cit.*, p. 430.

[145]Frutkin, *International Cooperation in Space, op. cit.*, pp. 92-95; *US International Space Programs*, Texts of Executive Agreements, Memoranda of Understanding and other International Arrangements, 1959-1965, Staff Report prepared for Committee on Aeronautics and Space Sciences, US Congress, Senate, 89th Congress, 1st session, Senate Document #44, July 30, 1965, p. 409.

[146]*Soviet Space Programs, 1962, op. cit.*, pp. 233-235; Text of letters in *Department of State Bulletin, 46* (1188), April 2, 1962, pp. 536-538; and *International Cooperation and Organization for Outer Space, 1965, op. cit.*, pp. 139-140.

[147]Whelan, "Soviet Attitude Toward International Cooperation in Space," *Soviet Space Programs, 1966-1970, op. cit.*, pp. 406-408.

[148]*Ibid.*, pp. 408-410.

[149]Leonard E. Schwartz, "International Space Organizations," in Odishaw, editor, *The Challenges of Space* (Chicago: University of Chicago Press, 1963), p. 258; and Van Dyke, *op. cit.*, p. 260.

[150]Donald E. Fink, "Soviets Block Space Data Exchange Pact," *Aviation Week and Space Technology, 81* (15), October 12, 1964, p. 25.

[151]Whelan, "Soviet Attitude Toward International Cooperation in Space," *Soviet*

Space Programs, 1962-1965, op. cit., pp. 473-474. The Soviets were right.

[152]Whelan, "Soviet Attitude Toward International Cooperation in Space," *Soviet Space Programs, 1966-1970, op. cit.,* pp. 419-420.

[153]*Ibid.,* pp. 420-422, 427-429.

[154]US Department of State, "US-USSR Cooperation in Space Research," Appendix E of *Soviet Space Programs, 1966-1970, op. cit.,* pp. 608-610.

[155]Whelan, "Political Goals and Purposes of the USSR in Space," *Soviet Space Programs, 1966-1970, op. cit.,* p. 42.

[156]NASA Press Release #70-210, December 9, 1970, in Appendix G of *Soviet Space Programs, 1966-1970, op. cit.,* pp. 615-616.

[157]NASA Press Release #71-57, March 31, 1971, in Appendix H of *Soviet Space Programs, 1966-1970, op. cit.,* pp. 617-622.

[158]Testimonies of U. Alexis Johnson, Under Secretary of State for Political Affairs, and George M. Low, deputy administrator of NASA, Before the Committee on Astronautics and Space Sciences, *Space Agreements with the Soviet Union, June 23, 1972, op. cit.,* pp. 13-15, 16-17, 53-57, 70-71. Text of executive agreement is on pp. 41-42.

[159]*Ibid.,* p. 57.

[160]*Space Cooperation Between the US and the Soviet Union, 1971, op. cit.,* p. 17.

[161]See Lovell, "The Great Competition in Space," *op. cit.,* pp. 132-133.

[162]See *Aviation Week and Space Technology,* 99 (19), November 5, 1973, pp. 20-21; *Aviation Week and Space Technology,* 98 (1), January 1, 1973, p. 13; and *Aviation Week and Space Technology,* 98 (2), January 8, 1973, p. 17.

[163]Testimony of Dr. George Low, *Space Agreements with the Soviet Union, op. cit.,* p. 26.

[164]Whelan, "Political Goals and Purposes of the USSR in Space," *Soviet Space Programs, 1966-1970, op. cit.,* p. 42.

[165]Every Driscoll, "The Soviet Space Program," *Science News* 99 (18), May 1, 1971, p. 304.

[166]Craig Covault, "Planet Flight Cooperation Sought," *op. cit.,* p. 12.

[167]Driscoll, *op. cit.,* p. 305.

[168]Testimony of Dr. George Low, *Space Cooperation Between the US and Soviet Union, op. cit.,* p. 14. Astronauts and cosmonauts have exchanged visits as part of the training for Apollo-*Soiuz.*

[169]Frutkin, *International Cooperation in Space, op. cit.,* p. 165; and Schwartz, *op. cit.,* p. 254.

[170]F. J. Krieger, *Behind the Sputniks, A Survey of Soviet Space Science* (Washington, D.C.: Rand Corporation, Public Affairs Press, 1958), p. 7.

[171]See Kash, *op. cit.,* pp. 58-60.

[172]Schwartz, *op. cit.,* p. 255.

[173]"Soviet Space Safety," *Aviation Week and Space Technology,* 71 (10), September 7, 1959, p. 28.

[174]"Soviets Seen Downgrading Space Congress," *op. cit.,* p. 15.

[175]Schwartz, *op. cit.,* p. 252; H. C. Van de Hulst, "COSPAR and Space Cooperation," in Hugh Odishaw, *op. cit.,* p. 293; see also Leonard E. Schwartz, *International Organizations and Space Cooperation* (Durham, N.C.: World Rule of Law Center, 1962).

[176]Jones, *op. cit.,* p. 99.

[177]Odishaw, "Science and Space," *op. cit.,* p. 87; Schwartz, "International Space Organizations," *op. cit.,* p. 249.

[178]Kash, *op. cit.,* p. 3.

[179]Frutkin, *International Cooperation in Space, op. cit.,* p. 37; Odishaw, "Science and Space," *op. cit.,* p. 86; Schwartz, "International Space Organizations," *op. cit.,* p. 243.

[180]Schwartz, *supra,* p. 244.

[181]*Loc. cit.*

[182]See Van de Hulst, "COSPAR and Space Cooperation," *op. cit.,* p. 293.

[183]*Ibid.,* pp. 293-294; Jones, *op. cit.,* pp. 100-101; Schwartz, "International Space Organizations," *op. cit.,* p. 245; see also H. C. Van de Hulst, "International Space Cooperation," *Bulletin of the Atomic Scientists, 17* (5-6), May-June 1961, pp. 233-236.

[184]Van de Hulst as cited by Schwartz, *supra,* p. 248.

[185]*Ibid.,* p. 250.

[186]Odishaw, "Space and Science," *op. cit.,* pp. 86-87; Sheldon, *Review of the Soviet Space Program, op. cit.,* p. 131.

[187]Jones, *op. cit.,* pp. 102-103.

[188]Frutkin, *International Cooperation in Space, op. cit.,* pp. 37-38, who cites the letter of March 14, 1959, from Dr. R. W. Porter to Professor Dr. H. C. Van de Hulst.

[189]Kash, *op. cit.,* p. 69.

[190]Lovell, "The Great Competition in Space," *op. cit.,* pp. 130-131.

[191]Whelan, "Soviet Attitude Toward International Cooperation in Space," *Soviet Space Programs, 1966-1970, op. cit.,* pp. 442-443.

[192]Schwartz, "International Space Organizations," *op. cit.,* p. 260. See also Schwartz, *International Organizations and Space Cooperation, op. cit.*

[193]Philip C. Jessup and Howard J. Taubenfeld, *Controls for Outer Space and the Antarctic Analogy* (New York: Columbia University Press, 1959), pp. 237-238; and Jones, *op. cit.,* p. 124.

[194]Kash, *op. cit.,* pp. 99-100. See UN General Assembly, "Questions of the Peaceful Use of Outer Space," UN Document A/C.1/L 219, November 7, 1958, 2 pp.; *International Cooperation in Space, 1965, op. cit.,* pp. 184-187; UN Document C.1/L220/ Rev. 1, November 14, 1958.

[195]Kash, *op. cit.,* pp. 100-102; and "Statement by Mr. Lodge in the UN," *Department of State Bulletin, 39* (1016), December 15, 1958, p. 976.

[196]Jessup and Taubenfeld, *op. cit.,* pp. 255-256; UN Document No. A/C.1/L219/ Rev. 1, November 18, 1958; UN Document No. A/C.1/SR. 995, November 24, 1958, p. 15.

[197]In his 1960 address to the UN as cited by Jones, *op. cit.,* p. 27.

[198]Kash, *op. cit.,* pp. 104-106; "Statement of Mr. Lodge in Committee I," *Department of State Bulletin, 42* (1072), January 11, 1960, pp. 65-66.

[199]Kash, *op. cit.,* pp. 107, 113.

[200]A. A. Blagonravov, "Space Research and International Collaboration," *Current Digest of the Soviet Press, 11* (18), June 3, 1959, p. 26, from *Izvestiia,* May 6, 1959, p. 2.

[201]Myres S. McDougal, Harold D. Lasswell and Ivan A. Vlasic, *Law and Public Order in Space* (New Haven: Yale University Press, 1963), pp. 572-573. According to the registry which describes the purpose of each satellite, no country has ever launched a space vehicle designed to accomplish a military mission!

[202]*International Cooperation in Space, 1965, op. cit.,* p. 194; UN Document A/C.1/ 857, November 15, 1961, p. 4; Whelan, "Soviet Attitude Toward International Cooperation in Space," *Soviet Space Programs, 1962-1965, op. cit.,* p. 430.

[203]See *Soviet Space Programs, 1962, op. cit.,* pp. 234-235.

[204]See McDougal, Lasswell and Vlasic, *op. cit.*, p. 580.

[205]Kash, *op. cit.*, pp. 115-117; Nicholas deB. Katzenbach, "The Law in Outer Space," in Levy, *op. cit.*, p. 76 (source of quote); James R. Killian, Jr., "Shaping a Policy for the Space Age," in Bloomfield, *Outer Space: Prospects for Man and Society, op. cit.*, pp. 239-240; and Frutkin, *International Cooperation in Space, op. cit.*, p. 62.

[206]See Donald G. Brennan, "Arms and Arms Control in Outer Space," in Bloomfield, *supra*, p. 171.

[207]Frutkin, *International Cooperation in Space, op. cit.*, p. 62.

[208]Schwartz, "International Space Organizations," *op. cit.*, p. 262.

[209]Odishaw, "Space and Society," *op. cit.*, pp. 88-89.

[210]Frutkin, *International Cooperation in Space, op. cit.*, p. 147.

[211]See *Aviation Week and Space Technology*, 98 (1), January 1, 1973, pp. 14-15.

[212]Jones, *op. cit.*, pp. 114-115.

[213]See Domas Krivickas and Armins Rusis, "Soviet Attitude Toward Law of Outer Space," *Soviet Space Programs, 1966-1970, op. cit.*, pp. 491-498; text in Appendix C of *ibid*, pp. 597-602; UN Document A/AC. 105/46, August 9, 1968.

[214]Whelan, "Soviet Attitude Toward International Cooperation in Space," *Soviet Space Programs, 1966-1970, op. cit.*, p. 438.

[215]Sheldon, *United States and Soviet Progress in Space* (1973), *op. cit.*, p. 68.

[216]Vladen Vereshchetin, *op. cit.*, pp. 38-39.

9   International Law and Outer Space

Law in outer space did not become a practical issue until the orbiting of
Sputnik in late 1957. Nonetheless, a certain precedence for space law is
available from previous international conventions in the areas of air law
and the law of the high seas. So, for example, international conventions
and practice generally extend "freedom of the seas" to all nations on the
high seas outside the territorial waters claimed by states, outlaw piracy and
unrestricted submarine warfare, and require national registry of vessels, aid
to the shipwrecked, etc.

The Paris Air Convention of 1919 and the Chicago Convention of 1944
both extend to states exclusive sovereignty over the *air space* above their
territories.[1] The Soviet Union did not participate in either convention, citing
in the case of the Chicago Convention the presence of the "pro-fascist"
states, Spain, Portugal and Switzerland, which did not recognize the Soviet
Union. The USSR, however, at least until Sputnik, championed the principle
of national sovereignty as set forth in the Chicago Convention.[2] In general,
despite its "proletarian internationalism," the USSR has taken the widest
interpretation of national sovereignty. Until Sputnik Soviet international
jurists would acknowledge no upper limits to sovereignty, defending as their
own everything *useque ad coelum*, or to the very heavens. This interpretation was applied to American high altitude balloons used to conduct photoreconnaissance and distribute political propaganda over Soviet territory.
The United States in the person of Secretary Dulles said the balloons flew
above the limits of sovereignty; the Soviets claimed that there were no such
limits.[3]

Almost immediately after Sputnik an article appeared by a Soviet legal
scholar which contended that the Sputnik did not violate international
sovereignty—not only because it was part of an international activity (IGY)

and because no nation voiced any objections, but because sovereignty does not extend past "air space" and into "outer space." The Soviet author held that Sputnik did not pass over the territory of states but that these territories passed, "so to speak," under Sputnik. He also applied to Sputnik the principle of freedom of the seas.[4]

Subsequent opinions by Soviet jurists have agreed that outer space is outside the boundaries of national sovereignty.[5] The Soviet view on this question largely coincides with that of American and other Western legal scholars. One can suspect that if only Western nations had mounted space programs, the traditional Soviet emphasis on sovereignty would have led to extensive legal disputes concerning alleged violations of sovereignty involving space activities.[6]

Soviet and other socialist legal scholars also agree with the predominant Western view that excludes surface-to-surface missiles such as ICBMs from the legal regime of space since they only pass through space in going from one point on the earth to another. Objects within the regime of space law should therefore include earth satellites and objects in deep space such as those on or in orbit around celestial bodies.[7] According to the Marxist interpretation, space law represents the consensus of ruling classes in existing states "having different social and economic systems." Sources of space law include international treaties, customary law and general principles of law such as sovereignty, etc.[8] Socialist jurists consider space to be a *res communis omnium*, i.e., belonging to all states and appropriable by none.[9] They have attempted to extend into the field of space law their own general theory of international law in the current period based on the Communist doctrine of peaceful coexistence. Peaceful coexistence applied to space law is said to include (a) sovereign equality of states, (b) abstention from threats of force or use of force, (c) prohibition of aggression and justification of individual and collective self-defense, (d) noninterference in the internal affairs of states, (e) good faith in the observance of space law, and (f) peaceful settlement of disputes.[10] According to one Soviet legal scholar, the 1961 General Assembly Resolution on the Peaceful Uses of Outer Space extends to outer space the principles of peaceful coexistence, which he says forbid military bases, orbital bombers, satellite reconnaissance and contamination.[11]

Legal delimitation of the boundary or boundaries between air space and outer space has not been agreed upon. This is not the result of differences between international jurists of the East and West, but of more general disagreements and technical difficulties arising from the fact that earth's atmosphere blends gradually into the vacuum of space and no easily identifiable boundary comparable to a shoreline exists. This is why the functional definition of space has been accepted by both socialist and capitalist states. Both sides have agreed that general principles of international law such as those included in the UN Charter require that the activities of states in outer space be "peaceful" in character. The major difference is that the USSR and other socialist states interpret peaceful to mean nonmilitary, while the United States and its allies interpret peaceful to mean nonaggressive.[12] Kash points out that the Soviet position encounters technical difficulties in distinguishing military from civilian activities, and the American position encounters difficulties in assessing whether the motives of decision-makers can be construed as aggressive.[13]

REGULATION OF SPACE ACTIVITIES: GENERAL PRINCIPLES

Generally speaking, the United States is opposed both in principle and in practice to international legal regulation of activities in outer space. American spokesmen object that such regulation could "impede scientific progress" and curb "a free society," and that space research should not be "hamstrung" by international restrictions.[14] The United States has contended that unregulated practice will build a body of precedents which will constitute a *customary law* of space.[15] Although the legal issues involved are outside the scope of this paper, the United States generally takes the position that international law is composed of treaties, agreements and practices which are in force at the will of states and that it can be abrogated at the will of states acting through their own legal mechanisms without any "violation" having occurred. At the same time the United States (and other nations) concede to international law a certain moral and political force in that they oppose international treaties and agreements which are very likely to be abrogated.

In principle and practice the United States has favored specific agreements affecting international space law and opposed more general agreements. In contrast, the USSR favors general regulations as a prerequisite to specific agreements.[16] In principle the USSR, unlike the USA, favors international regulation of space activities, but in practice it has supported only such regulations that would be of strategic benefit to itself, usually at the expense of the United States. So, for example, the Soviet Union has opposed activities by private capitalist enterprises in space, claiming that these bring about the "chaos" of competition and prevent international cooperation.[17] Soviet policy has favored the interpretation of the 1967 Space Treaty as *ius cogens*, that is, binding on states and superseding abrogation by individual states.[18] In distinction to its interpretation of the Space Treaty, the USSR has insisted that reporting and registration of space activities be voluntary, while the US has pressed for mandatory reporting, still maintaining, of course, that agreements to that effect would be abrogable by states.[19]

The most important aspect of international regulation of space, and the area that has engendered the greatest controversy between the two space powers, is that of arms control. The Soviet Union approaches the question of arms control from its traditional position favoring general and complete disarmament.[20] As such, the USSR has been reluctant to separate the issue of arms control in space or even arms control in general from the question of disarmament. The United States, by contrast, has pursued arms control in space as a *separate* issue of high priority.[21] The United States has put special emphasis on inspection as a means of monitoring and enforcing arms control and disarmament, while the USSR has resisted schemes for inspection which would obviate its strategic advantage emanating from the closed nature of Soviet society.[22]

Soviet policy, depending on the Soviet interpretation of "peaceful" as nonmilitary, favors complete demilitarization of space and claims that military activities in space violate the UN Charter and 1967 Space Treaty. It is the activities of reconnaissance satellites that have most strongly been opposed by the USSR on these grounds. American policy has upheld the legality of military reconnaissance from space and pressed only for the banning of instruments of mass violence from space.[23] Emphasizing the importance of enforcement, American spokesmen have claimed that it is impractical to completely bar space to military activities in that military missions are often indistinguishable from those of a scientific or other civilian purpose.[24] The Hungarian jurist Gyula Gal is basically correct in objecting that the development of space technology has produced a specialization of functions and missions such that most military space missions can be identified from orbital characteristics and distinguished from civilian missions.[25]

In contending the illegality of military reconnaissance spacecraft, authorities such as Gal seem to want to argue that such craft are illegal even under the Western contention that "peaceful" in the UN Charter and Space Treaty means nonaggressive. Gal claims that since they are used to select targets for ICBMs, reconnaissance satellites invoke an illegal "threat of force" against the country observed which is somehow (?) greater than peaceful impact of strategic reassurance provided by space-gleaned intelligence.[26]

Although the United States more or less acknowledges its military space activities and the USSR does not, both powers provide the basic legal justification for their military space activities through the principle of national and collective self-defense recognized in Article 51 of the UN Charter. The United States takes the position that it cannot refrain from undertaking military missions in space as long as its rivals are not prohibited from doing so. (The Soviets evidently consider themselves legally prohibited, but do it anyway in "self-defense.")[27] The leading Soviet text on military strategy even went so far in 1963 as to duplicate the claim, more familiar in the US, that Soviet security requires superiority to the United States in space defense systems.[28]

Although the Soviet international jurist G. P. Zhukov rejects the claim made by the American jurist J. C. Cooper that the right of self-defense extends even to *preemptive* "defensive" strikes,[29] Gyula Gal, presumably reflecting Soviet opinions, claims the legal right of a state to *intercept* attacking spacecraft, destroy foreign reconnaissance satellites and otherwise conduct self-defense measures outside its territory, in space, on the moon or on other celestial bodies.[30]

REGULATION OF SPACE ACTIVITIES: HISTORICAL RECORD

Although the Soviet Union has come to occupy a position which at least in principle favors broad international control of space activities, and the United States has resisted such control, the original impetus for general international regulation and control came from the United States. This occurred in January, 1957, when President Eisenhower in his State of the Union message to Congress proposed mutual international control of "outer space missile and satellite development." Within a week Eisenhower's suggestion was incorporated along with his "open skies" plan, which proposed that nations agree to permit aerial reconnaissance of one another as a means of preventing preparation of aggressive attacks, into the formal disarmament proposals of the United States. The proposals, submitted to the UN Political Committee and subsequently to the UN Disarmament Subcommittee, called for the use of outer space for *exclusively* peaceful and scientific purposes and the testing of intercontinental missiles and satellites under international inspection and control. In the succeeding months the USSR conducted the first ICBM tests and launched the first Sputnik. Nikita Khrushchev proclaimed that the USSR was willing to put earth satellites and missiles under international control *as a part of* a bilateral and general settlement of differences between the United States and the Soviet Union.[31]

The United States' disarmament proposals, including those affecting space, commanded majority support in the UN but did not meet with the approval of the Soviet bloc.[32] The one major stumbling block that had plagued the disarmament negotiations for years and turned them into a propaganda forum was the issue of inspection and controls. The Soviets viewed the proposal as legitimizing American espionage in a strategic area (missile and space technology) in which the USSR claimed a clear superiority.[33]

In January, 1958, the US separated the issue of space weaponry from the rest of its disarmament proposals. Eisenhower submitted a proposal in a letter to Soviet Premier Bulganin which called in effect for the banning of ICBMs from outer space. Again this was viewed by the Soviets as an attempt to neutralize their strategic advantage. In the words of Khrushchev, "They want to prohibit that which they do not possess."[34] At the UN Secretary Dulles proposed that an international commission be created to reserve the use of space for peaceful purposes.[35]

The Soviet response to these proposals was to propose a ban on the use of space for military purposes *combined with* the elimination of all military bases on foreign territories and the establishment of a UN agency for international cooperation in space research with all launchings by states to be part of an agreed international program.[36] Although the Soviet proposal could not be accepted by the US because of the clause on military bases, it was something of a propaganda coup for the USSR. The US, which had first injected the issue of disarmament into discussions of space in its proposal to ban the ICBM, now had to ask that the issue of peaceful uses of outer space be generally separated from that of disarmament. Moreover, the Soviet proposal for the conduct of space research under international auspices had great popular appeal and appeared to be a generous offer coming from a country that led in this field.[37]

Caught rather off balance, as it had been by Soviet disarmament proposals in 1955, the US retreated to a more cautious stance. It proposed that an ad hoc committee consider the role the UN might play "relating to the peaceful uses of outer space," future UN "organization arrangements" to facilitate such peaceful use, and the "nature of legal problems" which might arise from the exploration of space.[38] Revising its tactics, the USSR submitted a new proposal which dropped any mention of a UN space agency or the elimination of military bases but substituted a slate of ad hoc committee members composed of equal numbers of NATO members, Soviet-bloc countries and neutrals.[39] Evidently sensing that the General Assembly would take *some* action, the Soviet representatives must have decided that it was more important to raise the issue of parity than to angle for the greatest propaganda appeal.[40] The US proposal with some revisions passed the UN on November 24, 1958. The nine socialist states voted against it, and 18 countries abstained. Objecting to the composition of the ad hoc committee, the USSR, Czechoslovakia and Poland refused to participate. Subsequently, India and the UAR also withdrew their delegates.[41]

In early 1959 an article appeared written by the Soviet legal scholar E.

A. Korovin. Korovin repudiated the applicability of the 1919 Paris and 1944 Chicago conventions on air law to outer space, claiming that the infinite extensions of sovereignty would make space research impossible. He held that although sovereignty does not extend into outer space, international law does. This was the basis for his repudiation of the analogy between space and "freedom of the seas" put forward by J. C. Cooper and others. Korovin claimed that reconnaissance satellites are illegal under international law and that a state would be justified in taking diplomatic and other "reprisals and retaliation of a *nonmilitary* nature" against them.[42]

When the Soviet *Luna* 2 spacecraft landed on the moon in September, 1959, the USSR held to its *res communis omnium* (a thing which belongs to all nations in common) interpretation of outer space and celestial bodies by repudiating any national claim to the moon. The Soviet jurists did attempt to impute claims to the moon in the space plans of the United States despite the absence of any such claims by official sources.[43] Some Western legal scholars have held to the *res nullis* (a thing belonging to no one) interpretation of outer space, according to which space and celestial objects may be appropriated by persons or states.

In the course of 1959 the UN ad hoc committee had, without Soviet participation, divided itself into legal and technical subcommittees, dropped any consideration of a UN role in space exploration and recommended the creation of a permanent UN Committee on the Peaceful Uses of Outer Space (COPUOS). By late 1959 the US and the USSR were able to compromise on a slate for such a committee with 12 "western" members (in the sense that Japan and Iran can be considered "western"), seven members of the Warsaw Treaty Organization and five "neutrals" (in the sense that Lebanon and the UAR can be considered "neutral"). COPUOS was established as a permanent UN agency under the unanimous UN Resolution 1472/XIV. COPUOS was mandated to study the possibilities for international cooperation and legal questions in space.[44] The unanimity of the Resolution was not reflected in subsequent meetings of COPUOS in which the parity question reemerged in the form of Soviet demands that the committee operate on the basis of unanimity or two-thirds majority rather than simple majority rule. This effectively blocked COPUOS for two years.[45]

In 1960 the Commission on Space Law of the Soviet Academy of Sciences was established and chaired by E. A. Korovin, a corresponding member of the Academy.[46] The Soviet Union in the summer of that year proposed the destruction of the means of delivery of nuclear weapons and on-site inspection of all rocket launching facilities with prohibition of all military devices in space.[47] To the extent that either side takes the other's (or its own) proposals seriously, the United States may have been disinterested in the Soviet proposal in that its own military missions in space (such as Discoverer) were by 1960 considerably in advance of the Soviet Union, which in all probability did not launch a specialized military satellite until 1962. Nonetheless, President Eisenhower, addressing the UN in the fall, presumably had US military space in mind when he proposed that before it was too late, space be declared off limits to military activities, as Antarctica had been previously.[48] Eisenhower further proposed that it be agreed that (1) celestial bodies are not subject to national appropriation, (2) no "warlike" activities take place on them, (3) no nation launch into orbit or station in space weapons of mass destruction, (4) all launchings *be verified in advance* by the UN, (5) an international program be created under UN auspices.[49] The first three of Eisenhower's proposals were incorporated into a UN Resolution in 1961 and into the Space Treaty six years later; the others were blocked by rivalry on both sides and by the Soviet passion for secrecy.

Resolution 1721 (XVI) which passed the General Assembly unanimously on December 20, 1961, as the result of prior compromises by both the US and the USSR, declared that international law, including the UN Charter, applies in outer space and that outer space and all celestial bodies are free for exploitation and use by all states and not subject to national appropriation.[50] The Resolution also called on COPUOS to continue its work on legal questions arising from outer space. This it did, and in the course of its work several legal issues separating the United States and the USSR became abundantly clear in 1962.

Soviet popular literature kept up a steady stream of attack against US reconnaissance satellites. Basically, the Soviets claimed that satellite reconnaissance was no less illegal than aerial reconnaissance, attempting to confuse the issue of the illegality of espionage with that involved in violations of sovereign airspace.[51] Another paradox in the Soviet position was the claim on the one hand that space espionage was already illegal, while the USSR, nonetheless, demanded an *explicit* international legal ban on such espionage. Still another paradox: in the spring of 1962 the USSR launched its first observation satellite, *Kosmos* 4, thus apparently violating its own interpretations of international law.[52] Of course, the purpose of this launch and that of subsequent observation satellites has never been admitted by the USSR.

That same spring of 1962 the USSR began pressing in the legal subcommittee of COPUOS for a ban on all military space activities including reconnaissance.[53] At first the USSR insisted that peaceful exploration of space be tied to general disarmament.[54] Subsequently, the Soviet delegate to the legal subcommittee of COPUOS submitted a nine-point draft declaration on the peaceful uses of outer space. Although the declaration reiterated several points of UN Resolution 1721, it added several features which the United States found particularly objectionable.[55]

One stipulation of the Soviet draft would limit the exploration of space to activities carried on by states and international organizations. This was seen as an attempt at interference in the internal affairs of capitalist states and may have been explicitly aimed at Comsat, then in the process of formation.[56] Another stipulation would ban the use of space for propagating war, race hatred and national emnity. Previous experience in negotiating with the USSR on war propaganda turned the United States against this proposal.[57] Another plank in the Soviet proposal called for prior international discussion and agreement on any measures undertaken by states which might hinder space exploration by other countries. This was seen as an attempt by the Soviets to acquire a unilateral veto over such American military activities as Project West Ford, a military space-communications experiment involving the placing in orbit of millions of copper filaments as a radio reflector.[58] Finally, the Soviet proposal declared space reconnaissance to be "incompatible with the objectives of mankind in the conquest of space."[59] To the United States, space reconnaissance was not illegal any more than surveillance from the high seas, and it reduced the risk of war by penetrating Soviet secrecy and providing a possible basis for monitoring disarmament and arms control.[60] For the same reasons, the United States found noxious the Soviet proposal that states had no obligation to return spacecraft or astronauts to the launching state where evidence was found indicating these were employed to gather intelligence.[61] The negotiations broke down over these issues at about the time the Soviets launched *Kosmos* 7, their second photoreconnaissance satellite.[62] At least the USSR was no longer, to borrow a phrase from Khrushchev, seeking to "prohibit that which it does not possess."

As indicated in the previous chapter, the first breakthrough in Soviet-American space cooperation took place in 1962. Tension between the two countries reached a peak shortly afterward that year during the Cuban missile crisis. After the crisis both countries retreated a few paces from their previously competitive stances, paving the way for more fundamental agreements which took place during 1963. The Sino-Soviet dispute by the summer of 1963 reached a peak which it was not to surpass for at least another three years. Evidently both the failure of confrontation politics with the US in 1962 and the difficulties with the Chinese helped to prepare for a substantial shift in the Soviet stance by the late summer and fall of 1963. The change of stance was reflected in international agreements affecting space.

In the spring the USSR, continuing with its secret program of reconnaissance satellites, still held to a position which would make them illegal. This remained unacceptable to the United States and prolonged the impasse in COPUOS.[63] Similarly, the USSR held to its proposal to confine space exploration to activities carried on exclusively by states and international organizations.[64]

In July Khrushchev made a veiled reference to reconnaissance satellites by suggesting to the Belgian Foreign Minister, Paul Henri Spaak, that they should be employed instead of on-site inspection to police any agreement banning nuclear tests.[65] Implicit recognition of the monitoring function of satellites along with the more general political factors mentioned above probably helped bring about agreement in August on the Treaty Banning Nuclear Weapons Tests in the Atmosphere, in Outer Space and Under Water.[66] Agreements affecting space were also reached at about the same time for furthering space radio regulation by the ITU.[67]

The major shift in the Soviet posture on space became evident in September when the Soviet Foreign Minister, Gromyko, expressed his country's willingness to accept the American position of long standing to entertain separately from broader issues of arms control and space law a ban on nuclear weapons in space. Kennedy evidently was sufficiently impressed by Gromyko's remarks to make his own rejoinder in the form of the proposal for a joint lunar program.[68] The USSR quietly dropped its insistence on the outlawry of satellite reconnaissance and the prohibition of "private enterprise" from space.[69] This first resulted in the unanimous passage of UN Resolution 1884 (XVIII) banning nuclear weapons from outer space (October 17, 1963) and UN Resolution 1962 (XVIII) which (1) reaffirmed the *res communis omnium* character of outer space and the applicability of the Charter, (2) recognized the responsibility of states for national activities in outer space while acknowledging that such activities *could be carried out by nongovernmental entities*, (3) provided for previous international consultations (*not agreement*) by states which believe that some space activity of theirs might interfere with the space research of others, (4) recognized the jurisdiction of the launching state over space objects, and (5) provided for the international liability of launching states and for rescue, assistance and return of astronauts and other space objects.[70] Conjoined to Resolution 1963 (XVIII), which called for negotiations leading to an international treaty on outer space, Resolution 1962 provided the basis for the Space Treaty of 1967.[71] Instead of voting against the proposal, as it had done in 1959, and blocking the work of COPUOS, as it did subsequently, the Soviet Union in this case made substantial concessions and took positive initiatives on its own part to reach agreement.

After 1963, despite occasional exceptions in the popular press, the Soviet official attitude toward reconnaissance satellites became one of tacit approval. Their own program in this area evidently is of sufficient worth to them not to threaten its continuance.[72] They may have also reconsidered the strategic value of strict secrecy in that American misinformation, fed by Soviet boasts, had contributed to an American missile buildup, putting the USSR in a position of marked strategic inferiority at the time of the Cuban missile crisis.[73] Finally, as a part of the bilateral Strategic Arms Limitation Agreements of 1972, the Soviets more or less acknowledged their acceptance of satellite reconnaissance. Both the permanent agreement on ABMs and the interim agreement on offensive weapons signed by the US and the USSR pledge the signatories not to interfere with or use concealment to defeat the purpose of the other side's "national technical means of verification."[74]

Implementation of Resolutions 1962 and 1963 in treaty form proved to be a lengthy process. Changes of leadership in both the United States and the Soviet Union in 1963 and 1964 probably contributed to an atmosphere of guarded caution in negotiations affecting space. The Vietnam War further complicated Soviet policy toward the United States, as did the prospect between late 1964 and 1966 of effecting some kind of settlement between the CPSU and the Communist Party of China. By 1966 the CPSU leadership had stabilized sufficiently to hold a party Congress, one year after the deadline for such a congress according to the party rules. The summer of 1966 saw the first tete-a-tete of the new leaders of the US and the USSR when President Johnson conferred with Alexei Kosygin in Glassboro, New Jersey. The two countries were also able in 1966 to conclude an agreement, implemented two years later, for direct air flights between New York and Moscow.[75]

In May of 1966 President Johnson called for conclusion of an international treaty on outer space, probably considering that one or both of the space powers would soon be landing men on the moon and that existing legal agreements needed to be reasserted and further invigorated. According to *The New York Times*, Johnson also sought to isolate and show up the bellicosity of the People's Republic of China as compared to the United States (and the Soviet Union).[76] The Soviet permanent representative, Fedorenko, concurred in putting such a treaty on the UN agenda. A proposed text was agreed upon in June as the basis for discussion in the legal subcommittee of COPUOS.[77]

Discussions progressed, but a number of issues separated the positions of the two space powers. The Soviets wished to make reporting of national space activities to the UN voluntary; the United States wanted it to be mandatory. The Soviets wanted the treaty to recognize the equal access of all states to the tracking facilities of others, the United States maintaining that such access should be at the discretion of the state within whose boundaries the tracking facilities were located. The USSR also objected, in keeping with its established policies, to admitting any jurisdiction of the International Court of Justice in legal disputes related to space. The United States wanted the installations of any nation on celestial bodies to be open to all states, and the USSR wished to put this "on the basis of reciprocity." Finally, the USSR wanted the treaty to be considered *ius cogens* and open to all states, presumably including East Germany, China, etc., and not only to UN members.[78]

The issues separating the US and the USSR were compromised at the expense of introducing certain areas of vagueness into the treaty. Final agreement was reached on December 8, and the Treaty was opened for signature in January, 1967, entering into force on October 10, 1967.[79] Within less than a year the United States and the Soviet Union were able to arrive at an Agreement on the Rescue and Return of Astronauts and Space Objects, which was opened for signing in April, 1968, and went into effect in December of that year. Although the USSR attempted to include in the Agreement a clause to the effect that states were not obliged to return space objects employed for intelligence gathering, neither this Agreement nor the 1967 Space Treaty makes any mention of such space objects.[80]

Since the foundation of COPUOS negotiations have been under way to establish international agreements concerning liability incurred by states, international organizations and others in the conduct of space activities. Significant progress was made in agreement to certain principles through 1971, but disagreements on certain details in part relating to ideological issues involving socialism and private property prevented conclusion of an agreement suitable for signature by states. The USSR has objected to original liability for international organizations, possibly reflecting a reluctance to pay part of an immense settlement that might be adjudged against an international agency. It has also objected to compensation for "moral" damages (pain and suffering) and has rejected any role in assessing liability for the International Court of Justice in light of the national composition of the Court.[81]

When the legal subcommittee and a special working group of COPUOS began in 1968 to examine the problems of direct broadcast from space at the invitation of General Assembly Resolution 2260 (XXII), the USSR and other socialist countries reacted noticeably. After the working group issued its first report, they asked that it be discontinued and strove to establish legal barriers to any use of direct broadcasts that would involve reception by individuals in any country without prior and continuing consent of their governments. The USSR was particularly concerned about "ideological imperialism" carried out by private firms employing direct broadcast from space. Such broadcasts were viewed as a violation of sovereignty and a threat to peace.[82] They are, of course, a threat to the policies of the USSR and many other states which seek to isolate their citizens from and to filter through official agencies information from the world outside their frontiers.

In June, 1971, the USSR submitted to the UN the Draft Treaty for the Peaceful Uses of the Moon. The Treaty includes provisions from the Space Treaty, UN resolutions and other agreements, and applies them to the moon. Such provisions include a ban on any kind of activity on the moon serving military purposes, provisions for protecting the lunar environment, and provisions for rescue and assistance to persons on the moon. The proposed draft would also bar commercial appropriation of lunar material or property rights on the moon and be open to accession by all states.[83] As yet, no final action has been taken on an international treaty directed exclusively to the moon.

Future involvement of international law and international agreements in space will probably be in the form of actions in the areas of liability and legal regulations affecting the moon and planets. Just as the Soviets probably view with horror the prospects of American commercial exploitation of the moon, others view with trepidation the Soviet intention to subdue the planets to the uses of man.[84] Just as the 1972 agreements on strategic arms impacted the area of space reconnaissance, it is also possible that further negotiations on strategic armaments may have some impact on military uses of outer space. The Soviets have already tested such space weapons systems as FOBS and an orbital interceptor. It may be found to be in the interests of both the USSR and the United States to provide a legal ban to extensive competition in space-borne weapons systems. The Soviet Union ought to favor such action since it has for a long time been in favor (at least in principle) of the banning of *all* military activities in space.[85]

A brief consideration of Soviet legal policies in three separate areas is provided below: contamination, rescue and return of astronauts and space objects, and space-related liability.

CONTAMINATION OF SPACE

Both the Soviet Union and the United States conducted nuclear tests in space and in the upper atmosphere before the 1963 Test-Ban Treaty.[86] The Soviet Government took official note of international protest of a one-megaton nuclear test conducted at an altitude of 400 kilometers by the US in 1962. The test allegedly altered the earth's radiation belts (indeed this was one aim of the test). It was also contended that the test damaged or otherwise interfered with the satellites of three Western countries. The Soviet Union circulated at the UN the published protests against this test made by British scientists and others.[87]

In the United States the claim has been made that Soviet policies on the sterilization of spacecraft are considerably less thorough than those of the United States, may not meet the standards set down by COSPAR and probably have resulted in the transfer of terrestrial microorganisms to the surfaces of Venus and Mars.[88] The major difficulty occasioned by Soviet practices concerning the sterilization of spacecraft to avoid contamination is that in contrast to the United States, the USSR has not provided extensive information about its sterilization techniques, so that while it claims to follow COSPAR requirements, this is difficult or impossible for outsiders to verify.[89]

The two West Ford military experiments involving the launch of hundreds of millions of copper filaments into an eccentric orbit for communications tests, conducted in 1961 and 1963, occasioned protests by the Soviet Academy of Sciences and by the Government of the USSR at the UN. As was the case with the 1962 US nuclear test, the USSR was in these two cases able to capitalize on the protests of scientists from Western nations who alleged that West Ford interfered with radio astronomy and with communications with spacecraft and could endanger astronauts. The US denied such claims and was able to gather majority support for its position in COSPAR.[90]

The Soviet Union, as noted previously, has endeavored to establish legal requirements for prior international consultations and agreements on all potentially harmful space experiments. The United States has agreed to incorporate prior consultation into UN resolutions and the Space Treaty but views the Soviet position as an attempt to secure a veto on US space experiments, particularly those of a military nature.[91] One might object to the US stance that the Soviets seek to place no limitations on American activities which are not concurred in by the Soviet Union itself. This is not quite the case. The USSR would in fact be less subject to limitations in that it holds that its own reporting of its space activities is voluntary. It does not report its space activities in advance and sometimes not at all. Legal scholars from the Soviet bloc have indicated that they would like to clarify the restrictions on contamination in the 1967 Space Treaty and make them stricter.[92]

RESCUE AND RETURN OF ASTRONAUTS AND SPACE OBJECTS

Liability, rescue and recovery were from the first days of COPUOS considered to be the most pressing legal questions affecting outer space from the point of view of the United States. The Soviet Union in the early 1960s chose to stress instead *much more general* questions of space law such as bans on all military activities, harmful effects on the space environments and propaganda activities.[93] Nonetheless, the Soviet interest in rescue and return was voiced as early as 1962 in Nikita Khrushchev's second letter to President Kennedy on space cooperation.[94]

Khrushchev's suggestion for cooperation in rescue and recovery was probably prompted in part by the superiority at that time of US ground-based tracking facilities and naval recovery vessels. Despite Soviet advances in recent years, American naval resources, which can be employed in rescue and recovery of objects falling into the oceans, remain more widely dispersed and extensive than those of any other nation.

In the earliest proposals to COPUOS on rescue and return, the principal differences between the drafts of the two space powers were that the United States favored unconditional return, while the USSR wished to make return contingent upon identification markings of craft signifying nationality, proof that launching had been officially announced by the launching state, and the absence of on-board equipment for the collection of intelligence.[95] It will be recalled that by 1962 American military uses of space had proceeded much farther than those of the Soviet Union and that the USSR had begun to substitute use of the *Kosmos* label for nonannouncement of flights, while the US had recently experimented with a policy of nonannouncement in relation to the military Discoverer satellites. Thus, the USSR was, as is to be expected, advancing its own interests in these proposals. Any agreement that implied the illegality of reconnaissance satellites was then and continues to be unacceptable to the United States.[96] The US view has prevailed in UN resolutions, agreements and treaties affecting requirements of rescue and return. However, some, but not all, Soviet legal scholars have still held to the view that space reconnaissance is illegal and therefore not entitled to the legal protections of such agreements.[97] The accidental death of Soviet cosmonauts aboard *Soiuz* 1 and 11 in 1966 and 1971 has probably reinforced the interest of the USSR in making legal provisions concerning rescue, return and assistance to astronauts more effective. The USSR has defended the priority of an agreement on rescue, assistance and return of astronauts on the grounds of "humanitarian" interests involved. This made possible the conclusion of an international agreement covering these matters in 1968. Further progress has been made in the form of bilateral agreements between the US and the USSR on the docking of the spacecraft of the two countries.

SPACE LIABILITY

McDougal, Lasswell and Vlasic cite as the principal ways in which space liability could be incurred:

(1) impact of spacecraft or parts thereof on earth;
(2) collision or interference of spacecraft with other spacecraft and other manmade objects in space or in the earth's atmosphere;
(3) deprivations arising from contamination of earth, space or celestial bodies by space activities;
(4) interference with space communications by space activities;
(5) deprivations occasioned by observation of one country by the spacecraft of another.[98]

Space debris from craft of both the USA and the USSR have impacted earth outside the boundaries of the launching sites. No claims of liability have been made, although Cuba has claimed the undoing of a cow by US space debris, and protests have been made in connection with nuclear tests and Project West Ford.[99]

Both space powers have conceded that they are liable for any damages caused by their spacecraft.[100] The main legal principles have been concurred in by them in recent years, although the USSR refuses to admit the jurisdiction of the International Court of Justice and has differed with the United States over the proper means of recovering damages caused by the space activities of international organizations.[101]

NOTES

[1]*Survey of Space Law*, Staff Report of the Select Committee on Astronautics and Space Exploration, 86th Congress, 1st session, House Document #89, 1959, p. 17.

[2]Leon Lipson, *Outer Space and International Law*, Rand Corporation, Paper #P-1434, February 24, 1958, pp. 1-4.

[3]See A. Kislov and S. Krilov, "State Sovereignty in Air Space," *International Affairs*, No. 3, March 1956, pp. 35-44; and Lipson, *supra*, pp. 12-13.

[4]G. P. Zadorozhnyi, "Iskustvennii sputnik zemli i mezhdunarodnoe pravo," *Sovetskaia Rossiia*, October 17, 1957, translated by Anne M. Jonas as "The Artificial Satellite and International Law," Rand Corporation, Translation #T-78, November 12, 1957, pp. 3-4.

[5]See A. Galina (Onitskaia), "On the Question of Interplanetary Law," *Sovetskoe Gosudarstvo i Pravo*, No. 7, July 1958, translated by F. J. Krieger and J. R. Thomas, *Translations of Two Soviet Articles on Law and Order in Outer Space*, Rand Corporation, Rand Translation #T-98, Septemmber 25, 1958, p. 10; E. A. Korovin, "Inter-

national Status of Cosmic Space," *International Affairs*, No. 1, January 1959, pp. 53-59; and G. A. Onitskaia, "Mezdunarodnye pravovye voprosy osvoeniia kosmicheskogo prostranstva," *Sovetskii Ezhegodnik Mezhdunarodnogo Prava* (Moscow, 1959), pp. 64-65.

[6]See Lipson, *op. cit.*, pp. 14-16.

[7]See the study by the Hungarian jurist Gyula Gal, *Space Law* (Leyden: A. W. Sijthoff, and Dobbs Ferry, N.Y.: Oceana Publications, Inc., 1969), pp. 36-37.

[8]*Ibid.*, pp. 39, 41.

[9]*Ibid.*, pp. 128-129, 191; and Onitskaia, *op. cit.*, pp. 64-65.

[10]Gal, *supra*, pp. 131-132; and Domas Krivickas and Armins Rusis, "Soviet Attitude Toward Space Law," in *Soviet Space Programs, 1962-1965, op. cit.*, pp. 495-496.

[11]E. A. Korovin, Chairmman, Commission on Space Law, USSR, "Peaceful Cooperation in Space," *International Affairs*, No. 3, March 1962, p. 63.

[12]Deputy Assistant Secretary of State Richard N. Gardner, "Outer Space: A Breakthrough for International Law," *American Bar Association Journal*, 50, January 1964, pp. 32-33; Gyula Gal, *op. cit.*, pp. 165, 168; Erin B. Jones, *Earth Satellite Telecommunications Systems and International Law* (Austin: University of Texas, 1970), pp. 57-58; G. P. Zhukov, "Iadernaia demilitarizatsiia kosmosa," *Sovetskoe Gosudarstvo i Pravo*, No. 3, March 1964, pp. 78-79; and Nicholas deB. Katzenbach, "The Law in Outer Space," in Lillian Levy, editor, *Space: Its Impact on Man and Society* (New York: W. W. Norton and Co., Inc., 1965), p. 77.

[13]Don E. Kash, *The Politics of Space Cooperation* (Purdue University Studies, Purdue Research Foundation, 1967), p. 96.

[14]Katzenbach, *op. cit.*, p. 76; and Arnold W. Frutkin, "Patterns of International Space Applications and Their Extension," American Astronautical Society Reprint #67-146, May 2, 1967, pp. 379, 385, 389.

[15]Loftus Becker, legal advisor to US Department of State (1959), as cited by Lt. Commander Richard L. Fruchterman, Jr., "Introduction to Space Law," *Judge Advocate General of the Navy Journal*, 20 (1), July-August 1965, p. 12.

[16]Krivickas and Russis, "Soviet Attitude Toward Space Law," *Soviet Space Programs, 1962-1965, op. cit.*, p. 497.

[17]I. I. Cheprov, "Pravovoe regulirovanie deatelnosti v kosmicheskom prostranstve," in *Kosmos i Mezhdunarodnoe Sotrudnichestvo* (Moscow, 1963), p. 58; and E. A. Korovin and G. P. Zhukov, *Sovremennye Problemy Kosmicheskogo Pravo* (Moscow, 1963), p. 28.

[18]Krivickas and Rusis, "Soviet Attitude Toward Law of Outer Space," *Soviet Space Programs, 1966-1970, op. cit.*, pp. 457, 458-459.

[19]*Ibid.*, p. 467.

[20]G. P. Zhukov as cited by Gyula Gal, *op. cit.*, p. 182; and E. A. Korovin, "Peaceful Cooperation in Space," *op. cit.*, p. 62.

[21]Paul Kecskemeti, "Outer Space and World Peace," in Joseph M. Goldsen, editor, *Outer Space in World Politics* (New York: Frederick A. Praeger, 1963), p. 32.

[22]Edwin Diamond, *The Rise and Fall of the Space Age* (Garden City, New York: Doubleday and Co., Inc., 1964), pp. 28-29.

[23]Gyula Gal, *op. cit.*, p. 180, and Soviet works cited by him in notes 1-3. Also see Kecskemeti, *op. cit.*, pp. 34-35.

[24]Senator Albert Gore speaking at the UN in December 1962, as cited by Krivickas and Rusis, "Soviet Attitudes Toward Space Law," *Soviet Space Programs, 1962-1965, op. cit.*, p. 495.

[25]Gyula Gal, *op. cit.*, p. 179.

[26]*Ibid.*, pp. 169, 170, 179-180; Lt. General (retired) B. N. Teplinskii, "The Strategic Concepts of US Aggressive Policy," *International Affairs*, No. 12, December 1960, p. 39; and G. P. Zhukov, "Space Espionage and International Law," *International Affairs*, No. 10, October 1960, pp. 53-57.

[27]Kecskemeti, *op. cit.*, p. 33; Lipson and Katzenbach in *Legal Problems of Space Exploration: A Symposium*, Senate Document #26, 87th Congress, 1st session, 1961, p. 807; G. P. Zhukov, "Problems of Space Law at the Present Stage," Memorandum of the Soviet Association of International Law at the Brussels Conference of the International Law Association, August 1962, pp. 30, 35-36; and G. P. Zhukov in *International Affairs*, May 1963, p. 28, as cited by Thomas W. Wolfe, *Soviet Strategy at the Crossroads* (The Rand Corporation, Cambridge, Mass.: Harvard University Press, 1964), pp. 201-202; E. A. Korovin, "Outer Space Must Become a Zone of Real Peace," *International Affairs*, No. 9, September 1963, p. 92; and Gal, *supra*, p. 183.

[28]Cited by Gal, *loc. cit.*, note 192.

[29]G. P. Zhukov, "Problems of Space Law at the Present Stage," *op. cit.*, p. 36, as cited by Gal, *supra*, pp. 183-184.

[30]Gal, *supra*, pp. 184-185, 197.

[31]Philip J. Klass, *Secret Sentries in Space* (New York: Random House, 1971), p. 18; Leonard W. Schwartz, "International Space Organizations," in Hugh Odishaw, editor, *The Challenges of Space* (Chicago: University of Chicago Press, 1963), p. 255; J. Goldsen and L. Lipson, *Some Political Implications of the Space Age*, Rand Corporation, Paper #P-1435, February 24, 1958, pp. 1-2. See also US Congress, House Committee on Science and Astronautics, *US Policy on the Control of Outer Space*, Report #353, 86th Congress, 1st session, 1959, p. 2.

[32]Philip C. Jessup and Howard J. Taubenfeld, *Controls for Outer Space and the Antarctic Analogy* (New York: Columbia University Press, 1959), pp. 252-253.

[33]According to Kash, *op. cit.*, p. 97.

[34]*Loc. cit.*; and "Letter from Eisenhower to Premier Bulganin," *Department of State Bulletin, 38* (976), March 10, 1958, p. 373.

[35]Jessup and Taubenfeld, *op. cit.*, p. 254.

[36]*New York Times*, March 16, 1958, p. 34; and Gyula Gal, *op. cit.*, pp. 272-273.

[37]Kash, *op. cit.*, p. 100. See also I. Ermasev, "Cosmic Problems and Earth Reality," *New Times*, No. 14, April 4, 1958, pp. 11-14; and the Soviet proposal, UN Document A/C.1/L.219, November 7, 1958.

[38]Kash, *supra*, pp. 100-101; and "Statement by Mr. Lodge in the UN," *Department of State Bulletin, 39* (1016), December 15, 1958, p. 976.

[39]Jessup and Taubenfeld, *op. cit.*, pp. 255-256. See UN Document A/C.1/L219/ Rev. 1.

[40]Kash, *op. cit.*, pp. 101-103.

[41]*International Cooperation and Organization for Outer Space, 1965, op. cit.*, pp. 184-187; and Gyula Gal, *Space Law, op. cit.*, pp. 272-273.

[42]E. A. Korovin, "International Status of Cosmic Space," *International Affairs*, No. 1, January 1959, pp. 53-59 (emphasis added).

[43]*Soviet Space Programs, 1962, op. cit.*, pp. 206-207.

[44]Kash, *op. cit.*, pp. 104-106; and Gal, *Space Law, op. cit.*, p. 274.

[45]Kash, *op. cit.*, p. 107.

[46]See E. A. Korovin, "Outer Space and International Law," *New Times*, No. 17, April 25, 1962, pp. 13-14.

[47]*International Cooperation and Organization for Outer Space, 1965, op. cit.*, p. 168; and Donald G. Brennan, "Arms and Arms Control in Outer Space," in Lincoln P.

Bloomfield, editor, *Outer Space: Prospects for Man and Society* (New York: Frederick A. Praeger, Inc., 1968), p. 174.

[48]J. C. Cooper, *Explorations in Space Law*, essays edited by Ivan A. Vlasic (Montreal: McGill University Press, 1968), p. 287.

[49]Jones, *op. cit.*, p. 27.

[50]Kash, *op. cit.*, pp. 108-109.

[51]G. P. Zhukov, "Space Espionage and International Law," *International Affairs*, No. 10, October 1960, pp. 53-57; *Soviet Space Programs, 1962, op. cit.*, pp. 208-209; Zadorozhnyi as cited by Lincoln P. Bloomfield, "The Prospects for Law and Order," in American Assembly, *Outer Space, Prospects for Man and Society* (Englewood Cliffs, N.J.: Prentice-Hall, Inc., 1962), p. 173; E. A. Korovin, "Outer Space and International Law," *op. cit.*, pp. 13-14.

[52]Charles S. Sheldon II, "Program Details of Unmanned Flights," *Soviet Space Programs, 1966-1970, op. cit.*, pp. 180-181.

[53]Joseph M. Goldsen, "Outer Space and World Politics" in Goldsen, *op. cit.*, p. 20.

[54]Krivickas and Rusis, "Soviet Attitudes Toward Space Law," *Soviet Space Programs, 1962-1965, op. cit.*, p. 497; and UN Document A/AC.105/PV.8, May 15, 1962, pp. 43-45.

[55]UN Document A/AC.105/C.2/L.1 in A/AC.105/6, July 9, 1962, pp. 3-4; and Krivickas and Rusis, *supra*, pp. 498-509.

[56]Gyula Gal, *Space Law, op. cit.*, p. 142; and Kash, *op. cit.*, p. 117.

[57]Kash, *op. cit.*, p. 116; and Krivickas and Rusis, "Soviet Attitudes Toward Space Law," *Soviet Space Programs, 1962-1965, op. cit.*, p. 501.

[58]Kash, *loc. cit.*

[59]*Ibid.*, p. 117.

[60]Deputy Assistant Secretary of State Richard N. Gardner, "Cooperation in Outer Space," *Foreign Affairs, 41* (2), January 1963, pp. 5-6.

[61]Kash, *op. cit.*, pp. 116, 117.

[62]Klass, *op. cit.*, p. 124.

[63]*Ibid.*, p. 125.

[64]Krivickas and Rusis, "Soviet Attitudes Toward Space Law," *Soviet Space Programs, 1962-1965, op. cit.*, pp. 507-508; and E. G. Vasilevskaia, "Voprosy kosmicheskogo prava v novishi amerikanskoi literature," *Kosmos i mezhdunarodnoe sotrudnichestvo* (Moscow, 1963) pp. 231-232; and I. I. Cheprov, *op. cit.*, p. 58.

[65]Klass, *op. cit.*, p. 127.

[66]Treaty Banning Nuclear Weapon Tests in the Atmosphere, in Outer Space and Under Water, signed August 5, 1963, *American Journal of International Law, 57*, October 1963, pp 1026-1029.

[67]Gal, *op. cit.*, p. 42.

[68]Alton Frye, *The Proposal for a Joint Lunar Expedition: Background and Prospects*, Rand Corporation, Paper #P-2808, January 1964, pp. 12-14; and Arthur Schlesinger, Jr., *A Thousand Days* (Boston: Houghton-Mifflin, 1965), pp. 919-920.

[69]Alton Frye, "Soviet Space Activities: A Decade of Pyrrhic Politics," in Bloomfield, editor, *op. cit.*, p. 189.

[70]Krivickas and Rusis, "Soviet Attitude Toward Space Law," *Soviet Space Programs, 1962-1965, op. cit.*, p. 523; Gal, *op. cit.*, p. 276; UN Res. 1962 (XVIII), United Nations Resolution on Legal Principles Governing Activities in Outer Space, December 24, 1963, *International Legal Materials Current Documents, 3* (1), January 1964, pp. 157-159.

[71]Kash, *op. cit.*, p. 118.

[72]See Charles S. Sheldon II, "Soviet Military Space Activities," *Soviet Space Programs, 1966-1970, op. cit.*, p. 332.

[73]See Klass, *op. cit.*, xiv and pp. 215-216.

[74]Treaty of the United States of America and the Union of Soviet Socialist Republics on the Limitation of Anti-Ballistic Missile Systems (May 26, 1972), Article XII; and Interim Agreement between the United States of America and the Union of Soviet Socialist Republics on Certain Measures with Respect to the Limitation of Strategic Offensive Arms (May 26, 1972), Article V. The texts of the treaties appear in *Arms Control: Readings from Scientific American* (San Francisco: W. H. Freeman & Co., 1973), pp. 260-273.

[75]See Joseph G. Whelan, "Political Goals and Purposes of the USSR in Space," *Soviet Space Programs, 1966-1970, op. cit.*, p. 7.

[76]*New York Times*, December 9, 1966, pp. 18, 19.

[77]Krivickas and Rusis, "Soviet Attitude Toward Law of Outer Space," *Soviet Space Programs, 1966-1970, op. cit.*, pp. 456-457.

[78]Betty Goetz Lall, "Cooperation and Arms Control in Outer Space," *Bulletin of the Atomic Scientists*, 22 (9), November 1966, pp. 34-37; and Krivickas and Rusis, *supra*, pp. 456-468.

[79]Kash, *op. cit.*, p. 121; Krivickas and Rusis, *supra*, pp. 457-458, 461-468.

[80]Krivickas and Rusis, *supra*, pp. 468-472.

[81]*Ibid.*, pp. 473-490.

[82]*Ibid.*, pp. 499-501.

[83]TASS Release of June 8, 1971, *On Draft Treaty for Peaceful Uses of the Moon*, Appendix I of *Soviet Space Programs, 1966-1970, op. cit.*, pp. 623-627.

[84]Sir Bernard Lovell, "The Great Competition in Space," *Foreign Affairs, 51* (1), October 1972, p. 138.

[85]G. P. Zhukov (1970) as cited in Krivickas and Rusis, "Soviet Attitude Toward Law of Outer Space," *Soviet Space Programs, 1966-1970, op. cit.*, p. 462. See also Brennan, *op. cit.*, p. 174.

[86]Arnold W. Frutkin, *International Cooperation in Space* (Englewood Cliffs, N.J.: Prentice-Hall, Inc., 1965), p. 148; and Robert Salkeld, *War and Space* (Englewood Cliffs, N.J.: Prentice-Hall, Inc., 1970), p. 125.

[87]Gyula Gal, *op. cit.*, pp. 149-151; and UN Document A/AC.105/C.1/1, June 3, 1962.

[88]B. C. Murray, M. E. Davies and P. K. Eckman, *Planetary Contamination II: Soviet and US Practices and Policies*, Rand Corporation, Paper #P-3517, March 1967, pp. 1-11.

[89]*Ibid.*, p. 11; and Joseph G. Whelan, "Soviet Attitude Toward International Cooperation in Space," *Soviet Space Programs, 1966-1970, op. cit.*, pp. 410-413.

[90]Frutkin, *International Cooperation in Space, op. cit.*, p. 48; Gal, *op. cit.*, pp. 146-147; UN Document A/AC.1/SR.1210, December 4, 1961, p. 37; A/AC.105/13, May 28, 1963, pp. 2-5; UN Document A/AC.105/15, June 7, 1963, p. 5. See also the Commission on Legal Problems of Outer Space of the Soviet Academy of Sciences, "American Diversion in Space," *International Affairs*, No. 12, December 1961, pp. 117-118.

[91]Francis Plimpton, US Delegate to COPUOS, as cited by Krivickas and Rusis, "Soviet Attitude Toward Space Law," *Soviet Space Programs, 1962-1965, op. cit.*, p. 501 and pp. 498-503 of the same volume.

[92]G. P. Zhukov and Gyula Gal as cited by Krivickas and Rusis, "Soviet Attitude Toward Law of Outer Space," *Soviet Space Programs, 1966-1970, op. cit.*, p. 468.

[93]Frutkin, *International Cooperation in Space, op. cit.*, p. 149.

[94]Letter of March 20, 1962, in *International Cooperation and Organization for Outer Space, 1965, op. cit.*, pp. 139-140.

[95]Myres S. McDougal, Harold D. Lasswell and Ivan A. Vlasic, *Law and Public Order in Space* (New Hvaen: Yale University Press, 1963), p. 580; Jones, *op. cit.*, p. 64; and Krivickas and Rusis, "Soviet Attitude Toward Space Law," *Soviet Space Programs, 1962-1965, op. cit.*, p. 511.

[96]See Kash, *op. cit.*, pp. 116-117.

[97]Krivickas and Rusis, "Soviet Attitude Toward Law of Outer Space," *Soviet Space Programs, 1966-1970, op. cit.*, p. 471.

[98]McDougal, Lasswell and Vlasic, *op. cit.*, pp. 531-532.

[99]Frutkin, *International Cooperation in Space, op. cit.*, pp. 148-149; and Gal, *op. cit.*, p. 228.

[100]Gal, *loc. cit.*, citing UN Document A/AC.105/PV.14, pp. 56-60; Whelan, "Soviet Attitude Toward International Cooperation in Space," *Soviet Space Programs, 1966-1970, op. cit.*, p. 425; UN Resolution 1962 (XVIII); and the Treaty on Principles Governing the Activities of States in the Exploration and Use of Outer Space, Including the Moon and Other Celestial Bodies of January 24, 1967, in *Treaties in Force, A List of Treaties and Other International Agreements of the US in Force on January 1, 1973*, US Department of State, Publication #8697, 1973, p. 386.

[101]Krivickas and Rusis, "Soviet Attitude Toward Space Law," *Soviet Space Programs, 1962-1965, op. cit.*, pp. 518-519; Nicholas Daniloff, *The Kremlin and the Cosmos* (New York: Alfred A. Knopf, 1972), p. 188; and Krivickas and Rusis, "Soviet Attitude Toward Law of Outer Space," *Soviet Space Programs, 1966-1970, op. cit.*, p. 487.

10 Soviet Space Flight Activities of Practical Application to the Economy

This chapter will briefly examine the efforts of the Soviet space program in those areas of *direct* application to the Soviet economy. Some comparisons with this aspect of the American space program will be made. Indirect applications of space programs such as economic pump-priming, technological spinoffs, national inspiration, etc., were dealt with in Chapter 4.

Both the American and Soviet space programs began with their main goals as the launching of men into space and preliminary exploration of the moon and planets. At least until the advent of manned space stations like *Saliut* and Skylab, space missions whose main purpose is direct application to the economy have been accomplished by unmanned satellites in earth orbit. Figures for the proportion of Soviet space funding allocated to applications activities are not available. Between fiscal years 1960 and 1969, NASA devoted $5.8 billion to space science and applications activities representing about 16.8 per cent of total budget outlays of $34.5 billion. For 1969 the figure was $.6 billion, or about 14 per cent of the NASA budget planned for that year.[1] Factoring out the percentage of this amount allocable to general categories and space science, a report of the National Academy of Science put NASA spending on applications activities for 1968 at about $100 million, or roughly two per cent of the total for that year.[2] The Soviet Union began its applications programs much later than NASA and has accomplished less in general in the field of space applications than the United States. The greater number of Soviet applications satellites in recent years has undoubtedly increased the ratio of total space spending devoted to those programs, but the overall percentage and total for the period through 1969 is almost certainly much less than that of the United States.[3]

Table VII below provides a rough measure of the total efforts devoted to space applications activities by the United States and the Soviet Union as measured by the number of space payloads in each country attributable to direct applications activities. One main mission is attributed to payloads which serve several purposes, and in all categories, especially navigation/ferret, it is often difficult to distinguish civilian and military purposes. In the case of the Soviet Union, it may be presumed that most of all satellites in the navigation/ferret category serve military purposes, since no Soviet navigation (or ferret) satellites have been identified as such by Soviet sources.

TABLE VII

SPACE APPLICATIONS PAYLOADS
BY MISSION CATEGORY, 1957–1973

YEAR	COMMUNICATIONS		WEATHER**		NAVIGATION OR FERRET**		GEODESY**		EARTH RESOURCES		APPLICATIONS TOTAL		GRAND TOTAL*	
	US	USSR	US	USSR	US	USSR	US	USSR	US	USSR	US	USSR	US	USSR
1957													0	2
1958	1												5	1
1959													9	3
1960	2		2		2								16	3
1961	3		1		3								35	6
1962	3		3		1		1						54	20
1963	5		2										60	17
1964	3	1	1	1		3						5	69	35
1965	7	2	2	2		22						26	94	64
1966	11	2	6	2	4		4				25	4	100	44
1967	19	4	6	4	3	4	1				29	12	87	66
1968	11	4	4	2	1	6	1				17	12	64	77
1969	6	2	3	2	0	6	1				10	10	66	71
1970	5	5	5	5	1	16	1				13	26	39	91
1971	6	3	4	4	0	27	0				10	34	53	102
1972	4	56	4	3	1	-(23)	0		1		10	36	40	97
1973	4	32	2	2	1	9	0		0		7	43	27	115
Totals	91	111	55	27	27	70	17		1		190	206	831	814

*Includes mission categories not recorded in this table.
**Categories of US annual figures before 1966 are incomplete owing to classification of many payloads with military purposes since late 1961.
Sources: Charles S. Sheldon II, "Overview, Supporting Facilities and Launch Vehicles of the Soviet Space Program" and "Postscript," *Soviet Space Programs, 1966-1970, op. cit.,* pp. 120-121, 508; *Soviet Space Programs, 1971, op. cit.,* pp. 3, 5; Sheldon, *Review of the Soviet Space Program . . . , op. cit.,* pp. 34, 35; and Sheldon, *US and Soviet Progress in Space,* 1972, *op. cit.,* p. 42, and *US and Soviet Progress in Space,* 1973, *op. cit.,* p. 42, which either contain errors in the figures for Soviet communications and navigation/ferret satellites for 1972 or represent the result or reclassification of navigation/ferret satellites from previous years into other categories for 1972 including communications.

In published literature the USSR was quick to recognize the potential use-fulness of the application of space activities to the civilian economy. None-theless, the first Soviet applications flights did not take place until several years after the inception of the Soviet space programs, while American applications-oriented missions began as early as 1958, the first year in which a space satellite was successfully launched by the United States.[4] Since that time, Russian authorities have evidently become convinced that applications activities in such areas as space meteorology, communications, navigation, etc., are worth the costs in terms of benefits returned, and they have roughly matched the American effort in these areas.[5] In the years since Apollo 11, Soviet space spokesmen have put much heavier stress on the contributions of the space program to the national economy.[6] It has been estimated that the United States will have saved $38 billion by the achieve-ments of applications satellites by the end of the 1970s.[7] Moreover, the civilian uses of applications satellites are often connected with military bene-fits as well, and both these factors help keep applications programs going in both the USA and the USSR despite budgetary constraints.[8] The extent to which applications satellites pay for themselves is reflected by the fact that the bulk of NASA's launches in 1974 were applications satellites paid for by Comsat, Western Union, various US government agencies and foreign countries.[9]

COMMUNICATIONS SATELLITES

Communications along with satellite meteorology is one field in which contemporary costs and benefits make it possible to envision profitable applications of space systems. Just as the far-flung business, military and political connections of the United States have been a powerful impetus to the development of communications satellite systems, the Soviet Union, which is a more insular, regional and rather xenophobic power, is by reason of its very vastness also inclined to the use of communications satellites. Because Soviet settlements of population and technology are widely scat-tered, and because of the underdevelopment and climactic harshness of the intervening territories, communications satellites offer an excellent means for political, economic and social integration of the USSR.[10]

The first precursor of what has become the *Molniia* (Lightning) system of Soviet communications satellites was *Kosmos* 41, launched in August, 1964. The first of a series of 26 *Molniia* I satellites through 1973 was *Molniia* (I)-1, launched in April, 1965. Counting eight *Molniia* IIs (an improved model), there have been a total of 34 successful civilian com-munications satellites launched from 1964 through 1973,[11] about one-third of the US total.

Work began in 1965 on the *Orbita* system of ground stations linked by *Molniia* satellites. *Molniia* 1 was used in 1965 for a successful test of live television transmission from Vladivostok to Moscow. By 1970 there were 36 *Orbita* stations. The USSR has set the goal of linking the whole of the USSR through television via satellite and *Orbita* ground stations by 1980. The *Molniia-Orbita* system also provides communications linkages for telephone, telegraph, facsimile and weather data communications.[12]

As early as 1961 the Soviet Professor Ari Shternfeld had proposed a global communications network employing three communications satellites in 24-hour synchronous orbits. Communications and navigation satellites were also described that year by the Chief Designer (Korolev).[13] The lag in Soviet communications satellite development, however, left the leading role in development and operation of global communications satellites to the United States. In mid-1962 the US Congress passed the legislation establishing Comsat, and two years later Intelsat was set up. In 1973 Western Union was licensed to operate a system of domestic communications satellites. As described in the previous chapters, the Soviets objected to the "free-enterprise" aspects of Comsat and Intelsat as well as to the dominant American role in these organizations. Without the necessary technology and in an atmosphere of socialist versus capitalist competition, the USSR could not attract a world market for its own international communications system.[14]

In November, 1971, the USSR launched an improved communications satellite, *Molniia* II. *Molniia* II flies the same synchronous 12-hour eccentric orbit as the *Molniia* I satellite but employs higher frequencies for communications similar to those of the Intelsat satellites.[15] In 1968 Soviet scientists announced that the USSR would in 1970 launch communications satellites to 24-hour synchronous circular orbits as proposed by Shternfeld in 1961 and like those which have for several years been launched by NASA for Comsat and Intelsat and by the US Air Force for reconnaissance purposes. This new Soviet satellite known as *Statsionar* has yet to be put into 24-hour synchronous orbit, nor has any other Soviet satellite.[16] Not only does this represent a substantial and continuing lag in Soviet satellite communications technology, but it also is in part a reflection of the difficulties of launching satellites into (equatorial) synchronous orbits from the high latitude of the southernmost Soviet cosmodrome. One author has suggested that to help overcome this lag and to get around the lack of international response to the Soviet international communications system, *Intersputnik*, the USSR might join Intelsat and seek contracts with it for the launch of communications satellites.[17]

METEOROLOGICAL SATELLITES

As with communications, satellite meteorology promises to be another space activity that can pay for itself.[18] For the USSR space meteorology serves agriculture and other activities that benefit from improved weather forecasting. Around the world Soviet aircraft and naval, merchant and fishing fleets also are benefited by the observation and prediction of weather made possible by satellites. Here, as with communications satellites, early statements of Soviet spokesmen reflected great optimism and ambition, which were followed up by little real action for several years.[19]

The first satellite testing of systems later employed in the Soviet series of *Meteora* weather satellites occured in April and December, 1963, with the flights of *Kosmos* 14 and 23. The military satellites, *Kosmos* 45, 65 and 92, launched between September, 1964, and October, 1965, also aided in the development of *Meteora*. The first precursors of the *Meteora* satellite were *Kosmos* 44, 58, 100 and 118, and the announced weather satellites *Kosmos* 122, 144, 156, 184, 206 and 226.

In 1967 the Soviets announced the development of the *Meteora* system of satellites, stations for receiving and processing data, and services for operation and regulation of these satellites. Evidently, the system was not fully operational until March, 1969, when the name *Meteora* 1 was given to a Soviet satellite. The *Meteora* satellites have provided valuable services to the USSR, and their developers were the winners of the Lenin Prize for 1970 in science and technology.[20] Observations of weather have also been carried out by *Molniia, Soiuz* and *Kosmos* scientific and military satellites. Soviet scientists have proposed the launching of weather satellites in higher circular and synchronous orbits.[21]

The considerable lag in the development of Soviet meteorological satellites complicated cooperation between the US and the USSR in this field. Recent Soviet progress and the absence of competitive issues associated with other programs, such as those involved with satellite communications, have made possible a substantial program of cooperation between the United States and the USSR in this field. As with other applications satellites, the usefulness of meteorological satellites helps to maintain the viability of the Soviet space program among other programs demanding political attention and allocation of resources in the USSR. Through 1973 the USSR launched 27 meteorological satellites compared with 55 for the United States.

NAVIGATIONAL SATELLITES

Satellites are used to compute the position of ships and sometimes aircraft and to guide them to their destinations without their interfering with one another. Satellites can do this with greater accuracy than conventional astronavigational techniques. The US Navy launched its first Transit navigational satellite in 1960. As with other applications satellite systems, the USSR was considerably behind the United States and did not launch a navigational satellite until 1965. Both systems served military customers, although Transit was declassified in 1967 and has been offered to and used by the merchant navies of other countries.[22] As with other applications satellites, the need for such a system and the intention to build it were signaled by the Chief Designer as early as 1961, but the opening of the operational phase was delayed for several years.[23] The delay in Soviet development of navigational and other applications satellites is an indication that the Soviet space program operates under severe budgetary constraints and that, much like the American program, there has been a marked preference for space spectaculars of great political impact, especially during the tenure of Nikita Khrushchev.

The most curious and politically revealing aspect of Soviet navigational satellites, other than the relative tardiness of their development, is the Soviet announcement policy concerning them. On the one hand, Soviet spokesmen claim to have an operating navigational satellite system. On the other, they have never named this system or identified a single Soviet satellite as being part of it. On these grounds, it is probably safe to assume that Soviet navigational satellites serve mostly military purposes. It is probable that their main use, as with the Transit system, is to provide precise navigation for missile-launching submarines.[24]

The announcement that a Soviet navigational satellite system was under development came from Leonid Sedov in 1965. An operational system was claimed by Academy President Keldysh in early 1966. Although it is difficult to distinguish navigational satellites from those used for electronic ferreting, it may be that some Soviet satellites have served both purposes. The first Soviet satellites with the characteristics of navigational satellites (multiple payload, circular orbit at medium to high altitude, precise elements, small size) were *Kosmos* 38-40, launched by a single launch vehicle in August, 1964. The series of navigation/ferret satellites has continued, and this has become the second largest single category of Soviet applications satellites, indicating it serves multiple purposes. Through 1973 the USSR launched 70 of these satellites, and the United States 27.[25] This is the only category of applications satellites for which the Soviet total is substantially larger than the US. It is possible that a few of the satellites may have been scientific in purpose and that for some reason, possibly mission failure, no scientific data has been published from them.[26]

As in other applications categories, there have been published suggestions in the USSR that 24-hour synchronous satellites be employed for navigation. This remains a possibility for the near future as do fuller employment of navigation satellites by the Soviet merchant navy and possibly the subsequent lifting of some of the security wraps that now surround the system.[27]

EARTH RESOURCES SATELLITES

Location, measurement and monitoring of earth resources employing space satellites is one field in which there is a great promise in terms of potential economic benefits, but one in which neither the United States nor the Soviet Union has so far accomplished the progress already achieved in communications, meteorology and navigation.[28]

The United States now has under way an Earth Resources Technology Satellite (ERTS) system employing unmanned satellites to do earth resources work. The first satellite in this system, ERTS-1, was launched in July, 1972. This satellite has passed over the USSR and other socialist countries. Originally it was not known whether it would be programmed to shut down at such times or whether the data made available would be shared with the USSR or used by the United States for its own purposes, presumably as strategic intelligence.[29] Many countries are now participating in ERTS, and the USSR is among them. Published records indicate that no data is being collected from countries which do not participate in ERTS.

A certain amount of data from Soviet *Meteora* weather satellites and stereophotographs taken by *Zond* manned precursor satellites in 1968 and 1969 has been used in the USSR in connection with work on earth resources. The military reconnaissance satellite *Kosmos* 243, besides testing improved techniques for weather satellites in 1968, also provided data of use in mapping earth resources.[30] More extensive mapping of earth resources was conducted by *Soiuz* 7-9, *Soiuz* 11 and *Soiuz* 13. It appears that the Soviet Union will rely on manned orbital stations such as the *Saliut* space stations and associated *Soiuz* spacecraft to do earth resources work, while the United States will use both manned and unmanned spacecraft.[31]

Earth resources application of space systems has in the past and continues to be extensively discussed and very highly praised by Soviet scientists and Academy personnel. Possible benefits enumerated by Professor Kiril Ia. Kondratev, Academician Blagonravov and others include location of minerals, oil and gas, application to the extensive Soviet fishing industry, monitoring pollution, location of minerals in Antarctica, possible use of satellites to increase Soviet crop yields by 25 to 50 per cent, use in forestry, water resources and geology.[32]

Permanent and semipermanent manned space stations are now the major missions of both the American and the Soviet space programs as witnessed by continuing work by the Soviets after a two-year lapse with the *Soiuz* and *Saliut* spacecraft and, in the American case, by the Skylab program and the missions indicated for the space shuttle. The present decade will probably provide an indication of the usefulness of space stations for earth resources work and the cost savings that can be achieved by the possible use of shuttle vehicles, now in preliminary development in the United States and actively discussed in the Soviet Union.[33]

It should be noted in passing that although Table VII records 17 American geodetic satellites and none for the Soviet Union, geodesy and mapping have undoubtedly been done by Soviet satellites, and the data gleaned from them has been applied to both civilian geological purposes and military tasks, such as the targeting of ICBMs. Evidently, the military significance of geodetic satellites, as with that of navigation satellites, has inhibited Soviet authorities from permitting the identification of such satellites or publication of geodetic data.[34]

FUTURE POSSIBILITIES

Besides those possibilities for applications satellites already enumerated above, a number of others exist and will be briefly noted here. Khlebsevich, the designer of *Lunakhod* several years ago, proposed the stationing on the moon of a radio telescope for observation of earth.[35] One Soviet writer has proposed the creation of dust rings around the earth similar to those of Saturn in order to deflect sunlight to the northerly reaches of the USSR—an unlikely prospect for many reasons.[36]

Savings which might be accomplished through technological advances could reduce the break-even point for satellite versus conventional telephone communications from the 1,500 miles of 1967 to 100 miles by the year 2000.[37] Advances in communications satellites could also be used to secure conferences between distant businessmen or political leaders, reduce authority delegated to business and diplomatic personnel by establishing direct communications links, sample world public opinion and employ two-way television for international education and propaganda.[38] It is the possibility of such propaganda use (against the USSR) that has kept Soviet spokesmen ambivalent about occasional suggestions by Academician Keldysh and others for direct broadcast from space to home receivers.[39] Other uses of space are likely to emerge if, through such means as reusable shuttle vehicles, the cost of orbiting space payloads can be substantially reduced.

NOTES

[1]B. W. Augenstein, *Policy Analysis in the National Space Program*, Rand Corporation, Paper #P-4137, July 1969, pp. 29-30.

[2]Ralph E. Lapp, "Investing in Space," *New Republic, 160* (2), January 11, 1969, p. 12.

[3]Remarks of Reichelderfer, Friedman and Kantrowitz in Jerry and Vivian Grey, eds., *Space Flight Report to the Nation* (New York: Basic Books, 1962), pp. 47-48; and Alton Frye, "Soviet Space Activities: A Decade of Pyrrhic Politics," in Lincoln P. Bloomfield, editor, *Outer Space, Prospects for Man and Society* (New York: Frederick A. Praeger, Inc., 1968), p. 81.

[4]Barbara DeVoe, "Soviet Application of Space to the Economy," *Soviet Space Programs, 1966-1970, op. cit.*, p. 297.

[5]Charles S. Sheldon II, "Projections of Soviet Space Plans," *Soviet Space Programs, 1966-1970, op. cit.*, p. 395.

[6]For a recent example see Boris Petrov in *Selskaia Zhizn*, January 4, 1972, as cited in *Soviet Space Programs, 1971, op. cit.*, p. 58.

[7]According to Charles Sheldon II, "Peaceful Applications," in Bloomfield, *op. cit.*, p. 63.

[8]*Loc. cit.*; and Salkeld, *War and Space* (Englewood Cliffs, N.J.: Prentice-Hall, Inc., 1970), pp. 149-150.

[9]See *Space World*, K-4-124, April 1974, pp. 12-13.

[10]DeVoe, "Soviet Application of Space to the Economy," *op. cit.*, p. 299.

[11]*Ibid.*, pp. 298-301; and *Soviet Space Programs, 1971, op. cit.*, pp. 46-47, 67-70; and "Satellite Report," *Space World*, K-6-126, June 1974, p. 30.

[12]DeVoe, "Soviet Application of Space to the Economy," *op. cit.*, pp. 298, 301-303.

[13]*Soviet Space Programs, 1962, op. cit.*, pp. 89-90.

[14]Joseph G. Whelan, "Soviet Attitude Toward International Cooperation in Space," *Soviet Space Programs, 1962-1965, op. cit.*, pp. 473-475; *USSR Probes Space* (Moscow: Novosti Press Agency Publishing House), no date, no pagination; F. J. Krieger's remarks in Grey, *op. cit.*, p. 189; Erin B. Jones, *Earth Satellite Telecommunications Systems and International Law* (Austin: University of Texas, 1970), pp. 105, 114-115.

[15]*Soviet Space Programs, 1971, op. cit.*, pp. 46-47. A second *Molniia* II was launched in May 1972.

[16]DeVoe, "Soviet Application of Space to the Economy," *op. cit.*, pp. 305, 318; and Charles Sheldon II, *United States and Soviet Progress in Space: Summary Data Through 1972 and a Forward Look*, Library of Congress, Congressional Research Service, January 29, 1973, p. 39.

[17]Sheldon, *supra*, p. 67.

[18]Vernon Van Dyke, *Pride and Power, The Rationale of the Space Program* (Urbana: University of Illinois Press, 1964), p. 118.

[19]See the remarks of Khlebsevich quoted by Sergei Gouschev and Mikhail Vassiliev, editors, *Russian Science in the 21st Century* (New York: McGraw-Hill Book Co., Inc., 1960), p. 210.

[20]DeVoe, "Soviet Application of Space to the Economy," *op. cit.*, pp. 305-312.

[21]*Ibid.*, pp. 311-314; and *Soviet Space Programs, 1971, op. cit.*, pp. 35-37.

[22]Sheldon, "Peaceful Applications," *op. cit.*, p. 57.

[23]"Chief Designer," in *Izvestiia*, December 31, 1961, p. 3, as cited in *Soviet Space Programs, 1962, op. cit.*, p. 90.

[24]DeVoe, "Soviet Application of Space to the Economy," *op. cit.*, p. 314; and Shel-

don, "Soviet Military Space Activities," and "Projections of Soviet Space Plans," in *Soviet Space Programs, 1966-1970, op. cit.,* pp. 328-329 and 385.

[25]See Table VII above.

[26]*Soviet Space Programs, 1971, op. cit.,* pp. 13-14.

[27]See Charles S. Sheldon II, "Summary" in *Soviet Space Programs, 1966-1970, op. cit.,* xxxii; and DeVoe, "Soviet Application of Space to the Economy," *op. cit.,* p. 314.

[28]DeVoe, *supra,* p. 315.

[29]See "New Eye in Space," *Newsweek, 80* (6), August 7, 1972, p. 43; and Sheldon, *United States and Soviet Progress in Space* (1972), *op. cit.,* pp. 13, 40-41.

[30]DeVoe, "Soviet Application of Space to the Economy," *op. cit.,* pp. 311, 313-314; and *Soviet Space Programs, 1971, op. cit.,* p. 47.

[31]DeVoe, *supra,* pp. 315-317, 319-321; *Soviet Space Programs, 1971, supra,* pp. 35-37; Sheldon, "Summary" in *Soviet Space Programs, 1966-1970, op. cit.,* xxxii; and Sheldon, *United States and Soviet Progress in Space* (1972), *op. cit.,* p. 6.

[32]Sheldon, "Summary," *loc. cit.;* DeVoe, *supra,* pp. 316-317; *Soviet Space Programs, 1971, supra,* pp. 48-49 and notes 4-13 on p. 48. See also the comments by G. Petrov (p. 54), Kondratev and Blagonravov (p. 55) and B. Petrov (p. 58) in *ibid.*

[33]See Chapter VII: *Soviet Space Programs, 1971, supra,* p. 63. See also the comments of cosmonaut Beregovoi (p. 55), Academician Keldysh (p. 56) and correspondent V. Denisov (p. 57) in *ibid.*

[34]G. Petrov in *ibid.,* p. 54; and Sheldon, "Soviet Military Space Activities," *op. cit.,* pp. 328-329.

[35]Khlebsevich in Gouschev and Vassiliev, *op. cit.,* p. 209. Use of a lunar telescope for earth meteorology has been more recently (1966) suggested in the USSR by Kiril Kondratev (p. 319).

[36]V. Petrov in *Pravda Ukrainy,* as cited by Sheldon, "Peaceful Applications," *op. cit.,* p. 55.

[37]Francis A. Gicca, "Space Communication in the 21st Century," Raytheon Corporation, May 1, 1967, American Astronautical Society Reprint 67-092, pp. 3-4.

[38]J. L. Hult, "Satellites and Future Communication Including Broadcast," Rand Corporation, 1967, American Astronautical Society Reprint 67-091, pp. 11-12.

[39]DeVoe, "Soviet Application of Space to the Economy," *op. cit.,* pp. 318-319.

11 Domestic and International Impacts of Space Exploration

DOMESTIC IMPACTS

The social and technical innovations which have emerged from the national space programs of the US and the USSR are many. Gauging their impact on the economic and political life of the two countries is, however, an extremely complex task. In one sense, there is never anything new "under the sun." Social and technical inventions are built upon an existing foundation, and the functions they serve generally were and continue to be served by other social structures and techniques. Moreover, new organizations and techniques seldom, if ever, have a unidirectional impact. Their impact in some areas is likely to be positive, but given the scarcity of resources in any society, the positive impact in one area is likely to be tied to a negative impact elsewhere.[1]

Such impacts as innovations wrought through space are likely to develop in stages with differing effects over time. The effects of innovations which emerge from national space programs will in some cases be similar, but, in general, will not be the same in societies as different as those of the USA and the USSR.[2] It is impossible to say with certainty in most cases what the "alternatives" to space exploration were and will be. It is easy enough to say that the quanta of national commitment and resources drawn into space programs *might* have been directed elsewhere or might be directed elsewhere in the future, but just where these quanta would have gone or will go is difficult to say. Some of the impacts of space exploration that have been mentioned are so elusive and intangible that it is hard to weigh them in any meaningful way. For example, it is claimed that "the stimulus of deeply inspiring and commonly appreciated goals" of space feats inspires human society to rise "out of its lethargy to new levels of productivity."[3] Although general indices of labor productivity are available, the extent to which they are influenced by inspiration from space among a complex of other factors is difficult, if not impossible, to measure with certainty.

Space as a Technological Stimulus and Drain

Here the impact of space exploration as a form of pump-priming or general economic stimulus will be ignored. Not only is such impact largely irrelevent in the Soviet context, but even in the United States this impact is likely to be transient and relatively easily replaced by other kinds of general stimuli. This section will rather make a general examination of the impacts of transfer of technology from space programs into other sectors of the economy and the "drain" of resources away from other sectors into national space programs.

Proponents of an extensive space program in both the Soviet Union and the United States have claimed that innovations emerging from space will increase the productivity of labor and capital, creating a rise in real incomes. New products of space research, it is claimed, will stimulate demand and create new markets. New resources and new uses for old resources discovered by space research can be employed more efficiently, thus increasing total production.[4] Transfers of technology may take place in several ways. Vertical transfers of technology take place when general discoveries find specific application. Horizontal transfers take place when techniques employed in one field find uses in another. NASA's contracts within educational institutions and supply firms may generate discoveries and products of diverse application. Contacts among scientists help to promote such transfers. Organizations, public and private, may shift their interests from space to other fields, taking with them the knowledge and organizational resources they have acquired. Some transfers of this sort, perhaps the most important ones, are likely to be intangible, that is, in the form of general skills, organizational resources, problem-solving orientations, "mission mindedness," etc.[5]

Examples in which such transfers may have taken place include the nonaerospace work done by aerospace firms. In November, 1964, Aerojet General Corporation took a six-months, $100,000 government contract to study air and water pollution. In 1965 an Aerojet subsidiary, Space General Corporation, contracted for a study on crime; Lockheed Missile and Space Corporation was awarded a contract for a space data system, and North American Aviation got a contract for a state-wide transportation system.[6]

NASA has established special programs to facilitate transfers of space technology, has predicted the public benefits therefrom, and has attempted to focus public attention on spinoffs of technology from space exploration.[7] Such activities may be more important for NASA's public relations than for any impact on the US economy. No US agency exists the specific function of which is the widest sharing of technology throughout the economy.[8] The Soviet Union does have such an agency, the State Committee for Science and Technology, established in 1965. Its chairman, Vladimir A. Kirillin, is a deputy prime minister and member of the Central Committee of the CPSU. It has been suggested that the State Committee has a special responsibility for the Soviet space program. The Committee is the successor to the State Committee for the Coordination of Scientific Research created in 1961.[9]

Despite the opportunities and agencies that exist, there have been relatively few transfers of technology from space research to civilian sectors of the national economy. This is certainly true of the United States, and by implication, of the USSR as well.[10] One US House subcommittee reported that of nearly 1,200 prime research contracts and thousands of subcontracts funded by NASA, the agency received only 159 disclosures of inventions and authorized patent applications on only 23 developments.[11]

Space technology and civilian industrial and consumer technology are not generally and directly compatible. The Rockwell Corporation was in part motivated by the expectation that it could "get rich on the stuff in [North American's] wastebaskets" when it merged with North American Aviation. William F. Rockwell, chairman of the new corporation, North American Rockwell, reported that it turned out that the technology developed for defense and space was too complicated and too expensive to be profitably adapted for competitive civilian markets.[12] Awareness of space-developed innovations is not a sufficient condition for their use elsewhere. Considerable risks and expenses must often be borne before the transfer is made, and the results may be disappointing. Criteria of cost and performance are probably the areas of greatest disparity between the civilian and aerospace markets.[13]

The impetus for technological progress occasioned by space research must be diminished by the forces of technological drag that space research creates in any consideration of its total or net impact. Space research can draw resources away from other fields which also contribute to technological progress. This is particularly true of one resource, highly skilled and competent scientific, executive and technical manpower.[14] NASA and its contractors have employed some five to seven per cent of American scientists and engineers.[15] Sensitive to this effect, NASA in 1962 began its own program of grants aimed at producing 1,000 Ph.Ds. per year and endeavored to reduce its demands for fully qualified scientists and engineers by using and training technicians to take their places. NASA has also sponsored research in educational and other nonprofit organizations.[16]

Aside from human resources and affecting their distribution is the national allocation of funds for research and development. In the mid-1960s NASA accounted for about one-third of American spending on research and development, another third went to the American military, and the rest was divided between diverse civilian sectors.[17] The proportion of Soviet research and development money devoted to space for the same period is estimated at about one-fourth.[18] Presuming that the Soviet Government affords a proportion of research and development funds to military purposes similar to that of the United States, both then allocate at least half of national research to activities of which the technological impact on the civilian sector is at best indirect.

In both Great Britain and the United States extremely rapid growth rates in research and development spending in the fifties and early sixties were not matched by accompanying growth in general economic productivity. Vast expenditures on R & D in the United States also were not reflected by any corresponding growth in labor productivity. The economist Robert A. Solo offers an explanation for this by categorizing two kinds of R & D expenditures: those that go into civilian industry, which he calls "growth-oriented" R & D, and that portion which goes for military, space and medical research, which does not have a favorable impact on overall economic growth.[19] It is the latter category which accounts for the bulk of R & D growth in the postwar period in both the United States and the Soviet Union. Space and military R & D compete with growth-oriented R & D in the sense that both make demands on more or less scarce economic resources especially in that they compete for employees in the same manpower pools. Since such resources are relatively more scarce in the Soviet Union, the "drag" on growth effected by military and space spending is likely to be greater there than in the United States.[20] The two reasons why the USSR in recent years has failed to match the growth rates of many capitalist countries that have reinvestment rates only half or one-third of the Soviet Union are (1) the low productivity of labor and capital in the USSR, and (2) the fact that so much of the Soviet effort, including the best personnel and technology, has been diverted to military and space programs.

Solo calculates that while growth-oriented R & D in the US through the fifties and early sixties remained at or below 1953 levels, nongrowth R & D grew by 233 per cent.[21] Western Europe and Japan, without high military budgets and lacking any ambitious national space programs, were able to show higher growth rates than the United States (and in some cases higher than those of the USSR). These nations have produced commodities for civilian markets which compete favorably with those of the United States and are virtually unavailable in the Soviet Union. If economic scarcity has made the course of the Soviet Union more difficult in this respect, one suspects that *political* factors have had a greater impact in the United States than in the USSR. The former country compared to the latter cannot commit resources easily outside the framework of "national commitments," as in defense and space.[22] Even though the wealth of the United States makes it possible to have guns, butter and a man on the moon, the political sanctions necessary to mobilize these abundant resources are harder to come by than is the case in the Soviet political system where a consensus among a handful of oligarchs can set in effect greater changes. Campaigns for civilian technological advancement are also contained in the United States by a lingering *laissez-faire* philosophy and by an immense semicaptive home market, which the USSR also enjoys at home and in East-Central Europe.

American and Soviet scientists have complained that military and space research have had a distorting impact on scientific priorities and the development of science in both societies.[23] American observers have held that the USSR has suffered severe economic dislocations resulting from its investments in space.[24] Although space activities with direct economic applications *do* promise to have a positive impact on national production, it was shown in the last chapter that programs in this area have been on a minor scale within the national space programs of both the USA and the USSR, and that the lion's share of space budgets has gone into the military, manned, and planetary missions of both powers, which so far have had little economic application and relatively little spinoff of technology. The technology of Soviet manned missions is basically that of 1957, and that of Apollo was largely 1962 technology. The major mission goal orientation followed so far by the United States is generally inhospitable to technological advances. At its height, Project Apollo commanded three-fifths of NASA's budget and one-third of all national expenditures for research.[25]

One Soviet source claims that space and military research and development were commanding too great a share of Soviet technical resources, and that this was behind the mid-1960s shift of institutes, rubles and personnel away from the Academy of Sciences and into the various institutes of the industrial ministries of the USSR. The same source holds that a simultaneous shift occurred away from crash programs in space and missile development to a broader and more "rational" distribution of research effort designed to promote broad economic progress.[26] In support of this contention, one can note that the USSR has not successfully developed a new space booster since 1965 and, until lately, has continued to rely on early and mid-1960s technology for the bulk of its ample strategic missile force.

Impact on Politics

The Soviet space program derives political support from and helps to maintain the considerable influence of Soviet scientists in general and the Soviet Academy of Sciences in particular. Space programs have undoubtedly increased the political influence of scientists in both countries.[27] As one might expect, dissension about the national space program has been more prominent in the United States. In the USSR the principal public spokesmen for the space program and the leading representatives of the Soviet Union in international negotiations affecting space research are members of the Academy. In this sense the Academy is as much responsible for the space program as any other organization. Although the CPSU and government leadership are ultimately responsible for space and all other major fields of policy in the USSR, they have, for whatever reason, evidently decided to yield a very substantial amount of authority to the Academy in this one field. Of all groups in the USSR, excepting the cosmonauts, the Academy is the main beneficiary of the Soviet space program.

Since most Soviet satellites can be classified as performing some kind of military mission, one can also identify the Soviet Armed Forces and especially the Strategic Rocket Troops and Red Air Force as at the same time important supporters and beneficiaries of Soviet space activities.

A similar role of benefits and support for space activities is also played by the Armed Forces of the United States, especially the US Air Force. As with the USSR, military satellites comprise the largest share of American payloads. The American scientific community is much less important than the Soviet scientific community as a supporter or beneficiary of the American space program. Although the Space Science Board of the National Academy of Sciences serves in an advisory capacity to NASA, neither the Board nor the Academy as a whole has any special responsibility for the American space program. The representatives of the United States in international space dealings are NASA bureaucrats and State Department personnel. NASA executives have been chosen from the worlds of business, science and engineering. Besides the executives of NASA, prominent spokesmen on space programs include national political leaders and interested scientists, scholars and journalists.

Excepting the astronauts and NASA employees, the aerospace industry and areas impacted by it are the principal beneficiaries and supporters of the American space program. Experiments with the corporate form are just beginning in the Soviet Union, so there is no corresponding institution to exert political influence in the USSR. Certain Soviet industrial ministries, however, do the bulk of their work on military and space hardware and presumably defend the importance and priority of their efforts. In the United States the two-party system and the "politics of pluralism" contribute to the political influence of aerospace firms. As noted previously, 90 per cent of NASA's budget is ultimately spent in private industry.[28]

The influence of the military services and the aerospace industry as supporters and beneficiaries of the national space programs of the US and the USSR is substantially diminished by the interchangeability of space and military production and spending. Although US firms may have some preference for the positive image that attaches to space contracts, the stigma attached to defense contracts never reached great proportions and has probably declined with the phasing out of the war in Indochina. Generally, aerospace firms are likely to favor more and larger contracts rather than those with good public images anyway.

NASA's budget for 1974 was only about $3.2 billion. Aerospace contractors therefore favor business in general, not space in particular, and the bulk of that spending is military. A few dozen veterans of the astronaut corps cannot compete with the political influence of millions of veterans of the armed services. Similarly, the impact of military spending is greater and of wider scope in areas impacted by military bases and military contractors and on labor unions and professions connected with aerospace. Thus the military and the industries associated with it, while providing a modicum of support to the national space programs of both countries, are likely also to be their main competitors. Especially in the United States, the national space program lacks a distinct basis of support apart from the military-industrial complex. In neither country has the national space program emerged as a viable political alternative to military efforts.

What some call the major mission approach in the American space program and what others might characterize as the "boondoggle" approach reveals the greater, almost exclusive, influence of the military-industrial complex in the American space program as compared to the Soviet program. NASA is now in between one giant project (Apollo) and the next (the space shuttle). This is reflected in the steady drop in the NASA budget since Apollo spending peaked in 1966. Major missions like Apollo and major technological efforts like the shuttle call for immense expenditures of concentrated economic and political impact.

Although the original impetus for such missions may come from the commitment to space exploration in NASA or from the political goals of the national executive, ultimate support is based in large measure on the efforts of prime contractors. North American Aviation, now Rockwell International, was the largest prime contractor for Gemini and Apollo and has been selected as the largest prime contractor for the shuttle.

Although subcontractors, NASA personnel, and areas in which NASA facilities are located stand to benefit by funding of the shuttle project, North American Rockwell, which has fallen on hard days, and those dependent on the company are faced with the alternatives of boom or bust.

Boom and bust are not features of the Soviet economy. Not only was the "storming" of certain sectors of the Soviet economy enforced on *reluctant* managers by political leaders, but it is no longer characteristic of the Soviet system. No stockholders or "tycoons" are dependent for their livelihoods on immense government contracts, but managers and workers alike are more or less guaranteed the "right" to the steady enjoyment of their stations in life. Differential rewards are afforded to those who are reliable rather than innovative and who perform dependably without pulling off any major economic coups. Political leaders are shifted from one geographical and functional economic base to another without acquiring fixed attachments to the well-being of particular constituencies.

As suggested earlier in this study, the Soviet space program since Khrushchev has proceeded in the manner of the tortoise, in a more broadly aimed and steady fashion with an ultimate goal. Lately it has incorporated a greater and growing interest in applications of space to the economy, an area well suited to this pace and approach. The political impact of the American program, however, has helped prepare it for the role of the hare, moving from one dashing sprint to the next whenever a single project of astounding proportions can be pushed through the political system. Steady development, as in the field of communications satellites, has been largely the result of turning the program over to private industry. Comparing the American to the Soviet space program in this regard, a critic could view it as either spectacularly wasteful (Apollo) or narrowly pragmatic (Comsat). The Soviets lately seem to plod along somewhere in between, moderating the virtues and vices of both extremes.

Applications Activities

Applications activities in space, such as meteorology, communications and earth resources, promise to have a positive net impact on the economies of the space powers; indeed, at least in the area of communications, they already have. Because applications activities have operated at a much lower cost than more spectacular space feats, there is less likelihood of their imposing great *opportunity costs*. Moreover, the returns are direct and profitable enough to enable Comsat, Intelsat and Western Union, which pay for launches by NASA and for equipment from private firms, to turn a healthy profit. Still, communications satellites probably will not secure any great windfall for their owners and operators that will change the face of world or domestic politics and economics in the foreseeable future.[29] Nonetheless, this promises to be an expanding field, and the 1962 Communications Satellite Act is a landmark of sorts. For the present at least this one space activity can continue to be carried out on a pay-as-you-go basis without massive involvement by government. Comsat and Intelsat maintain an ongoing means for continuing at least one aspect of the American space program, if all else fails, and it will be in their interest to encourage developments in space technology, like the space shuttle, which promise to lower their costs of operation.

For the foreseeable future Western Europe will lack the technological capability to launch communications satellites for Comsat and Intelsat. The Soviet Union also lags in communications satellite technology, and there are political barriers that will probably prevent it from becoming a launcher and supplier for these organizations. It is therefore in the interest of these organizations and their principal owners to keep up the commitment to space by both NASA and American aerospace contractors. Comsat has set an important precedent in the American space program that can be applied when and if other space activities approach the position of net profitability.

Space meteorology promises substantial benefits to agriculture and in avoiding crop damage caused by storms. Because the services provided by space meteorology are largely informational, it is not a field which is particularly adaptable to private enterprise. Satellite meteorology probably offers greater benefits to the Soviet Union than to the United States in that agricultural insufficiency is a greater problem in the USSR, and the climate is more severe and variable there. Many of the problems of Soviet agriculture, however, will not be solved by the most perfect weather forecasting. By itself satellite meteorology can probably incur net savings for both countries, but its impact on agriculture and other sectors of the economy is not likely to be revolutionary.[30]

The first Earth Resources and Technology Satellite, ERTS-1, was launched in July, 1972, by the United States. Reports indicate that it has recorded data on the Soviet Union and other countries which have specifically requested such information from the United States. Earth resources work is the one application of space activities to earth problems which promises to have the greatest impact in the foreseeable future. If satellites doing earth resources and meteorological work could make the Soviet Union dependably self-sufficient in food production (agriculture and fishing), this would have a dramatic domestic and international impact. If earth resources satellites can be cheaply and reliably used to locate and aid in the exploitation of the vast resources of the country, the USSR could become the wealthiest nation on earth. It is by no means possible to say at this time whether such an outcome is certain or even likely. Yet, if the USSR could effectively exploit its immense resources, the impact would be staggering. The situation in agriculture and the declining growth rates of the USSR are the biggest problems currently faced by the Communist regime of that country. Both problems confound the grounds on which that regime claims legitimacy for its rule.

The United States lacks the untapped fossil fuel resources of the Soviet Union, except in the cases of coal and oil-bearing shale, and is not self-sufficient in many minerals which are abundant but difficult to locate and exploit in the USSR. Earth resources satellites will not be of any great benefit in locating and exploiting what the US does not possess in the first place. Nonetheless, earth resources might make it possible for the United States to employ its resources more efficiently, especially in areas like Alaska where the sparseness of the population and climactic characteristics pose similar barriers to those in the Soviet north and east. The greatest potential benefit to the US from space research would then be not the discovery and exploitation of existing resources but the discovery of new and cheaper forms of energy. It is possible, but not likely, that such discoveries would emerge from space research as a spinoff rather than as a result of applications satellites. Nuclear propulsion systems and solar batteries employed in space research might lead to such a breakthrough.[31] One suspects, however, that as is generally the case, it is easier, cheaper and quicker to make important discoveries by looking for them rather than waiting for them to "spin off" discoveries in other fields.

Space activities in general, especially communications activities, have an *integrating* impact on national societies. By making instantaneous communications from one end of the country to the other available more cheaply, problems and events in given localities assume a national character. The United States is already so thoroughly linked by national communications media that the impact of space in this respect is likely to be less than in the Soviet Union, which is larger, has more scattered and isolated population centers and is more ethnically and culturally diverse. The United States is, of course, ethnically diverse as well, but in *most* cases it has accommodated and moderated the effect of ethnic diversity more than has the USSR. Sovietization, or cultural homogenization of the USSR, has been a long-term goal of the political authorities there and one which they are using satellite technology to achieve. Communications satellites will not necessarily make it possible to achieve this goal, but they will help.

INTERNATIONAL IMPACTS

The degree to which cooperation and/or competition in space might become an alternative to war and the arms race was discussed in Chapter 3. The conclusion was largely negative. It is the opinion of this author that "non-lethal" competition[32] is not an alternative but an adjunct to lethal competition as witnessed by the coincidence of the arms race and the space race from Sputnik to the man on the moon.[33] Space cooperation between the US and the USSR, which is only now approaching significant levels, seems more promising than nonlethal competition. This is so because it helps to destroy much of the rationale for emnity by reinforcing interdependence and, where it succeeds, trust.

Understanding may not bring about peace, as witnessed by the endemic nature of wars among well-acquainted peoples, but misunderstanding is deliberately inculcated, especially by the Soviet leaders, as part of rationalizing the structure of conflict. Space technology will not in itself break down barriers to communications. If it is deliberately used to do so by one or both space powers against the will of the other, it will only exacerbate tensions.[34] Given the views of the Soviet leaders, one can suspect that the development of a space interceptor may have been intended more for use against propaganda satellites than against reconnaissance satellites or space weapons. The USSR has not shot down American reconnaissance satellites and there are no orbiting American weapons systems to be shot down. Such an incident would certainly help to promote a war in space.

If space research will help to achieve a stable peace, it will do so by instilling cooperative habits, by providing safeguards for arms control,[35] by distributing mutual economic benefits, and through dramatic cooperative endeavors such as the Apollo-*Soiuz* mission, thereby weakening the rationale for emnity in the popular mind. This last effect may be dulled somewhat by the declining psychological salience of space spectaculars.[36]

Whatever peaceful impact the improvement of communications between the American and Soviet peoples might have, it will have to await actual improvements in the relations between the two countries to which space cooperation can contribute.[37]* Improvement in relations through satellite communications will still be inhibited by the competitive nature of the two international satellite communications systems, Intelsat and *Intersputnik*. The existing systems are likely to have a greater impact on solidifying existing power blocs and increasing the means whereby both the USA and the USSR can assist and influence third countries.[38] If communications satellites are themselves used to promote peace and a relaxation of tensions, an agreement fusing or somehow combining Intelsat and *Intersputnik* would be a

*While this was being written the Soviet authorities shut off American television facilities in Moscow during the Nixon visit of July, 1974, when the correspondents attempted to report on Soviet dissidents.

necessary prerequisite. Such an agreement would also require active co-operation to insure the compatibility of satellites systems, which would contribute in the manner of other cooperative endeavors to the reduction of tensions.[39] The likelihood of such an agreement is very slight.

If earth applications activities could possibly solve the problems of Soviet agriculture and economic growth as previously suggested, the international impact would be as great or greater than that on Soviet society itself. The USSR would no longer be reliant upon the West and its good will to secure the necessary minimum of foodstuffs for the Soviet people. Not only would this reduce the interdependence of the US and the USSR, but it would also reduce American influence over the USSR, such as that evidenced by the recent announcement of the Soviet government to eliminate special taxes on those who leave the USSR.

If an abundance of mineral wealth and fossil fuels were made readily and cheaply available in the USSR during a period of acute energy crisis elsewhere in the world, the impact would be even greater. In comparison to the United States with its global political, military and economic connections, the USSR is still very much a *regional* power within the closely tied, relatively isolated system of Communist states, cut off from China with few interests outside Eurasia and conducting little more than one-fourth the total volume of trade of the United States.[40] The USSR could in such circumstances acquire a much more powerful voice in world affairs as a *global* power and perhaps end the decline in its economic growth which contributed to the ouster of Khrushchev and continues to confront the existing political authorities. The means employed would be a rapid increase in the trade of Soviet minerals and fossil fuels for Japanese and Western technology.

The regional character of contemporary Soviet power is in part a reflection of Soviet economic lag behind the United States and of a long history of Russian-Communist xenophobia toward the outside world. Both of these characteristics, incidentally, are mirrored in the space programs of the two countries as previous chapters have indicated, particularly with regard to cooperation with and openness toward other countries.

Khrushchev used to claim that the USSR was going to defeat the West in a great "economic competition" in which Soviet growth would make it possible for the USSR to outstrip the United States in a variety of economic indicators during the 1970s. Rapid Soviet economic growth is a bygone along with Mr. Khrushchev. The Communist movement as realized in the USSR has at least temporarily lost much of its claim to what many assert is its principal attraction to political elites and masses in the underdeveloped world—the rapid economic modernization and social and political transformation of a society. It is to be expected that the USSR will at least attempt to reinvigorate Soviet economic development employing applications satellites as one means of bolstering its strategic might in the international political arena and as a means of renewing the economic promise of Marxist-Leninist communism.

The above developments are suggested as possibilities. Neither the US, which has gone farther in this field, nor the USSR has yet reaped major, politically significant returns from the application of satellite technology to earth resources. Instead the most politically significant application of satellite technology to economic problems has been the American success in capturing the leading role in the field of satellite communications. It is simply too early to foresee the full impact of satellite technology elsewhere.

Beyond the largely speculative impact of satellite applications are other possible space activities, the impact of which is also speculative. Future space operations outside the vicinity of the earth promise to be sufficiently expensive so as to offer a strong impetus for international, meaning Soviet-American or multilateral, cooperation, as in the case of ESRO's commitment to build a space laboratory for the American space shuttle. NASA currently contends that if men are to visit the planets in this century, it will have to be accomplished through Soviet-American and/or American-European cooperation. Such cooperative endeavors in peacetime are unprecedented in the history of modern nations. If the Apollo-*Soiuz* mission is completed successfully, it could pave the way toward a cooperative mission to Mars. Both missions, especially the latter, necessitate close cooperation and could have a profoundly beneficial effect on relations between the two countries.

Both space powers are now working toward the creation of a permanent manned space station in earth orbit. This is currently the principal technological goal of the Soviet space program. As a cooperative endeavor, it could open the way for manned exploration of the planets and for extensive use of earth applications. Implemented unilaterally by either space power, such a station would dramatically affect the relative positions of the two space powers, destabilize the 1967 Space Treaty, and perhaps occasion a new episode in the space race.[41]

NOTES

[1]Raymond A. Bauer, with Richard S. Rosenbloom and Laure Sharp et al., *Second-Order Consequences, A Methodological Essay on the Impact of Technology* (Cambridge, Mass.: MIT Press, 1969), pp. 37-38; and Bruce Mazlish, "Historical Analogy: The Railroad and the Space Program and Their Impact on Society," in Mazlish, editor, *The Railroad and the Space Program: An Exploration in Historical Analogy*

(Cambridge, Mass.: The MIT Press, 1965), pp. 34-35.

[2]Bauer et al., *supra*, pp. 38-39; and Mazlish, *loc. cit.*

[3]Lloyd V. Berkner, Chairman of the Space Science Board of the National Academy of Science, "The Compelling Horizon," *Bulletin of the Atomic Scientists, 19,* May 1963, pp. 8-9.

[4]Leonard S. Silk, "The Impact on the American Economy," in American Assembly, *Outer Space Prospects for Man and Society* (Englewood Cliffs, N.J.: Prentice-Hall, Inc., 1962), pp. 76-77; and Alexei Kosygin as cited by James Reston in *Review of the Soviet Space Program, November 10, 1967, op. cit.,* p. 84.

[5]Bauer et al., *op. cit.,* pp. 166-169, 174.

[6]John McHale, "Big Business Enlists for the War on Poverty," *Transaction, 2* (4), May-June 1965, pp. 3-9.

[7]Bauer et al., *op. cit.,* p. 156.

[8]*Ibid.,* p. 180.

[9]John D. Holmfeld, "Organization of the Soviet Space Program," *Soviet Space Programs, 1966-1970, op. cit.,* pp. 94-101.

[10]F. S. Nyland, *Space: An International Adventure?,* Rand Corporation, Paper #P-3043, December 1964, p. 4; and B. W. Augenstein, *Policy Analysis in the National Space Program,* Rand Corporation, Paper #P-4137, July 1969, pp. 60-61.

[11]Cited by Amitai Etzioni, *The Moon-Doggle, Domestic and International Implications of the Space Race* (Garden City, N.Y.: Doubleday and Co., Inc., 1964), p. 80.

[12]Forbes, July 1, 1968 as cited by Erlend A. Kennan and Edmund H. Harvey, *Mission to the Moon, A Critical Examination of NASA and the Space Program* (New York: William Morrow and Co., 1969), p. 248.

[13]Bauer et al., *op. cit.,* pp. 171-172.

[14]Etzioni, *op. cit.,* ix and pp. 30, 80; and Killian as cited by Vernon Van Dyke, *Pride and Power, The Rationale of the Space Program* (Urbana: University of Illinois Press, 1964), p. 97.

[15]Van Dyke, *supra,* pp. 97-98.

[16]Bauer et al., *op. cit.,* p. 123; and Augenstein, *op. cit.,* p. 42.

[17]Etzioni, *op. cit.,* pp. 8, 87.

[18]Leon M. Herman, "Soviet Economy Capabilities for Scientific Research," *Soviet Space Programs, 1962-1965, op. cit.,* p. 419.

[19]See Etzioni, *op. cit.,* pp. 73-74.

[20]*Ibid.,* p. 72.

[21]Cited in *ibid.,* p. 74.

[22]*Ibid.,* p. 108.

[23]See *ibid.,* pp. 7, 8, 21; Dr. Philip Abelson as cited by Lillian Levy, "Conflict in the Race for Space," in Levy, editor, *Space: Its Impact on Man and Society* (New York: W. W. Norton & Co., Inc., 1965), p. 205; and Herman, *op. cit.,* p. 414.

[24]James R. Killian, Jr., "Shaping a Policy for the Space Age," and Alton Frye, "Soviet Space Activities: A Decade of Pyrrhic Politics," in Lincoln P. Bloomfield, editor, *Outer Space: Prospects for Man and Society* (New York: Frederick A. Praeger, Inc., 1968), p. 234 and pp. 178-179; and Herman, *supra,* p. 420-423.

[25]Kennan and Harvey, *op. cit.,* p. 87; and Augenstein, *op. cit.,* pp. 62-68.

[26]Unpublished address by Dr. Sergei P. Fedorenko at Illinois State University, December 6, 1972.

[27]See Don E. Kash, *The Politics of Space Cooperation* (Purdue University Studies, Purdue Research Foundation, 1967), p. 45.

[28]Augenstein, *op. cit.,* p. 48; and Van Dyke, *op. cit.,* p. 213.

[29]Klaus Knorr, "International Implications of Outer Space Activities," in Joseph M. Goldsen, editor, *Outer Space in World Politics* (New York: Frederick A. Praeger, 1963), pp. 120-121.

[30]*Ibid.*, p. 120.

[31]*Ibid.*, p. 121.

[32]See John Kenneth Galbraith, *The New Industrial State* (Boston: Houghton-Mifflin Co., 1967), p. 342; Christopher Wright, "The United Nations and Outer Space," in Hugh Odishaw, editor, *The Challenges of Space* (Chicago: University of Chicago Press, 1963), p. 289; and Leonid Sedov, "Interplanetary Travel Soon a Reality," *New Times*, No. 38, September 1959, pp. 5-6.

[33]Claims similar to the above are made by Levy, "Conflict in the Race for Space," *op. cit.*, p. 210; and Frank Gibney and George Feldman, *The Reluctant Spacefarers. A Study in the Politics of Discovery* (New York: New American Library, 1965), p. 162.

[34]The suggestion that the United States employ synchronous satellites to "recapture" the peoples of Eastern Europe through televised propaganda is made by William R. Kintner, "The Problem of Opening the Soviet System," in Frederick J. Ossenbeck and Patricia C. Kroeck, editors, *Open Space and Peace. A Symposium on Effects of Observation* (Stanford: Stanford University, The Hoover Institute, 1964), pp. 121-122. The contention that satellite propaganda would serve largely to exacerbate international tensions is made by Donald N. Michael, "Prospects for Human Welfare: Peaceful Uses," in American Assembly, *Outer Space Prospects for Man and Society*, *op. cit.*, p. 40.

[35]Philip J. Klass, *Secret Sentries in Space* (New York: Random House, 1971), xv and pp. 200-201, 209-210.

[36]See Frye, "Soviet Space Activities: A Decade of Pyrrhic Politics," *op. cit.*, pp. 194-196.

[37]See Van Dyke, *op. cit.*, p. 76; and Tunis A. M. Craven, Commissioner, Federal Communications Commission, "International Cooperation," in Jerry and Vivian Grey, eds., *Space Flight Report to the Nation* (New York: Basic Books, Inc., 1962), p. 143.

[38]See Michael, *op. cit.*, p. 39.

[39]*Ibid.*, p. 41.

[40]Figures for 1970 from Luman H. Long, editor, *The 1972 World Almanac and Book of Facts* (New York: Newspaper Enterprises Association Inc., 1971), pp. 103, 568.

[41]See Charles S. Sheldon II, "An American 'Sputnik' for the Russians?" in Eugene Rabinowitch and Richard S. Lewis, editors, *Man on the Moon—The Impact on Science, Technology and International Cooperation* (New York: Basic Books, Inc., 1969), p. 64.

Bibliography

BOOKS AND MONOGRAPHS

Aleksandrov, V., "Rocket-plane, Aircraft of the Future," in *Soviet Writings on Earth Satellites and Space Travel*, pp. 196-199.

American Assembly, Columbia University, *Outer Space: Prospects for Man and Society* (Englewood Cliffs, N.J.: Prentice-Hall, Inc., 1962).

Amme, Carl H., Jr., "The Implications of Satellite Observation for US Policy," in Ossenbeck and Kroeck, editors, *Open Space and Peace, A Symposium on Effects of Observation*, pp. 105-111.

Arms Control: Readings from Scientific American (San Francisco: W. H. Freeman & Co., 1973).

Astashenkov, P. T., *Akademik S. P. Korolev* (Moscow, 1969).

Augenstein, B. W., *Policy Analysis in the National Space Program*, Rand Corporation, Paper #P-4137, July 1969.

Bardin, I. P., "USSR Rocket and Earth Satellite Program for the IGY," Submitted to the GSAGI at Brussels, June 10, 1957, in Krieger, *Behind the Sputniks*, pp. 282-287.

Barghoorn, Frederick C., *Politics in the USSR* (Boston: Little, Brown and Co., Inc., 1972).

Bauer, Raymond A. with Richard S. Rosenbloom and Laure Sharp *et al.*, *Second-Order Consequences, A Methodological Essay on the Impact of Technology* (Cambridge, Mass.: MIT Press, 1969).

Berkner, L. V., *The Scientific Age, The Impact of Science on Society*, Based on the Trumbull Lectures delivered at Yale University (New Haven: Yale University Press, 1965).

Bloomfield, Lincoln P., editor, *Outer Space: Prospects for Man and Society* (New York: Frederick A. Praeger, Inc., 1968).

———, *The Peaceful Uses of Space*, Public Affairs Pamphlet No. 331, 1962.

———, "The Prospects for Law and Order," in American Assembly, *Outer Space: Prospects for Man and Society*, pp. 150-180.

———, "The Quest for Law and Order," in Bloomfield, editor, *Outer Space: Prospects for Man and Society*, pp. 114-144.

Brennan, Donald G., "Arms and Arms Control in Outer Space," in Bloomfield, *Outer Space: Prospects for Man and Society*, pp. 145-177.

Buchheim, Robert W. and the staff of Rand Corporation, *New Space Handbook: Astronautics and Its Applications* (New York: Vintage Books, 1963).

291

Caidin, Martin, *Red Star in Space* (New York: Crowell-Collier Press, 1963).

Cheprov, I. I., "Pravovoe regulirovanie deatelnosti v kosmicheskom prostranstve," in Akademiia Nauk SSSR, *Kosmos i Mezhdunarodnoe Sotrudnichestvo* (Moscow, 1963), pp. 49-76.

Cooper, John C., *Explorations in Aerospace Law*, Selected Essays edited by Ivan A. Vlasic (Montreal: McGill University Press, 1968).

Crankshaw, Edward, *Khrushchev, A Career* (New York: The Viking Press, 1966).

Craven, Tunis A. M., "International Cooperation," in Grey and Grey, editors, *Space Flight Report to the Nation*, pp. 141-143.

Daniloff, Nicholas, *The Kremlin and the Cosmos* (New York: Alfred A. Knopf, 1972).

Deutsch, Karl W., "Outer Space and International Politics," in Goldsen, editor, *Outer Space in World Politics*, pp. 139-174.

Diamond, Edwin, *The Rise and Fall of the Space Age* (Garden City, N.Y.: Doubleday and Co., Inc., 1964).

Emme, Eugene M., *Aeronautics and Astronautics* (Washington, D.C.: US Government Printing Office, 1961).

———, editor, *The Impact of Air Power* (Princeton: Van Nostrand, 1959).

Etzioni, Amitai, *The Moon-Doggle, Domestic and International Implications of the Space Race* (Garden City, N.Y.: Doubleday and Co., Inc., 1964).

Fainsod, Merle, *How Russia Is Ruled* (Cambridge, Mass.: Harvard University Press, 1963).

Feodosev, V. I. and G. B. Siniarev, *Vvedenie v Raketnuiu Tekniku* (Moscow: 1956).

Fortune editors, *The Space Industry, America's Newest Giant* (Englewood Cliffs, N.J.: Prentice-Hall, Inc., 1962).

Frutkin, Arnold W., *International Cooperation in Space* (Englewood Cliffs, N.J.: Prentice-Hall, 1965).

Frye, Alton, *The Proposal for a Joint Lunar Expedition: Background and Prospects*, Rand Corporation, Paper #P-2808, January, 1964.

———, "Soviet Space Activities: A Decade of Pyrrhic Politics," in Bloomfield, editor, *Outer Space: Prospects for Man and Society*, pp. 178-205.

Gal, Gyula, *Space Law* (Leyden: A. W. Sijthoff, and Dobbs Ferry, N.Y.: Oceana Publications, Inc., 1969).

Galbraith, John K., *The New Industrial State* (Boston: Houghton-Mifflin Co., 1967).

Gardner, Trevor, "Military Effects," in Grey and Grey, editors, *Space Flight Report to the Nation*, pp. 127-131.

Gavin, Lt. General James M., *War and Peace in the Space Age* (New York: Harper and Bros., 1958).

Gibney, Frank and George J. Feldman, *The Reluctant Spacefarers. A Study in the Politics of Discovery* (New York: New American Library, 1965).

Glennan, T. Keith, "The Task for Government," in American Assembly, *Outer Space: Prospects for Man and Society*, pp. 84-104.

Goldsen, Joseph M., editor, *International Political Implications of Activities in Outer Space; A Report of a Conference*, Rand Corporation, Report #R-362-RC, May 1961.

———, *Outer Space and the International Scene*, Rand Corporatioon, Paper #P-1688, May 6, 1959.

———, editor, *Outer Space in World Politics* (New York: Frederick A. Praeger, 1963).

———, "Outer Space in World Politics," in Goldsen, editor, *Outer Space in World*

Politics, pp. 3-24.

————, *Public Opinion and Social Effects of Space Activity,* Rand Corporation, Research Memorandum #RM-2417, July 20, 1959.

————, *Research on Social Consequences of Space Activities,* Rand Corporation, Paper #P-3220, August 1965.

———— and Leon Lipson, *Some Political Implications of the Space Age,* Rand Corporation, Paper #P-1435, February 24, 1958.

Golovine, Michael N., *Conflict in Space, A Pattern of War in a New Dimension* (London: Temple Press, Ltd., 1962).

Gouschev, Sergei and Mikhail Vassiliev, editors, *Russian Science in the 21st Century* (New York: McGraw-Hill Book Co., Inc., 1960).

Grey, Jerry and Vivian, editors, *Space Flight Report to the Nation* (New York: Basic Books, Inc., 1962).

Harvey, Mose L., "The Lunar Landing and the US-Soviet Equation," in Rabinowitch and Lewis, *Man on the Moon—The Impact of Science, Technology and International Cooperation,* pp. 65-84.

Hoffman, Stanley, *Gulliver's Troubles or the Setting of American Foreign Policy* (New York: McGraw-Hill Book Co., 1968).

Holmes, Jay, *America on the Moon: The Enterprise of the Sixties* (Philadelphia: J. B. Lippincott Co., 1962).

Horelick, Arnold L., "The Soviet Union and the Political Uses of Outer Space," in Goldsen, editor, *Outer Space in World Politics,* pp. 43-70.

Huntington, Samuel P., *The Common Defense: Strategic Programs in National Politics* (New York: Columbia University Press, 1961).

Hyman, Sidney, "Man on the Moon. The Columbian Dilemma," in Rabinowitch and Lewis, editors, *Man on the Moon—The Impact of Science, Technology and International Cooperation,* pp. 39-52.

International Council of Scientific Unions, *Annals of the International Geophysical Year* (London: 1958).

International Institute of Strategic Studies, *The Military Balance, 1968-1969* (London: 1968).

————, *The Military Balance, 1969-1970* (London: 1969).

————, *The Military Balance, 1970-1971* (London: 1970).

————, *The Military Balance, 1971-1972* (London: 1971).

————, *The Military Balance, 1972-1973* (London: 1972).

————, *The Military Balance, 1973-1974* (London: 1973).

————, *Strategic Survey, 1971* (London: 1972).

International Programs, Pamphlet prepared by the Office of International Programs, NASA, January 1963.

International Programs, Pamphlet prepared by the Office of International Programs, NASA, July 1967.

Jessup, Philip C. and Howard J. Taubenfeld, *Controls for Outer Space and the Antarctic Analogy* (New York: Columbia University Press, 1959).

Johnson, Lyndon B., "The Politics of the Space Age," in Levy, editor, *Space: Its Impact on Man and Society,* pp. 3-9.

Jones, Erin B., *Earth Satellite Telecommunications Systems and International Law* (Austin, Texas: University of Texas, 1970).

Kash, Don E., *The Politics of Space Cooperation* (Purdue University Studies, Purdue Research Foundation, 1967).

Katzenbach, Nicholas deB., "The Law in Outer Space," in Levy, editor, *Space: Its*

Impact on Man and Society, pp. 69-81 .

Kecskemeti, Paul, "Outer Space and World Peace," in Goldsen, editor, *Outer Space in World Politics*, pp. 25-42.

Kennan, Erlend A. and Edmund H. Harvey, *Mission to the Moon, A Critical Examination of NASA and the Space Program* (New York: William Morrow and Co., 1969).

Killian, James R., Jr., "Shaping a Public Policy for the Space Age," in American Assembly, *Outer Space: Prospects for Man and Society*, pp. 181-192.

———, "Shaping a Policy for the Space Age," in Bloomfield, editor, *Outer Space: Prospects for Man and Society*, pp. 230-242.

Kintner, William R., "The Problem of Opening the Soviet System," in Ossenbeck and Kroeck, editors, *Open Space and Peace, A Symposium on Effects of Observation*, pp. 112-125.

Kiselev, A. N. and M. F. Pebrov, *Ships Leave for Space* (Moscow: 1967, in Russian).

Klass, Philip J., *Secret Sentries in Space* (New York: Random House, 1971)

Knorr, Klaus, "The International Implications of Outer Space Activities," in Goldsen, editor, *Outer Space in World Politics*, pp. 114-138.

Korovin, E. A. and G. P. Zhukov, editors, *Sovremennye Problemy Kosmicheskogo Prava* (Moscow: 1963).

Khrushchev, N., *For Victory in Peaceful Competition with Capitalism* (New York: E. P. Dutton and Co., 1960).

Kosmodemianskii, A. A., *Konstantin Tsiolkovskii, His Life and Work* (Moscow: 1956 and 1960, in Russian).

Krieger, Fermin J., *Behind the Sputniks, A Survey of Soviet Space Science* (Washington, D.C.: The Rand Corporation, Public Affairs Press, 1958).

———, *Soviet Space Experiments and Astronautics*, Rand Corporation, Paper #P-2261, March 31, 1961.

———, *The Space Programs of the Soviet Union*, Rand Corporation, Paper #P-3632, July 1967.

Leonard, Wolfgang, *The Kremlin Since Stalin* (New York: Frederick A. Praeger, Inc., 1962).

Levy, Lillian, "Conflict in the Race for Space," in Levy, editor, *Space: Its Impact on Man and Society*, pp. 188-211.

———, editor, *Space: Its Impact on Man and Society* (New York: W. W. Norton and Co., Inc., 1965).

Lipson, Leon, *Outer Space and International Law*, Rand Corporation, Paper #P-1434, February 24, 1958.

Logsdon, John M., *The Decision To Go to the Moon: Project Apollo and the National Interest* (Cambridge, Mass.: MIT Press, 1970).

Mailer, Norman, *Of a Fire on the Moon* (Boston: Little, Brown and Co., 1970).

Mazlish, Bruce, "Historical Analogy: The Railroad and the Space Program and Their Impact on Society," in Mazlish, editor, *The Railroad and the Space Program: An Exploration in Historical Analogy*, pp. 1-52.

———, editor, *The Railroad and the Space Program: An Exploration in Historical Analogy* (Cambridge, Mass.: MIT Press, 1965).

McDougal, Myres S., Harold D. Lasswell and Ivan A. Vlasic, *Law and Public Order in Space* (New Haven: Yale University Press, 1963).

Medaris, Major General John B., *Countdown for Decision* (New York: G. P. Putnam's Sons, 1960).

Michael, Donald N., "Prospects for Human Welfare: Peaceful Uses," in American

Assembly, *Outer Space: Prospects for Man and Society*, pp. 31-63.

Morgernstern, Oskar, "Political Effects," in Grey, editors, *Space Flight Report to the Nation*, pp. 132-136.

Murray, Bruce C. and Merton E. Davies, *A Comparison of US and Soviet Efforts to Explore Mars*, Rand Corporation, Paper #P-3285, January 1966.

Murray, B. C., M. E. Davies and P. K. Eckman, *Planetary Contamination II: Soviet and US Practices and Policies*, Rand Corporation, Paper #P-3517, March 1967.

Nyland, F. S., *Space: An International Adventure?*, Rand Corporation, Paper #P-3043, December 1964.

Odishaw, Hugh, editor, *The Challenges of Space* (Chicago: University of Chicago Press, 1963).

――――, "Science and Space," in Bloomfield, editor, *Outer Space: Prospects for Man and Society*, pp. 75-93.

Onitskaia, G. A., "Mezdunarodnye pravovye voprosy osvoeniia kosmicheskogo prostranstva," in *Sovetskii Ezhegodnik Mezhdunarodnogo Prava* (Moscow: 1959), pp. 64-65.

Organization for Economic Cooperation and Development, *The Research and Development Effort in Western Europe, North America and the Soviet Union* (OECD, Paris: 1965).

Ossenbeck, Frederick J. and Patricia C. Kroeck, editors, *Open Space and Peace, A Symposium on Effects of Observation* (Stanford, Calif.: Stanford University, The Hoover Institute, 1964).

Payne, Seth T. and Leonard S. Silk, "The Impact on the American Economy," in Bloomfield, editor, *Outer Space: Prospects for Man and Society*, pp. 94-113.

Penkovsky, Oleg, *The Penkovsky Papers*, translated by P. Denatin (London: Collins Clear Type Press, 1966).

Pokrovskii, G. I., "Intercontinental Ballistic Missiles and Aviation," in *Soviet Writings on Earth Satellites and Space Travel*, pp. 199-200.

Puckett, Robert H., *The Military Role in Space—A Summary of Official, Public Justifications*, Rand Corporation, Paper #P-2681, August 1962.

Rabinowitch, Eugene and Richard S. Lewis, editors, *Man on the Moon—The Impact on Science, Technology and International Cooperation* (New York: Basic Books, Inc., 1969).

Report from Iron Mountain on the Possibility and Desirability of Peace (New York: Dial Press, 1967).

Retterer, R. W., "Career Opportunities in the Space Age," in Levy, editor, *Space: Its Impact on Man and Society*, pp. 92-102 .

Riabchikov, Evgeny, *Russians in Space,* edited by Colonel General Nikolai P. Kamanin, translated by Guy Daniels, prepared by the Novosti Press Agency Publishing House, Moscow (Garden City, N.Y.: Doubleday and Co., Inc., 1971).

Ritvo, Herbert, annotator, *The New Soviet Society* (New York: New Leader, 1962).

Romanov, Alexander P., *Designer of Cosmic Ships* (Moscow: 1969, in Russian).

――――, *Kosmodrom, Kosmonavty, Kosmos. dnevnik specialnogo korrespondenta TASS* (Moscow: Izd. DOSAFF, 1966).

Rosholt, Robert, *An Administrative History of NASA, 1958-1963* (Washington, D.C.: NASA, 1966).

Salkeld, Robert, *War and Space*, foreword by General Bernard A. Schriever, USAF (retired) (Englewood Cliffs, N.J.: Prentice-Hall, Inc., 1970).

Schlesinger, Arthur, Jr., *A Thousand Days* (Boston: Houghton-Mifflin Co., 1965).

Schriever, General Bernard A., "Does the Military Have a Role in Space?" in Levy, editor, *Space: Its Impact on Man and Society*, pp. 59-68.

Schwartz, Leonard E., *International Organizations and Space Cooperation* (Durham, N.C.: World Rule of Law Center, 1962).

———, "International Space Organizations," in Odishaw, editor, *The Challenges of Space*, pp. 241-266.

Shapiro, Leonard, *The Government and Politics of the Soviet Union* (New York: Vintage Books, 1965).

Sheldon, Charles S., II, "An American 'Sputnik' for the Russians?" in Rabinowitch and Lewis, editors, *Man on the Moon—The Impact on Science, Technology and International Cooperation*, pp. 53-64.

———, "Peaceful Applications," in Bloomfield, editor, *Outer Space: Prospects for Man and Society*, pp. 37-74.

———, *Review of the Soviet Space Program, With Comparative US Data* (New York: McGraw-Hill Book Co., Inc., 1968).

Shelton, William, *Soviet Space Exploration* (New York: Washington Square Press, Inc., 1968).

Sidey, Hugh, *John F. Kennedy, President* (New York: Atheneum Press, 1964).

Silk, Leonard S., "The Impact on the American Economy," in American Assembly, *Outer Space: Prospects for Man and Society*, pp. 64-83.

Sloss, Leon, "Unilateral Space Observation and the Atlantic Alliance," in Ossenbeck and Kroeck, editors, *Open Space and Peace, A Symposium on Effects of Observation*, pp. 155-162.

Smart, Ian, *Advanced Strategic Missiles: A Short Guide*, International Institute for Strategic Studies, Adelphi Papers, No. 63, December 1969, London.

Sokolovskii, V. D., editor, *Soviet Military Strategy*, translated by Herbert S. Dinerstein, Leon Goure and Thomas W. Wolfe, A Rand Corporation Research Study (Englewood Cliffs, N.J.: Prentice-Hall, Inc., 1963).

Soviet Writings on Earth Satellites and Space Travel (New York: Citadel Press, 1958).

Stoiko, Michael, *Soviet Rocketry: Past, Present and Future* (New York: Holt, Rinehart & Winston, 1970).

Sturm, Thomas A., *The USAF Scientific Advisory Board: Its First Twenty Years, 1944-1964* (Washington, D.C.: US Government Printing Office, 1967).

Sullivan, Walter, editor, *America's Race for the Moon. The New York Times Story of Project Apollo* (New York: Random House, 1962).

Swenson, Loyd S., Jr., James M. Grimwood and Charles C. Alexander, *This New Ocean, A History of Project Mercury* (Washington, D.C.: NASA, 1966).

Tikhonravov, M. K., "Interplanetary Communications," *Bolshaia Sovetskaia Entsiklopediia*, second edition, Volume 27, June 18, 1954, in Krieger, *Behind the Sputniks*, pp. 28-34.

Tokaty-Tokaev, G. I., *Comrade X*, translated by Alec Brown (London: Harvill Press, 1956).

Triska, Jan F. and David D. Finley, *Soviet Foreign Policy* (New York: Macmillan and Co., Inc., 1968).

Ulam, Adam B., *Expansion and Coexistence, A History of Soviet Foreign Policy, 1917-1967* (New York: Frederick A. Praeger, Inc., 1971).

USSR Probes Space (Moscow: Novosti Press Agency Publishing House, undated).

Van de Hulst, H. C., "COSPAR and Space Cooperation," in Odishaw, editor, *The Challenges of Space*, pp. 291-298.

Van Dyke, Vernon, *Pride and Power, The Rationale of the Space Program* (Urbana: University of Illinois, 1964).

Vasilevskaia, E. G., "Voprosy kosmicheskogo prava v novishi amerikanskoi literature," in Akademiia Nauk SSSR, *Kosmos i Mezhdunarodnoe Sotrudnichestvo* (Moscow: 1963), pp. 231-232.

Vladimirov, Leonid, *The Russian Space Bluff* (New York: Dial Press, 1973).

Witkin, Richard, "Pros and Cons," in Sullivan, editor, *America's Race for the Moon. The New York Times Story of Project Apollo*, pp. 146-152.

Wolfe, Thomas W., *Soviet Strategy at the Crossroads*, Rand Corporation (Cambridge, Mass.: Harvard University Press, 1964).

Wright, Christopher, "The United Nations and Outer Space," in Odishaw, editor, *The Challenges of Space*, pp. 277-290.

Yarmolinsky, Avrahm, *Road to Revolution, A Century of Russian Radicalism* (New York: Collier Books, 1962).

DOCUMENTS

Beck, Leonard N., "Recent Developments in the Soviet Space Program: A Survey of Space Activities and Space Science," in *Soviet Space Programs, 1962-1965*, pp. 147-352.

———, "Soviet Space Plans and Economic Capabilities, Part I: Soviet Projections of Space Plans," in *Soviet Space Programs, 1962-1965*, pp. 353-423.

Block, Herbert, "Value and Burden of Soviet Defense," in *Soviet Economic Prospects for the Seventies*, 1973, pp. 175-204.

Bourgin, Simon, "Impact of US Space Cooperation Abroad," in *International Cooperation in Outer Space, Symposium*, pp. 163-172.

Cohn, Stanley H., "Economic Burden of Defense Expenditures," in *Soviet Economic Prospects for the Seventies*, 1973, pp. 147-162.

Congressional Record

Department of State Bulletin

DeVoe, Barbara M., *Astronaut Information: American and Soviet (revised)*, Congressional Research Service, Library of Congress, No. 73-78SP, February 12, 1973.

———, *Biographies of Soviet Cosmonauts*, Appendix D of *Soviet Space Programs, 1966-1970*, pp. 603-605.

———, "Soviet Application of Space to the Economy," in *Soviet Space Programs, 1966-1970*, pp. 297-321.

Frutkin, Arnold W., "NASA's International Space Activities," in *International Cooperation in Outer Space, Symposium*, pp. 13-53.

Herman, Leon M., "Soviet Economic Capabilities for Scientific Research," in *Soviet Space Programs, 1962-1965*, pp. 397-423.

Holmfeld, John D., "Organizations of the Soviet Space Program," in *Soviet Space Programs, 1966-1970*, pp. 69-105.

———, "Resource Allocation and the Soviet Space Program," in *Soviet Space Programs, 1966-1970*, pp. 107-114.

Interim Agreement Between the United States of America and the Union of Soviet Socialist Republics on Certain Measures with Respect to the Limitation of Strategic Offensive Arms, May 26, 1972, in *Arms Control: Readings from Scientific American*, pp. 264-265.

Johnson, John A., "The International Activities of the Communications Satellite Corporation," in *International Cooperation in Outer Space, Symposium*, pp. 195-217.

Krivickas, Domas and Armins Rusis, "Soviet Attitude Toward Law in Outer Space," in *Soviet Space Programs, 1966-1970*, pp. 453-505.

————, "Soviet Attitudes Toward Space Law," in *Soviet Space Programs, 1962-1965*, pp. 493-528.

McCullough, James M., "Soviet Bioastronautics: Biological, Behavioral and Medical Problems," in *Soviet Space Programs, 1966-1970*, pp. 265-295.

NASA Press Release #70-210, December 9, 1970, in Appendix G of *Soviet Space Programs, 1966-1970*, pp. 615-616.

NASA Press Release #71-57, March 31, 1971, in Appendix H of *Soviet Space Programs, 1966-1970*, pp. 617-622.

Plummer, K. L., "Soviet-French Cooperation in Space," in Appendix F of *Soviet Space Programs, 1966-1970*, pp. 611-614.

Pollack, Herman, "Impact of the Space Program on America's Foreign Relations," in *International Cooperation in Outer Space, Symposium*, pp. 595-605.

Sheldon, Charles S., II, "The Challenge of International Competition," Address to AIAA/NASA, Houston, November 6, 1964, in *International Cooperation and Organization for Outer Space, 1965*, pp. 427-477.

————, "Illustrations of Soviet Launch Vehicles and Spacecraft," Appendix B of *Soviet Space Programs, 1966-1970*, pp. 559-596.

————, "Overview, Supporting Facilities and Launch Vehicles of the Soviet Space Program," in *Soviet Space Programs, 1966-1970*, pp. 115-157.

————, "Postscript," in *Soviet Space Programs, 1966-1970*, pp. 507-530.

————, "Program Details of Manned Flights," in *Soviet Space Programs, 1966-1970*, pp. 223-263.

————, "Program Details of Unmanned Flights," in *Soviet Space Programs, 1966-1970*, pp. 157-221.

————, "Projections of Soviet Space Plans," in *Soviet Space Programs, 1966-1970*, pp. 351-398.

————, "Soviet Military Space Activities," in *Soviet Space Programs, 1966-1970*, pp. 323-350.

————, "Summary," in *Soviet Space Programs, 1966-1970*, xxi-xl.

———— and Barbara M. DeVoe, "Table of Soviet Space Launches, 1957-1971 (June 6)," Appendix A of *Soviet Space Programs, 1966-1970*, pp. 531-538.

————, *United States and Soviet Progress in Space: Summary Data Through 1972 and a Forward Look,* Library of Congress, Congressional Research Service, January 29, 1973.

————, *United States and Soviet Progress in Space: Summary Data Through 1973 and a Forward Look,* Library of Congress, Congressional Research Service, January 8, 1974.

TASS Release of June 8, 1971, *On Draft Treaty for Peaceful Uses of the Moon,* Appendix I of *Soviet Space Programs, 1966-1970*, pp. 623-627.

Treaty Banning Nuclear Weapon Tests in the Atmosphere, in Outer Space and Under Water, signed by USA, United Kingdom and USSR at Moscow, August 5, 1963, *American Journal of International Law*, 57, October 1963, pp. 1026-1029.

Treaty on Principles Governing the Activities of States in the Exploration and Use of Outer Space, Including the Moon and Other Celestial Bodies of January 24, 1967 in *Treaties in Force*, p. 386.

Treaty of the United States of America and the Union of Soviet Socialist Republics on the Limitation of Anti-Ballistic Missile Systems, May 26, 1972, in *Arms Control: Readings from Scientific American*, pp. 260-263.

Treaties in Force, A List of Treaties and Other International Agreements of the US in Force on January 1, 1973, US Department of State, Publication #8697, 1973.

UN Document A/AC.105/13, May 28, 1963.

UN Document A/AC.105/46, August 9, 1968.

UN Document A/AC.105/15, June 7, 1963.

UN Document A/AC.105/C.1/1, June 3, 1962.

UN Document A/AC. 105/C.2/L.1 in A/AC.105/6, July 9, 1962.

UN Document A/AC.105/C.2/SR., April 1963.

UN Document A/AC.105/PV.8, May 15, 1962.

UN Document A/C.1/857, November 15, 1961.

UN Document A/C.1/L.219, November 7, 1958.

UN Document A/C.1/L.219/Rev. 1, November 18, 1958.

UN Document A/C.1/L.220/Rev. 1, November 14, 1958.

UN Document A/C.1/SR.995, November 24, 1958.

UN Document A/C.1/SR.1210, December 4, 1961.

UN Resolution 1962 (XVIII), United Nations Resolution on Legal Principles Governing Activities in Outer Space, December 24, 1963, *International Legal Materials Current Documents*, 3(1), January 1964, pp. 157-159.

US Congress, Joint Economic Committee, *Soviet Economic Prospects for the Seventies*, 93rd Congress, 1st session, June 27, 1973.

US Congress, House, Committee on Appropriations, *Independent Offices Appropriations for 1963*, Hearings Before a Subcommittee, 87th Congress, 2nd session, Part 3, 1962.

US Congress, House, Committee on Appropriations, *Independent Offices Appropriations for 1964*, Hearings Before a Subcommittee on Appropriations, 88th Congress, 1st session, Part 3, 1963.

US Congress, House, Committee on Appropriations, Subcommittee on Department of Defense Appropriations for 1960, Hearings, *Project Vanguard, A Scientific Earth Satellite Program for the IGY*, Report to the Committee on Appropriations by Surveys and Investigations Staff, 86th Congress, 1st session, 1959.

US Congress, House, Committee on Appropriations, Subcommittee on Department of Defense Appropriations, *Department of Defense Appropriations for 1962*, Hearings, 87th Congress, 1st session, 1961.

US Congress, House, Committee on Science and Astronautics, *Panel on Science and Technology*, Fifth meeting, 88th Congress, 1st session, 1963.

US Congress, House, Committee on Science and Astronautics, *Review of the Soviet Space Program*, 90th Congress, 1st session, November 10, 1967.

US Congress, House, Committee on Science and Astronautics, *US Policy on the Control of Outer Space*, Report #353, 86th Congress, 1st session, 1959.

US Congress, House, Committee on Science and Astronautics, Subcommittee on Applications and Tracking and Data Acquisition, *Hearings on NASA Authorization, FY 1964*, 88th Congress, 1st session, Part 4, 1963.

US Congress, House, Committee on Science and Astronautics, *Hearings on NASA Authorization, FY 1971*, 91st Congress, 2nd session, February 1970.

US Congress, House, "State-of-the-Union Address of the President of the United States delivered before a Joint Session of Congress," House Document #251, 88th Congress, 2nd session, January 8, 1964.

US Congress, House, Select Committee on Astronautics and Space Exploration, *Hearing on HR 11881, Astronautics and Space Explorations*, 85th Congress, 2nd session, 1958.

US Congress, House, Select Committee on Astronautics and Space Exploration, *The National Space Project*, House Report #1758, 85th Congress, 2nd session, 1958.

US Congress, House, Select Committee on Astronautics and Space Exploration, *Survey of Space Law*, Staff Report, House Document #89, 86th Congress, 1st session, 1959.

US Congress, House, Select Committee on Astronautics and Space Exploration, *The United States and Outer Space*, House Report #2710, 85th Congress, 2nd session, 1959.

US Congress, Senate, Committee on Aeronautics and Space Sciences, *Documents on International Aspects of the Exploration and Use of Outer Space*, 1954-1962, Senate Document #18, 88th Congress, 1st session, 1963.

US Congress, Senate, Committee on Aeronautics and Space Sciences, *International Cooperation and Organization for Outer Space*, Staff Report, Senate Document #56, 89th Congress, 1st session, 1965.
(Abbreviated as *International Cooperation and Organization for Outer Space, 1965*).

US Congress, Senate, Committee on Aeronautics and Space Sciences, *International Cooperation in Outer Space, Symposium*, Senate Document #59, 92nd Congress, 1st session, December 9, 1971.
(Abbreviated as *International Cooperation in Outer Space, Symposium*).

US Congress, Senate, Committee on Aeronautics and Space Sciences, *Legal Problems of Space Exploration: A Symposium*, Senate Document #26, 87th Congress, 1st session, 1961.

US Congress, Senate, Committee on Aeronautics and Space Sciences, *Manned Space Flight Programs of the NASA: Projects Mercury, Gemini and Apollo*, 87th Congress, 2nd session, 1962.

US Congress, Senate, Committee on Aeronautics and Space Sciences, *NASA Authorization for FY 1964*, Hearings, 88th Congress, 1st session, Part 2, 1963.

US Congress, Senate, Committee on Aeronautics and Space Sciences, *NASA Authorization for FY 1970*, Hearings, 91st Congress, 1st session, Part 1, 1969.

US Congress, Senate, Committee on Aeronautics and Space Sciences, *National Space Goals for the Post-Apollo Period*, 89th Congress, 1st session, 1965.

US Congress, Senate, Committee on Aeronautics and Space Sciences, *Scientists' Testimony on Space Goals*, Hearings, 88th Congress, 1st session, 1963.

US Congress, Senate, Committee on Aeronautics and Space Sciences, *Soviet Space Programs: Organization, Plans, Goals and International Implications*, 87th Congress, 2nd session, May 31, 1962.
(Abbreviated as *Soviet Space Programs, 1962*),

US Congress, Committee on Aeronautics and Space Sciences, *Soviet Space Programs, 1962-1965, Goals and Purposes, Achievements, Plans and International Implications*, Staff Report, 89th Congress, 2nd session, December 30, 1966.
(Abbreviated as *Soviet Space Programs, 1962-1965*).

US Congress, Senate, Committee on Aeronautics and Space Sciences, *Soviet Space Programs, 1966-1970, Goals and Purposes, Organization, Resources, Facilities and Hardware, Manned and Unmanned Flight Programs, Bioastronautics, Civil and Military Applications, Projections of Future Plans, Attitudes Toward International Cooperation and Space Law*, Staff Report prepared for the Use of the Committee, 92nd Congress, 1st session, Senate Document #92-51, December 9, 1971.
(Abbreviated as *Soviet Space Programs, 1966-1970*).

US Congress, Senate, Committee on Aeronautics and Space Sciences, *Soviet Space Programs, 1971, A Supplement to the Corresponding Report Covering the Period 1966-1970*, Staff Report prepared for the Use of the Committee by the Science

Policy Research Division, Congressional Research Service, Library of Congress, 92nd Congress, 2nd session, April 1972.

(Abbreviated as *Soviet Space Programs, 1971*).

US Congress, Senate, Committee on Aeronautics and Space Sciences, *Space Agreements with the Soviet Union*, Hearing, 92nd Congress, 2nd session, June 23, 1972. (Abbreviated as *Space Agreements with the Soviet Union*).

US Congress, Senate, Committee on Aeronautics and Space Sciences, *Space Cooperation Between the US and the Soviet Union*, Hearing, 92nd Congress, 1st session, March 17, 1971.

(Abbreviated as *Space Cooperation Between the US and USSR*).

US Congress, Senate, Committee on Aeronautics and Space Sciences, *US International Space Programs, Texts of Executive Agreements, Memoranda of Understanding and other International Arrangements, 1959-1965*, Staff Report prepared for Use by the Committee, Senate Document #44, 89th Congress, 1st session, July 30, 1965.

US Congress, Senate, Committee on Appropriations, Hearings Before a Subcommittee, *Independent Office Appropriations for 1963*, 87th Congress, 2nd session, Part 3, 1962.

US Department of State, "US-USSR Cooperation in Space Research," Appendix E of *Soviet Space Programs, 1966-1970*, pp. 607-610.

US Information Agency, Research and Reference Service, Survey Research Studies, "The Image of US Versus Soviet Science in West European Public Opinion. A Survey in Four West European Countries," *WE-3*, October 1961.

US President, *Public Papers of the Presidents of the US, Dwight D. Eisenhower, January 1 to December 31, 1957* (Washington, D.C.: 1958).

US President, *Public Papers of the Presidents of the US, Dwight D. Eisenhower, 1960-1961* (Washington, D.C.: 1961).

US Statutes, *Public Law 87-624*, August 31, 1962, volume 76, pp. 419-427 (Washington, D.C.: US Government Printing Office, 1963).

Whelan, Joseph G., "Political Goals and Purposes of the USSR in Space," in *Soviet Space Programs, 1962-1965*, pp. 19-146.

————, "Political Goals and Purposes of the USSR in Space," in *Soviet Space Programs, 1966-1970*, pp. 1-67.

————, "Soviet Attitude Toward International Cooperation in Space," in *Soviet Space Programs, 1962-1965*, pp. 425-492.

————, "Soviet Attitude Toward International Cooperation in Space," in *Soviet Space Programs, 1966-1970*, pp. 399-452.

————, "Western Projections of Soviet Space Plans," in *Soviet Space Programs, 1962-1965*, pp. 369-396.

PERIODICALS

Air Force and Space Digest

Aldoshin, V., "Outer Space Must Be a Peace Zone," *International Affairs*, No. 12, December 1968, pp. 38-41.

Alexandrov, N. A., "Why the USA Is Straining To Get into Outer Space," from *Soviet Fleet*, December 25, 1958, translated by J. R. Thomas, Rand Corporation, Translation #T-112, February 26, 1959.

Alsop, Joseph, "Facts About the Missile Balance," *Washington Post*, September 25, 1961, p. A-13.

Aviation Week and Space Technology

Baritz, J., "The Military Value of the Soviet Sputniks," *Bulletin of the Institute for the Study of the USSR*, 7(12), December 1960, pp. 30-38.

Bauer, Raymond A., "Keynes Via the Back Door," *Journal of Social Issues*, 17(2), 1961, pp. 50-55.

Berkner, L. V., "The Compelling Horizon," *Bulletin of the Atomic Scientists*, 19(5), May 1963, pp. 8-10.

———, "Earth Satellites and Foreign Policy," *Foreign Affairs*, 36(2), January 1958, pp. 221-231.

Blagonravov, A. A., "Space Research and International Scientific Collaboration," *Current Digest of the Soviet Press*, 11(18), June 3, 1959, p. 26, from *Izvestiia*, May 6, 1959, p. 3.

Christian Science Monitor

"Commission on Interplanetary Communications," *Vecherniaia Moskva*, April 16, 1955, in Krieger, *Behind the Sputniks*, Appendix B, pp. 329-330.

Commission on Legal Problems of Outer Space of the Soviet Academy of Sciences, "American Diversion in Space," *International Affairs*, No. 12, December 1961, pp. 117-118.

Covault, Craig, "Planet Flight Cooperation Sought," *Aviation Week and Space Technology*, 100(8), February 25, 1974, pp. 12-13.

———, "Soviet Plan to Buy Apollo Suits for Comparative Study Purposes," *Aviation Week and Space Technology*, 98(13), March 26, 1973, pp. 17-18.

Cowles, Gardner, "An Editor's Report on Europe and Russia," *Look*, 26, June 5, 1962, pp. 32-37.

DeWitt, Nicholas, "Reorganization of Science and Research in the USSR," *Science*, 133, June 23, 1961, pp. 1981-1991.

Driscoll, Every, "The Soviet Space Program," *Science News*, 99(18), May 1, 1971, pp. 303-305.

Dubovsky, M., "What For and Against Whom?" *New Times*, No. 30, July 29, 1970, pp. 9-11.

"Earth Moon," (anonymous), *New Times*, No. 38, September 1959, pp. 3-4.

Eisenhower, Dwight D., "Are We Headed in the Wrong Direction?" *Saturday Evening Post*, 235, August 11-18, 1962, pp. 19-25.

Ermasev, I., "Cosmic Problems and Earth Reality," *New Times*, No. 14, April 4, 1958, pp. 11-14.

Erskine, Hazel G., editor, "The Polls, Defense, Peace and Space," *Public Opinion Quarterly*, 25(3), Fall 1961, pp. 478-489.

Fink, Donald E., "Europeans See Wide Use of Spacelab," *Aviation Week and Space Technology*, 99(19), November 5, 1973, pp. 42, 43ff.

———, "Soviets Block Space Data Exchange Pact," *Aviation Week and Space Technology*, 81(15), October 12, 1964, p. 25.

Fruchterman, Lt. Commander Richard L., Jr., "Introduction to Space Law," *Judge Advocate General of the Navy Journal*, 20(1), July-August 1965, pp. 9-14.

Frutkin, Arnold, "International Programs of NASA," *Bulletin of the Atomic Scientists*, 17(5-6), May-June 1961, pp. 229-232.

Galina, A. (Onitskaia), "On the Question of Interplanetary Law," *Sovetskoe Gosudarstvo i Pravo*, No. 7, July 1958, translated by F. J. Krieger and J. R. Thomas, *Translations of Two Soviet Articles on Law and Order in Outer Space*, Rand

Corporation, Translation #T-98, September 25, 1958, pp. 1-14.

Gardner, Richard N., "Coöperation in Outer Space," *Foreign Affairs, 41*(2), January 1963, pp. 344-359.

———, "Outer Space: A Breakthrough for International Law," *American Bar Association Journal, 50*(1), January 1964, pp. 30-33.

Gartoff, Raymond L., "Red War Sputniks in the Works?" *Missiles and Rockets, 3*, May 1958, pp. 134-136.

Glasov, V., "Cannibals in Space," *New Times*, No. 24, June 19, 1963, pp. 12-13.

"Go for the Space Shuttle," *Newsweek, 80*(6), August 7, 1972, p. 43.

"Gold Medal Established for Outstanding Work in the Field of Interplanetary Communications," *Pravda*, September 25, 1964, in Krieger, *Behind the Sputniks*, Appendix A, p. 329.

Greenberg, D. S., "Space Accord: NASA's Enthusiasm for East-West Cooperation Is Not Shared by Pentagon," *Science, 136*, April 13, 1962, pp. 137-139.

Greenwood, Ted, "Reconnaissance and Arms Control," *Arms Control: Readings from Scientific American*, pp. 223-234.

Gvishiani, D., "Soviet Scientific and Technical Cooperation with Other Countries," *International Affairs*, No. 2-3, February-March, 1970, pp. 46-52.

Holloman, J. Herbert, "The Brain Mines of Tomorrow," *Saturday Review, 46*, May 4, 1963, pp. 46-47.

Hotz, Robert, "A Better Aerospace Budget," *Aviation Week and Space Technology, 100*(6), February 11, 1974, p. 7.

International Affairs (Moscow)

"International Cooperation in Science and Technology," an interview with D. M. Gvishiani, *New Times*, No. 3, January 20, 1970, pp. 7-9.

Ivanov, G., "Space Business and US Policy," *International Affairs*, No. 6, June 1964, pp. 64-68.

———, "Eurospace: Rocket Race," *International Affairs*, No. 12, December 1964, pp. 101-102.

Izvestiia

Karpenko, A. G., "Cosmic Laboratory," *Moskovskaia Pravda*, August 14, 1955, in Krieger, *Behind the Sputniks*, pp. 240-242.

Khozin, G., "Pentagon Seeking Control of Space Research in Asia and Africa," *International Affairs*, No. 2, February 1968, pp. 30-35.

Killian, James R., Jr., "The Crisis in Research," *Atlantic Monthly, 211*(3), March 1963, pp. 69-72.

———, "Making Science a Vital Force in Foreign Policy," Exerpts from a Speech to MIT Club of New York, December 13, 1960, *Science, 133,* January 6, 1961, pp. 24-25.

Kislov, A. and S. Krilov, "State Sovereignty in Air Space," *International Affairs*, No. 3, March 1956, pp. 35-44.

Kohler, Foy D. and Dodd C. Harvey, "Administering and Managing the US and Soviet Space Programs," *Science, 169*, September 11, 1970, pp. 1049-1056.

Kokum, Edward H., "Budget Steadies Aerospace Outlook," *Aviation Week and Space Technology, 100*(6), February 11, 1974, pp. 12-13.

———, "USSR's Space Effort Hits Economic Snags," *Aviation Week, 80*(11), March 16, 1964, pp. 137-139.

Komsomolskaia Pravda

Korovin, E., "International Status of Cosmic Space," *International Affairs*, No. 1, January 1959, pp. 53-59.

———, "Outer Space and International Law," *New Times*, No. 17, April 25, 1962, pp. 13-14.

———, "Outer Space Must Become a Zone of Real Peace," *International Affairs*, No. 9, September 1963, pp. 92-93.

———, "Peaceful Cooperation in Space," *International Affairs*, No. 3, March 1962, pp. 61-63.

Krasnaia Zvezda

Lall, Betty Goetz, "Cooperation and Arms Control in Outer Space," *Bulletin of the Atomic Scientists*, 22(9), November 1966, pp. 34-37.

Lapp, Ralph E., "Investing in Space," *New Republic*, 160(2), January 11, 1969, pp. 12-13.

Larionov, V., "The Doctrine of Military Domination in Outer Space," *International Affairs*, No. 10, October 1964, pp. 25-30.

"Leading on Earth and in Space," (anonymous), *International Affairs*, No. 9, September 1962, pp. 3-5.

Listvinov, T., "Space Research American Style," *International Affairs*, No. 11, November 1968, pp. 46-50.

Lovell, Sir Bernard, "The Great Competition in Space," *Foreign Affairs*, 51(1), October 1972, pp. 124-138.

———, "Russian Plans for Space," *Survey*, 52(52), July 1964, pp. 78-82.

———, "Soviet Aims in Astronomy and Space Research," *Bulletin of the Atomic Scientists*, 19(8), October 1963, pp. 36-39.

Lubell, Sam, "Sputnik and American Public Opinion," *Columbia University Forum*, 1(1), Winter 1957, pp. 15-21.

Marshak, Robert E., "Reexamining the Soviet Scientific Challenge," *Bulletin of the Atomic Scientists*, 19(4), April 1963, pp. 12-17.

Missiles and Rockets

Nesmeianov, A. N., "The Problem of Creating an Artificial Earth Satellite," *Pravda*, June 1, 1957, in Krieger, *Behind the Sputniks*, pp. 276-281.

Nevskii, S., "Apollo 12," *New Times*, No. 48, December 3, 1969, pp. 9-10.

"New Eye in Space," *Newsweek*, 80(6), August 7, 1972, p. 43.

"New Stage in Space Research," an interview with Academician L. I. Sedov, *New Times*, No. 9, March 3, 1971, pp. 23-24.

New Times (Moscow)

New York Times

Newsweek

Normyle, William J., "NASA Molding Post-Apollo Plans," *Aviation Week and Space Technology*, 89(26), December 23, 1968, pp. 16-17.

Observer, "Space, Science and Peace," *New Times*, No. 35, August 24, 1962, pp. 1-3.

"On Flights into Space," an Interview with L. I. Sedov, *Pravda*, September 26, 1955, in Krieger, *Behind the Sputniks*, pp. 112-115.

"On the Way to the Stars," *Tekhnika-Molodezhi*, July 1954, in Krieger, *Behind the Sputniks*, pp. 35-49.

Orberg, Jim, "Korolev," *Space World*, K-5-125, May 1974, pp. 17-24.

"The Outlook for Space Exploration," interview with Academician L. I. Sedov, *New Times*, No. 33, August 18, 1965, p. 5.

Pechorkin, V., "The Pentagon Theoreticians and the Cosmos," *International Affairs*, No. 3, March 1961, pp. 32-36.

Perkins, C. D., "Man and Military Space," *Journal of the Royal Aeronautical Society*, 67(631), July 1963, pp. 397-412.

Petrov, Boris, "Space Exploration: Progress and Trends," *World Marxist Review,* *14*(4), April 1971, pp. 82-89.

Petrovich, G. V. (pseudonym), "Pervii isskustvennii sputnik solntsa," *Vestnik Akademii Nauk SSSR, 29*(3), March 1959, pp. 8-14.

Pokrovskii, G. I. "Artificial Earth Satellite Problems," *Izvestiia*, August 19, 1955, in Krieger, *Behind the Sputniks*, pp. 243-245.

———, "Competition in Space," *New Times*, No. 20, May 17, 1961, pp. 3-4.

———, "Crime in Space," *New Times*, No. 25, June 20, 1962, pp. 9-11.

———, "Profitability of the Peaceful Uses of Outer Space," *International Affairs*, No. 9, September 1962, pp. 117-118.

———, "On the Problem of the Use of Cosmic Space," *International Affairs*, No. 7, July 1959, pp. 105-107.

Posev (Munich)

Pravda

"Probing the Secrets of the Universe," (anonymous), *New Times*, No. 33, August 15, 1962, pp. 1-2.

Rubel, J. H., "The Military in Space," *Bulletin of the Atomic Scientists, 19*(5), May 1963, pp. 20-22.

Sedov, Leonid, "Interplanetary Travel Soon a Reality," *New Times*, No. 38, September 1959, pp. 5-6.

———, "The Outlook for Space Exploration," *New Times*, No. 26, July 3, 1963, pp. 16-17.

———, "Outlook for Space Exploration," *New Times*, No. 8, February 23, 1966, pp. 1-3.

Sejnin, J. Y., "Space and Earth," *New Times*, No. 14, April 1, 1960, pp. 26-29.

Sevastianov, Vitalii, "Opening Up Outer Space," *New Times*, No. 25, June 1971, pp. 8-9.

Sheldon, Charles S., II, "Soviet Circumlunar Flight Seen," *Aviation Week, 80*(20), May 18, 1964, p. 77.

Shelton, William, "Neck and Neck in the Space Race," *Fortune, 76*(5), October 1967, pp. 166-168ff.

"Soviet Space Safety," (anonymous), *Aviation Week, 71*(10), September 7, 1959, p. 28.

"Soviets Planning Extensive IQSY Effort," (anonymous), *Aviation Week, 81*(8), August 24, 1964, pp. 65-67.

"Soviets Seen Downgrading Space Congress," (anonymous), *Aviation Week and Space Technology, 97*(16), October 16, 1972, p. 15.

"Soviets Suffer Setbacks in Space," (anonymous), *Aviation Week and Space Technology, 91*(20), November 17, 1969, pp. 26-27.

"Space Crime," editorial, *New Times*, No. 21, May 24, 1963, p. 2.

Space World

Stanford, Neal, "US Space Officials Say 'Race' Still On," *Christian Science Monitor, 56*(154), May 26, 1964, p. 3.

Staniukovich, K., "Artificial Earth Satellite Principles," *Krasnaia Zvezda,* August 7, 1955, in Krieger, *Behind the Sputniks*, pp. 237-239.

TASS, "Report on Intercontinental Ballistic Rocket," *Pravda*, August 27, 1957, in Krieger, *Behind the Sputniks*, pp. 233-234.

Teplinskii, Lt. General (retired) B. N., "Pentagon's Space Programme," *International Affairs*, No. 1, January 1963, pp. 56-62.

————, "The Strategic Concepts of US Aggressive Policy," *International Affairs*, No. 12, December 1960, pp. 36-41.

————, "Space Maniacs," *New Times*, No. 35, September 1, 1965, pp. 7-8.

Time

Trilling, Leon, "Soviet Astronautical Scientists: How They Work and Where They Publish," *Aerospace Engineering*, 20, July 1961, pp. 12-13ff.

Triska, Jan and David Finley, "Soviet-American Relations: A Multiple Symmetry Model," *Journal of Conflict Resolution*, 9(1), March 1965, pp. 37-53.

US News and World Report

Van de Hulst, H. C., "International Space Cooperation," *Bulletin of the Atomic Scientists*, 17(5-6), May-June 1961, pp. 233-236.

Vereshchetin, Vladen, "Intercosmos: Results and Prospects," *Space World*, I-1-97, January 1972, pp. 38-39.

Walsh, John R., "NASA: Talk of Togetherness with Soviets Further Complicates Space Politics for the Agency," *Science*, 142 (3588), October 4, 1963, pp. 35-38.

Weaver, Warren, "Dreams and Responsibilities," *Bulletin of the Atomic Scientists*, 19(5), May 1963, pp. 10-11.

Wolfe, T. W., "Are the Generals Taking Over?" *Problems of Communism*, 18, July-October 1969, pp. 106-110.

Young, Hugo, Bryan Sikock and Peter Dunn, "Why We Went to the Moon: From the Bay of Pigs to the Sea of Tranquility," *The Washington Monthly*, 2(2), April 1970, pp. 28-58.

Zadorozhnii, G., "The Artificial Satellite and International Law," from *Sovetskaia Rossiia*, October 17, 1957, translated by Anne M. Jonas, Rand Corporation, Translation #T-78, November 12, 1957, pp. 1-5.

Zhukov, G. P., "Iaderhaie demilitarizatsiia kosmosa," *Sovetskoe Gosudarstvo i Pravo*, No. 3, March 1964, pp. 79-89.

————, "Space Espionage and International Law," *International Affairs*, No. 10, October 1960, pp. 53-57.

Speeches and Occasional Papers

Frutkin, Arnold W., "Patterns of International Space Applications and Their Extension," American Astronautical Society Reprint #67-146, May 2, 1967, pp. 379-389.

Gicca, Francis A., "Space Communications in the 21st Century," Raytheon Corporation, American Astronautical Society Reprint #67-092, May 1, 1967, pp. 1-11.

Hult, J. L., "Satellites and Future Communication Including Broadcast," Rand Corporation, American Astronautical Society Reprint #67-091, 1967, pp. 1-22.

Zhukov, G. P., "Problems of Space Law at the Present Stage," International Law Association Memorandum, Brussels, 1962.

Unpublished Material

Fedorenko, Dr. Sergei P., Address delivered at Illinois State University, December 6, 1972.

General Reference

Long, Luman H., editor, *The 1972 World Almanac and Book of Facts* (New York: Newspaper Enterprises Association, Inc., 1971).

Appendices

A. TASS ANNOUNCEMENT OF *Kosmos* SERIES*

A series of artificial Earth satellites will be launched from different cosmodromes of the Soviet Union during 1962. Another launching of an artificial Earth satellite was carried out in the Soviet Union on 16 March 1962. . . .

The launching of the artificial Earth satellite continues the current program of studying the upper layers of the atmosphere and outer space in fulfillment of which a series of satellite launchings will be effected under this program from different cosmodromes of the Soviet Union in the course of 1962. The scientific program includes: the study of the concentration of charged particles in the ionosphere for investigating the propagation of radio waves; study of corpuscular flows and low energy particles; study of the energy composition of the radiation belts of the Earth for the purpose of further evaluating the radiation danger of prolonged space flights; study of the primary composition and intensity variations of cosmic rays; study of the magnetic field of the Earth; study of the shortwave radiation of the Sun and other celestial bodies; study of the upper layers of the atmosphere; study of the effects of meteoric matter on construction elements of space vehicles; and study of the distribution and formation of cloud patterns in the Earth's atmosphere.

Moreover, many elements of space vehicle construction will be checked and improved. The launchings of Sputniks of this series will be announced in separate reports. This program will give Soviet scientists new means for studying the physics of the upper atmospheric layers and outer space.

*TASS, March 16, 1962, 1701 GMT as cited in Charles S. Sheldon II, "Program Details of Unmanned Flights," *Soviet Space Programs, 1966–1970, op. cit.*, p. 173.

B. GLOSSARY OF ABBREVIATED TERMS

A-4	German short-range ballistic missile developed during World War II, better known as the V-2.
ABM	Anti-ballistic missile.
ABMA	The US Army Ballistic Missile Agency, which included the German scientists from the von Braun team, transferred to NASA in 1958.
ARPA	Advanced Research Projects Agency, the Pentagon agency which managed missile projects before the formation of NASA
CEUS	Commission on the Exploration and Utilization of Space, has general responsibility for space matters within the Soviet Academy of Sciences. Formerly ICIC.
Comsat	Communications Satellite Corporation, US-based private corporation for satellite communications.
COPUOS	UN Committee for the Peaceful Uses of Outer Space.
COSPAR	Committee on Space Research of the International Council of Scientific Unions.
CPSU	Communist Party of the Soviet Union.
CSAGI	French initials for the Special Committee for the International Geophysical Year (IGY).
DoD	Department of Defense (US).
ELDO	European Launcher Development Organization.
ERTS	Earth Resources Technology Satellite (US).
ESRO	European Satellite Research Organization.
FOBS	Fractional Orbital Bombardment System.
GDL	Gas Dynamics Laboratory (USSR).
GIRD	Group for the Study of Reactive Motion (USSR).
IAF	International Astronomical Federation.
ICBM	Intercontinental ballistic missile.
ICIC	Interdepartmental Commission on Interplanetary Communications (USSR). See CEUS.
IGY	International Geophysical Year.
Intelsat	International Telecommunications Satellite Consortium.
Intersputnik	International satellite communications organization largely confined to the Soviet bloc.
IRBM	Intermediate-range ballistic missile.
ITU	International Telecommunications Union.
Lengird	Leningrad branch of GIRD.
MODS	Military Orbital Development System, planned USAF manned satellite. Succeeded by MOL.
MOL	Manned Orbiting Laboratory, planned USAF manned satellite. The project was canceled.
Mosgird	Moscow branch of GIRD.
MRBM	Medium-range ballistic missile.
NAS	National Academy of Sciences (US).
NASA	National Aeronautics and Space Administration (US).
NASC	National Aeronautics and Space Council (US).
OIMS	All-Union Society to Study Interplanetary Communications (USSR).

OSOAVIAKHIM	All-Union Society for Air and Chemical Defense (USSR).
R & D	Research and development.
RNII	Jet Scientific Research Institute (USSR).
SALT	Strategic Arms Limitation Talks.
SS-4	Soviet MRBM used to launch small satellites, also known as "Sandal." Several were shipped to Cuba in Fall, 1962.
SS-5	Soviet IRBM, used to launch small to medium satellites. Also known as "Skean."
SS-6	Standard Soviet launch vehicle employed to launch the Sputniks, all Soviet manned satellites, etc. Also known as the "Sapwood" ICBM and *Vostok*.
SS-9	Soviet combat launch vehicle employed as an ICBM, as a launcher for FOBS and for a satellite interceptor. Also known as the "Scarp" ICBM.
SS-10	Soviet ICBM for which an orbital capability was claimed, also known as "Scrag."
TsAGI	Central Aero-Hydrodynamics Institute (USSR).
TsBIRP	Central Bureau for the Study of the Problems of Rockets (USSR).
USAF	United States Air Force.
USIA	United States Information Agency.

INDEX